GRAY GOLD

Gray Gold

Lead Mining
and Its Impact on the
Natural and Cultural
Environment,
1700–1840

MARK MILTON CHAMBERS

The University of Tennessee Press / Knoxville

Copyright © 2021 by The University of Tennessee Press / Knoxville.
All Rights Reserved. Manufactured in the United States of America.
First Edition.

Library of Congress Cataloging-in-Publication Data

Names: Chambers, Mark Milton, author.
Title: Gray gold : lead mining and its impact on the natural and cultural environment, 1700–1840 / Mark Milton Chambers.
Description: First edition. | Knoxville : The University of Tennessee Press, [2021] | Outgrowth of the author's thesis (doctoral)—Stony Brook University, 2012, under the title: River of gray gold : cultural and material changes in the land of ores, country of minerals, 1719–1839. | Includes bibliographical references and index. | Summary: "This book explores Native American and Euro-American lead mining in the Midwest. As Europeans flooded North America and moved westward, their own mining practices were greatly informed by Native American mining methods already in place. While many researchers have explored gold, silver, and copper mining and smelting, lead has not received much scholarly attention despite a long history of Native American and European desire for the ore. Chambers reflects on how early mining techniques affected the culture clash between Native Americans and European colonists, all the while tracking the impact increased mining had on the environment of what would become the states of Illinois and Missouri"—Provided by publisher.
Identifiers: LCCN 2021038182 (print) | LCCN 2021038183 (ebook) | ISBN 9781621906988 (hardcover) | ISBN 9781621906995 (pdf)
Subjects: LCSH: Lead mines and mining—Middle West—History. | Lead mines and mining—Environmental aspects—Middle West. | Indians of North America—Middle West. | Cultural landscapes—Middle West.
Classification: LCC TN453.A5 C49 2021 (print) | LCC TN453.A5 (ebook) | DDC 338.2/744097709033—dc23
LC record available at https://lccn.loc.gov/2021038182
LC ebook record available at https://lccn.loc.gov/2021038183

To Mom and Dad,

while always remembering Paul and Stacie

CONTENTS

Preface xi

INTRODUCTION. Lead Ore and Surveying *le Pays des Illinois* 1

CHAPTER 1. Early Mining and Smelting in North America 15

CHAPTER 2. Tracing Eighty Years of Early Mining Associations 58

CHAPTER 3. Early Mineralogical Assessments and Emerging Science 101

CHAPTER 4. Unhealthy Spaces Fitted Up with Furnaces 137

Conclusion 165

Notes 171

Bibliography 213

Index 251

ILLUSTRATIONS

FIGURE 1. Louisiana Map, 1765   25

FIGURE 2. French and Spanish Period Villages and
Land Routes in the early 1700s   38

FIGURE 3. French and Spanish Period Villages and
Land Routes in the late 1700s   72

FIGURE 4. View of Herculaneum and the Shot Tower   126

FIGURE 5. View of Mine La Motte Village   150

FIGURE 6. View of Mine La Motte with Three Miners   150

PREFACE

I started working on this project as an undergraduate living and working in St. Louis, Missouri. It all began after learning that I required one additional history course to complete my application to a master's program in education, which lead me back to the course catalog and eventually to enrolling in an early colonial history class on Ste. Geneviève. Dr. Carl Ekberg, a professor at Illinois State University, jumped in his car every Wednesday in Bloomington and arrived in St. Louis in time to teach class. Throughout the semester, Dr. Ekberg continually mentioned the need for more scholarly attention to the *country full of mines* in the colonial era. Beginning with that course, my first environmental history course taught by Dr. Conevery Bolton Valenčius at Washington University in St. Louis, and under the guidance of Dr. Christopher Sellers at Stony Brook University, this project has grown, with many changes, into a book.

Before endeavoring on this path of scholarship as a nontraditional student at Washington University in St. Louis, a number of professional experiences helped support my education, for which I am very grateful. As a flight service manager flying between six and twelve hours, I was grateful for the long domestic or international layovers that allowed me to be regenerated for the return trip back home or to write papers, but also, I am very grateful for all the institutions and people who have made this a most valuable and worthwhile journey. I owe more debts than I can count to the many friends, coworkers, generous family members, and scholars who made this book what it is today. What follows is only an unfinished list. Those whom I could not include here know my appreciation for their kindness and support.

What follows is a description of moments in the classroom and the immeasurable time with friends immersed in wonderful cross-cultural experiences that prepared me to believe that anything is possible. This book is a result of the amalgamation of those numerous life experiences. Reaching beyond my North American academic circles across the Pacific to the "Land of the Morning Calm," it is important to thank my friends David and Ellen Ross, with whom I had the pleasure of working while attending the Korean Language Institute at Yonsei University in Seoul, Korea. I would also like to acknowledge the friends and coworkers of the former Trans World Airlines who encouraged me to continue my education. As I studied to earn my degrees, numerous colleagues made it possible for me to manage both my onboard flight and in-class instructor schedules so that I could attend college at night. I did not know during those many years of employment in the Charles A. Lindbergh Training Center, at the end of the longest runway at St. Louis International Airport, instructing numerous flight service managers and flight attendants on in-flight procedures, that I was developing the skills needed to become a fulltime historian and educator. Three immensely extraordinary people at the Charles A. Lindbergh Training Center are Sharon McCullough, Lucille Collins, and Marilyn Taber; thank you for assisting me in my journey toward teaching and historical research. Most significantly, thank you American Airlines for making it possible for me to fully embrace my goal.

The Marguerite Ross-Barnett Scholarship from Washington University in St. Louis supported my undergraduate work. My initial years of graduate work at Washington University in St. Louis were supported by a number of research assistantships and tuition scholarships. Multiple graduate assistantships and tuition scholarships from the History Department supported my doctoral studies and a major part of my research for this project at Stony Brook University. In addition, a Dissertation Research Grant from the Early American Industries Association helped with numerous trips to the Missouri Historical Society and the Western Historical Manuscripts Collection in St. Louis. More recently, while writing, revising, and editing the book, I am deeply grateful for the opportunity I had to develop my teaching skills and for all the financial assistance provided to me in the History Department, the Department of Africana Studies, and the Educational Opportunity Program (EOP) at Stony Brook University.

I appreciate the help of the staffs of the many libraries and archives at which I gathered material: the archives of the University of Pennsylvania, Olin Library of Washington University in St. Louis, the Bernard Becker Medical Library, the St. Louis University Library and Medical Archives, the New York Public Library Research Division, and the New York Academy of Medicine Research Library. In particular, I applaud the staff of the Missouri Historical Society, the Western Historical Manuscripts Collection, and the Missouri Department of

Natural Resources for consistently providing excellent help over the course of my initial research visits and in the years since by phone conversations. Lynn Morrow, the former director of the Local Records Division of the Missouri State Archives, gifted me with his own materials and maps from his private collection that have not only aided my research for this book, but for the next book as well. I am also grateful to Stony Brook University for awarding me the 2017 President's Distinguished Travel Grant for Faculty in the Fine Arts, Humanities, and Lettered Social Sciences and the 2019 Research Initiative Grant for Faculty in the Arts, Humanities, and Lettered Social Sciences in support of revising the manuscript. Finally, I am deeply honored to have been selected as a 2021–2023 Civic Science Fellow with the Environmental Data and Governance Initiative with support from the Rita Allen and David and Lucile Packard Foundations during the final stages of the book's completion.

After many years away from New York, I returned to enroll and be trained in the Department of History at Stony Brook University, where I was welcomed by a humble and collegial faculty. Susan Grumet, Domenica Tafuro, Roxanne Fernandez, and Ann Berrios consistently eased my fears while listening and encouraging me, trying to meet my practical needs, and contributing to my intellectual growth. For passing conversations, advice, and insights shared, I particularly wish to thank Tracey Walters. At Stony Brook University, able and generous scholars guided me. I would like to thank Kathleen Wilson, Shirley Lim, Michael Barnhart, Paul Gootenberg, Donna Rilling, Ned Landsman, Brooke Larson, and Jennifer Anderson for creating a valuable intellectual community. A number of scholars have contributed commentary on each chapter. My thanks to Eugene Hammond, the director of the University writing program, who helped me present a well-written argument. I very much appreciate the suggestions and comments I have gained from presenting sections of this project to audiences at Washington University in St. Louis, the University of Texas at Arlington, the American Society for Environmental History, the History of Science Society, and the American Historical Association conferences. Many thanks to the staff at the University of Tennessee Press, especially Thomas Wells, Jon Boggs, and Linsey Perry. I am grateful as well to the anonymous readers of the press, and Gerry Krieg of Krieg Mapping for his attention to detail and patience preparing the maps. Thank you to the Muséum d'histoire naturelle–Le Havre for granting permissions for the reproduction of the nineteenth-century Charles LeSueur sketches.

My advisor, Christopher Sellers, faithfully read every chapter a number of times and advised my PhD thesis with fine attention to writing and offered new perspectives on my historical argument and research methods. In similar ways, during the writing of my dissertation and this book, Conevery Bolton Valenčius offered feedback and advice on every aspect of this work. Nancy Tomes's analysis and advice from the dissertation prospectus class through the finished product have shaped my thinking. Donna Rilling provided advice on a number of chapters, for which I am forever grateful. In addition, from the time I arrived at Stony Brook, April Masten offered guidance and writing advice on every chapter; she has been instrumental in shaping my arguments and shaping who I am as a writer.

I am honored to offer great thanks to my graduate advisor, Christopher Sellers. I deeply appreciate his guidance as my adviser and his example of engaged scholarly life. His comments on my arguments and on my writing, I hope, are reflected on each page; with discerning care, he has helped hone and refine the essence of this work. Christopher Sellers sets a model of understandable and humane scholarship, which I hope to sustain and follow for the rest of my career.

I am blessed with wonderful family. My older brother Kenny, his wife Kathy, and my younger brother Paul, while quietly wondering "when will he be finished," have voiced their support throughout my new journey. They have each been for me a continual source of encouragement. I am grateful for the unfailing support of my McCray cousins Stacie, Arthur, and his wife, Rosie. I am honored to acknowledge my Uncle Leroy and Aunt Marjorie McCray who watered my vision during each and every conversation, and whose many reminiscences of a long life well lived are a constant reminder regarding the importance of education.

Finally, I owe my parents my deepest and most lasting gratitude. Milton Henry Chambers, who I am sure will be looking for the French miners poems that he so faithfully translated, has quietly nurtured me with patience and constant support. Delores Ann Chambers has sustained me throughout this project with her unstinting encouragement, her pragmatic perspective, reciting of scriptures, and quiet prayers. Trained as a nurse, she made a risky but calculated decision to follow her other passion a number of years earlier. Witnessing her creative energy and willingness to take the next leap in my own efforts, I am grateful for that inheritance. In these and so many other ways, my parents have launched and sustained me throughout and into this new chapter. It is to them that this work is dedicated.

INTRODUCTION

Lead Ore and Surveying
le Pays des Illinois

By the 1803 Louisiana Purchase, Thomas Jefferson had already developed a long relationship with natural philosophy. His knowledge of natural science came from Comte de Buffon's voluminous *Natural History*, and Jefferson referred to him as "the best informed of any naturalist who has ever written." Jefferson's own foray into natural history is his voluminous *Notes on the State of Virginia*, first published in 1785. Inspired by a diplomatic questionnaire and formatted after Denis Diderot and Jean le Rond d'Alembert's eighteenth-century *Encyclopedia, or a Systematic Dictionary of the Sciences, Arts, and Crafts*, the content of *Notes on Virginia* includes sections on the natural world and humans and their artifacts. Actually, the term "Virginia" is a bit misleading, for at the time, the Atlantic states' boundaries were poorly defined or were nonexistent. "Virginia" included parts of western Pennsylvania, all of West Virginia, and parts of Kentucky and Tennessee. In *Notes on Virginia*, Jefferson outlines Virginia's mineral resources and mining operations. Although he mentions deposits of gold, copper, and iron, Jefferson also discusses the layout, the operation, and the workers of a particular lead mine. He suggests that if the pounding mill and the furnace were on the same side of the river, the mining procedures would be more efficient. He reveals that the lead mine employed thirty laborers who cultivated corn and produced anywhere between twenty and twenty-five tons of lead a year. Finally, Jefferson describes how the workers transported refined lead 130 miles by canoe and wagon along a good road to the local ferry on the James River, and then it was carried another 130 miles on water to its destination.[1] *Notes on Virginia* introduces the importance of the scientific study of minerals as an element of the emerging sciences' influence on the Enlightenment in the United States. Jefferson's survey of mining in Virginia, in particular his example of a lead mining operation combined with corn cultivation,

1

illustrates his recognition of settlements where mineral extraction and farming occurred as an important part of American national formation.

In his late eighteenth-century *Notes on Virginia*, Jefferson addressed a variety of subjects in addition to minerals, such as Virginia's rivers, agriculture, manufacturing, customs, history, animals, plants, and Native Americans. He along with other natural philosophers—those interested in studying nature through observation and scientific speculation—viewed nature as a subject more closely allied to fulfilling human needs. In addition, not only did some natural philosophers like Jefferson want to understand human interactions with minerals and plants, but because many understood how Native Americans used fire to manage the forest, these same naturalists desired to learn more about how Native Americans interacted with the environment.[2]

Jefferson not only deals with the mineral resources of Virginia in *Notes on Virginia*, he also broadens his geographical scope noting the occurrence of lead ore mining near Kaskaskia, Illinois. Of these mines he wrote, "The greatest, however, known in the western country. . . . The lead used in that country being from the banks on the Spanish side of the Mississippi, opposite to Kaskaskia."[3] By the late seventeenth century, the French had settled one small piece of the total of French-claimed territory that stretched from the Saint Lawrence River to New Orleans, with much of the territory lying in close proximity to the Mississippi River in the North American interior. In the fall of 1700, the French Jesuit missionary Father Jacques Gravier wrote about the same lead mines near Kaskaskia that Jefferson mentioned. Just as Jefferson recognized a particular lead mine in eighteenth-century Virginia as an important factor in American national formation, Gravier also understood the significance of metals to the French court. When Father Gravier wrote about the occurrence of lead ore near Kaskaskia, he was unsure what the officials in Paris would "decide with reference to the Mississippi settlements, if no silver mines [were] found there . . . but little heed [was] paid to the lead mines, which [were] very plentiful toward the Illinois Country" with ore that would possibly "yields three-fourths metal."[4] Father Gravier not only established a missionary outpost in the New World, his knowledge of natural philosophy, acquired at the Collège Louis-le-Grand in Paris, equipped him to survey *le pays des Illinois* for the romantic metals silver and gold, both valuable commodities to the French court. In this period of eighteenth-century emerging science, Father Gravier and Thomas Jefferson each developed a relationship with natural philosophy. Both understood the importance of sharing their knowledge about a mineral deposit like lead ore with others to reveal the power of this valuable metal, which they believed could influence national formation.[5]

This book is about lead, which still remains as important today as it has for thousands of years; lead was extracted and used from the very inception of metallurgy and indeed may have been the first metal to be smelted.[6] The story of lead ore is not only important for its value to societies, but the story of lead is also significant because it reveals how Native Americans mined and smelted lead ore long before French explorers and settlers colonized lead mines near Kaskaskia, Illinois. For example, Mississippians exchanged galena in the form of crystals and nodules. These shiny lead cubes were used as ornaments in rituals and were buried with the dead. Native Americans produced a metallic blue-gray pigment by crushing galena and grinding it into powder with a handheld stone. They also combined crushed galena with plants, like beeweed or tansy mustard, and with hematite and manganese to make glaze paints.[7]

Throughout the eighteenth century, metals were being employed in a multitude of products. Iron hollowware—pots, pans, and kettles—was cast directly from blast furnaces, as were anchors, fireplace andirons, and pokers. In addition, blacksmiths fashioned iron into horseshoes, scythes, sickles, hay rakes, knives, razors, shears, axes, chains, and plow irons. Coppersmiths sold kettles, boilers, and roofing cooper as well as stills for the manufacture of beer and whiskey. Tin was alloyed with copper to produce bronze statues and bells.[8] Lead was retained to make ribs for windowpanes, to mold musket balls in settlers' homes, and to make sheet lead for roofing homes, sheathing ships, and rolling into pipe. Eventually, following the Louisiana Purchase, American settlers mined, smelted, and shipped lead to Philadelphia where it was used to manufacture red and white lead. Red lead became an important product for the production of flint and crystal glass. White lead continued to be an important chemical product for making paint.[9] Additionally, lead was most useful to Americans in the form of musket balls and eventually shot for protection of personal property.[10]

Toward the end of the seventeenth century, René-Robert Cavelier, Sieur de La Salle, after descending the Mississippi River, claimed this entire basin in 1682. The term *le pays des Illinois* (Illinois Country) best describes the area where Jefferson noted the greatest lead mines known in the Louisiana Territory on the Spanish side of the Mississippi, just opposite Kaskaskia where Native Americans and French settlers extracted and smelted lead ores. In the seventeenth century, prior to the arrival of the Kaskaskia Indians and French settlers, the Illinois inhabited these lands on the east side of the Mississippi in what is now Illinois; they were a loose confederacy of Algonquian-speaking tribes who later gave their name to the state. In the years following La Salle's and Gravier's reporting on their travels along the Mississippi River, both received the assistance of Native Americans who guided them along the banks of the rivers and who guided Father Gravier to their lead mines. Eventually, French immigrants and Native Americans would settle Kaskaskia in 1703. The community was not only a significant colonial fur trading outpost but would also play an important role in the early development of French and Native American colonial lead mining.[11]

Similar to the Spanish explorers Álvar Núñez Cabeza de Vaca and Hernando de Soto, and the French priest Father Gravier, who searched for New World minerals for their respective nations, when the English stepped off their ships onto the Virginia shores, they too searched for minerals—of a shinier sort than lead—gold and silver. However, unlike the New Spain explorers, the English and French explorers and settlers did not locate vast quantities of gold and silver in North America. However, drawn to the possibility of gaining wealth from their new environment, those French, and later Americans, who settled in *le pays des Illinois* found the less romantic galena. Though not shiny like silver and gold, lead ore was still a very valuable commodity. While Thomas Jefferson highlighted Virginia's mineral resources of gold, copper, and iron, he clearly understood that the less notable metal, lead, sold by the pound would also contribute to America's national wealth. Like Cabeza de Vaca, De Soto, and Gravier, historians too have gravitated toward telling stories about settlers' interactions with gold, silver, iron, and copper, which has proved alluring, as evidenced by the forty-niners and other nineteenth-century prospectors who went in search of these metals throughout the United States.[12]

The general body of literature on mining in the United States touches on early (pre-1830) iron production, copper mining, and anthracite coal mining, but only starts broadly focusing on metal mining histories with the Georgia gold rush and Michigan copper rushes of the 1830s and 40s. This book will add to the understanding of mining from early Native Americans through the arrival of French and American settlers to the lead ore terrain in what became the state of Missouri. This is not only a story about the surveying of a territory in search of more galena or lead ore, but also a narrative about a forgotten region that contributed to the national formation of the United States. Placing one particular metal at center stage also adds to the broader understanding of a mining history where Native Americans, French settlers, African American slaves, Americans, and English immigrants prospected, extracted, and smelted lead ore as the scientific study of minerals continued to emerge. Historians have pointed to the development of mining in the United States by noting the significance of the California gold rush, and then examining late nineteenth-century mining expeditions, camps, and industrialization. However, examining the emergence and evolution of early eighteenth-century mining will highlight the importance of one nonprecious metal, lead.[13]

Early eighteenth-century lead mining in North America has been largely overlooked, as much of the early mining history is dominated by western developments. Rodman Wilson Paul and Elliott West's *Mining Frontiers of the Far West* focuses on the well-known nineteenth-century gold rushes.[14] Paul identifies a series of rushes that transpired bringing the first permanent settlers to western America. Beginning in California (1848–1858), moving to Nevada and the Comstock Lode (1859–1880), to Colorado (1859–1880), and on to the Pacific Northwest and Southwest (1860–1880), and ending in the Black Hills in the late 1870s, Paul's emphasis is on the enduring phases of western mining and on the men who made significant contributions to mining history. Paul conveys stories about Americans who were the ingenious contrivers of new techniques or machinery, the capitalists who subsidized the development of the most promising mines, and the builders who constructed railroad routes linking mining camps with markets, all of which eventually transpired in the lead mining regions along the Mississippi River. Lead ore mining in *le pays des Illinois* has similar stories about Native Americans, French, and Americans who devised new methods; it's about capitalists who created mining associations to develop some of the most promising mines; and it's about miners and merchants who built overland and water routes to ship their lead from mining spaces to towns and cities in the Mississippi valley and to the east.

Elliot West revisits Paul's work in light of Western scholarship today. West's chapters contrast sharply with Paul's original work. West writes a general social history of the region based in the scholarship of the forty years since *Mining Frontiers of the Far West* was first published. For example, in "Breaking and Building Communities," West presents the devastation of Native American communities and the development of mining not simply as counterpoints, but as closely interconnected relationships that were created between early French settlers and Native American lead mines. Calling for additional research on these "contact points," West encourages historians to acknowledge the existence of early North American mining that took place in the Mississippi valley more than a century prior to Native American and European contact.[15] Viewing mining through the material and cultural lens can highlight encounters between colonists and indigenes, as well as the syncretic residuals of meetings that predate primarily Euro-American mining practices and technologies.

To describe these lead mining contacts in this book, Native Americans guides are paired with Jesuit priests, naturalists, and miners to shine a bright light on the eighteenth-century mining narratives highlighting early mutual intersections between Native Americans who learned to prospect, extract, and smelt lead ores and the French settlers who carried with them their own ideas and practices about mining. Along the way they also established a set of syncretic methods that were neither wholly French nor wholly Native American, but rather, a complicated amalgamation (or hybridization) of techniques. One work, Susan Martin's *Wonderful Power: The Story of Ancient Copper Working in the Lake Superior Basin* focuses on Native American metalworking and early coordination with French, British, and American occupiers; however, any recognition of mining amalgamations between settlers and Native Americans is absent. Such an alternative formulation will offer a sense of the complexity of the North American scene as colonial miners developed vital relationships to extract and smelt the valuable resource lead.[16] Lucy Eldersveld Murphy's in-depth study of lead mining and frontier exchange economy does a wonderful job of investigating the social and economic history of Native American and immigrant mining communities in Southern Wisconsin and northwestern Illinois. Murphy presents the relationships and conflicts with various Native American neighbors, while simultaneously noting that Native American and European lead mining technology was separate, and not the result of a hybridization of practices.[17]

Just like Thomas Jefferson conducted a written survey about the lead mining operations in Virginia, this book will survey the terrain of the lead mines to further understand how the early Native Americans and French miners, and ultimately American and British miners, amalgamated their prospecting, extracting, and smelting skills and knowledge to contribute to the international and local economy of *le pays des Illinois*.[18] French priests and explorers who experienced lead mining with Native Americans realized that their indigenous guides had developed longstanding extracting and smelting methods that began earlier than they otherwise understood. It is possible that French settlers arrived expecting to see a primitive culture and Native Americans who passively interacted with their environment. Instead they learned to depend on Native Americans not just for geographical information as guides but also for their contributions to the settlers' and miners' wellbeing. The first period of lead mining in this area was initiated by Native Americans. The second period is marked by Native American miners and French settlers or miners hungry to locate a valuable metal joining their techniques to extract lead ores. The third period sees Euro-Americans continuing the syncretic mining methods established during the second period, but it also marks stages of departure from intercultural interactions to more European technologies and commercial connections to mine and sell lead from this small part of North America.[19]

Therefore, understanding how Native American mining practices and knowledge encouraged French miners to reimagine how to conduct mining in a colonial outpost without financial support from Paris opens a window on how colonial–Native American history and mining created an alternative "middle ground," similar to the core ideas advanced by Richard White's *The Middle Ground*. White argues for a *le pays des Illinois* middle ground where peoples intermixed through trade, agriculture, and marriage. The middle of North America was not the frontier that Frederick Jackson Turner would have recognized with Native American and French mining traditions converging. Beginning in the early seventeenth century, as explorers, fur traders, and missionaries adopted Native American communication and cultural practices, so too did the Kaskaskia and French miners learn to work with lead ore using their respective environmental knowledge and practices together. The "middle ground" they forged at the mines was replete with syncretism and alliances; as a by-product of cohabitation, the French and Kaskaskia miners' practices became blended.[20] Indeed, in a different context, the research of Daniel Usner on the cultural milieu of early Louisiana complements White's research and represents a concerted effort to recast the customary American frontier narrative.[21] White and Usner, and others, have pointed out that the convergence of Native American and European practices carries the potential for highlighting frontiers and borderlands not as places of straightforward conquests but as places where invaders and indigenes constructed accommodations.

This amalgamation of techniques changed the trajectory and timing of the emergence of mining and smelting work that began with indigenous and French ideas and systems.[22] The meeting between Native American and French miners in the middle Mississippi valley fits, in some sense, in the story of American mining long before the western gold rushes. Faced with increasing numbers of French settlers searching for gold and silver during initial contact, the Kaskaskia set about to accommodate French miners by sharing when and where they mined lead and how they extracted and smelted lead ores. In mutual exchange, French miners agreed to share their tools, and together they created a new alliance centered on mining lead ore. These partnerships represent an alternative middle ground, akin to but also distinct from that described by Richard White in connection with the fur trade.[23] Along this mining frontier, the Native American and French miners employed a variety of customs to forge longstanding cross-cultural mining practices to exploit their environment in their quest for mineral extraction.

In 1763, with the transfer of the eastern bank to Britain and the western bank to Spain, the Mississippi River became an international boundary. Thereafter, the western side of the river would be called Spanish Illinois, and the eastern side became known as British (and later American) Illinois. The syncretic mining techniques were still being applied until the arrival of the American Moses Austin, which began the transition away from amalgamation of techniques to more European methods. Within a few years of these significant changes and Thomas Jefferson's publishing of *Notes on Virginia*, wherein he briefly mentions the lead district across the river from the village of Kaskaskia, Jefferson's actions would mark a change and the anticipation of a new century. Expectations of new growth and expansion spread from the East Coast of the United States to the banks of the Mississippi River. As the American population grew and traveled westward, settlers looked for affordable land. Simultaneously, dramatic events were unfolding that would change the geography of the United States. Ultimately, the nation's landscape, economic future, and new technological practices to extract and smelt lead ores in the Louisiana Territory would contribute to supplantation of the old practices.

Like many Americans, Jefferson believed that Louisiana would someday belong to the United States.[24] It was thought that control of Louisiana was a natural extension of the country and critical to access valuable natural resources. Although the lucrative fur trade was on Jefferson's mind, he continued to be interested in the lead mines just across the river from the village of Kaskaskia where approximately one hundred years earlier the Kaskaskia, French settlers, Creoles, African slaves, and now the American settler Moses Austin worked the ores "on the Spanish side of the Mississippi river," not far from present day Ste. Geneviève, Missouri.[25] Within a year of the Louisiana Purchase, the newly appointed commandant Captain Amos Stoddard traveled to St. Louis to establish the American government in Upper Louisiana. He subsequently reached out to Moses Austin of Mine à Breton requesting a summary of all the local mines.[26] The mineralogical assessments and experiments Austin had compiled were sent to Jefferson who noted in his 1804 message to Congress that "the lead mines in that territory (Louisiana) offer so rich a supply of that metal as to merit attention. The report now communicated will inform you of their state, and the necessity of immediate inquiry into their occupation and titles."[27]

The late eighteenth and early nineteenth century were not only a time of burgeoning mercantile growth and expansion of the United States but also a period marked by attention to the role of scientific and technological developments forged by natural philosophers whose writings reveal the messy world of eighteenth- and nineteenth-century scientific advances. Leaving their European or American homes, natural philosophers witnessed environmental transformations and produced some of the first detailed accounts of the state of nature in America.[28] Richard Judd argues that these early explorer-scientists' writings outlined the regions' rocks, climates, plants, animals, and most important, its minerals. As the early development of mineralogical studies in the United States continued, it coincided with the emergence of scientific mineralogy in Europe.[29] Like British and European mineralogists, the early American mineralogist Moses Austin was curious about what we now call "science." Before sending his report to the president, he meticulously observed, collected data, and experimented on minerals—metal ores—to produce knowledge for national leaders and his local business associates. British and European mineralogists would have characterized their observations as either natural philosophy or natural history, but their real motive was in pursuit of economically useful knowledge. Mineralogists collected specimens from mines and quarries, carrying what they had gathered home for further examination in their studies or laboratories.

As the scientific study of minerals increasingly became intertwined in elements of emerging science, natural philosophers searched for meaning in nature through observation, scientific speculation, and imagination.[30] Long considered by historians, and rightly so, to be the work of priests, public officials, fur traders, miners, engineers, glassmakers, and naturalists, the writings of eighteenth- and early nineteenth-century naturalists deserve more attention. For example, the French Intendant (government official) Marc Antonine de la Loire des Ursin, the mining engineer Philippe de La Renaudière, and the young French officer Antoine Valentin de Gruy all used observation, scientific speculation, and imagination to record their scientific encounters with nature and lead ores. Most of them derived a wealth of geographical information from Native American guides on their way to the lead mines. They each had good reason to acknowledge the Native American presence in their explorations. These guides displayed their environmental knowledge and technological skills when prospecting, extracting, and smelting lead ore, all of which the explorer-scientists noted and documented. Additionally, while Henry Rowe Schoolcraft and Henry Marie Brackenridge are two early nineteenth-century explorer-naturalists known for their ethnographic descriptions, they also surveyed and wrote about the lead mining terrain. Schoolcraft and Brackenridge introduced their concerns about forest conservation and protection of species living in the vicinity of lead ore furnaces that emitted harmful toxic lead fumes. For example, beginning with Spanish occupation of the Upper Louisiana lead mines, there was a significant impact on the local landscape resulting from an abundance of fuelwood consumption needed to fire stone furnaces similar to those used in Europe. To meet the demand for lead, large bundles of timber were required to simply smelt lead, which resulted in deforestation. As new settlers poured into the mining district, they observed a local landscape completely denuded of trees.[31]

By the eighteenth century, the term *mineral* referred to all the naturally occurring inorganic solid objects. Mineral also referred to any valuable ore mass. Toward the century's end, Abraham Gottlob Werner defined three aspects of mineralogy: identifying and cataloging minerals, the distribution of rocks and minerals, and the structure of rocks and minerals.[32] If early American mineralogists like Thomas Jefferson or Moses Austin were asked to define what encompassed the earth's solid parts, they would reply metals, earths and stones, salts, and sulfurs, and "minerals" organized in rock masses, strata, and veins. Like Georgius Agricola in the sixteenth century and Abraham Gottlob Werner in the eighteenth century, the work of the mineralogist was to describe the major mineral classes. Werner and other mineralogists of the Enlightenment all looked back in time to the German physician and natural historian Agricola as the founder of modern mineralogy. In *De Natura Fossilium* (1546), Agricola identified a metal like lead as a solid that melted when heated and returned to its original form once cooled. Aristotle also described "metals" in *Meteorologica III, Part 6* "as things mined" in essentially the same terms as Agricola.[33]

By closely following Agricola's method of observation and detailed documentation of local miners' techniques, the natural philosophers Antonine de la Loire des Ursin, Philippe de La Renaudière, and Antoine Valentin de Gruy went to the lead mines near Kaskaskia as mineralogists looking for lead or for their utilitarian value. Like Agricola, they not only observed Native American miners' techniques, they also worked the mines to study, to understand, and ultimately, to exploit minerals for the promotion of local and national wealth.[34] Similarly, the American mineralogist Moses Austin most likely developed his mineral analysis skills by reading Agricola and other European and British mining manuals. In addition, like many mining engineers, Austin and the British miners who helped him to establish Mine à Breton in Upper Louisiana developed their skills working as mining apprentices, where they learned mineralogy and smelting before arriving in Upper Louisiana.[35] Also referred to as "practical men," mineralogists understood the nature and distribution of minerals in their mining areas. However, lacking institutional support and the ability to publish results, these practical men learned to depend on wealthy professionals, or as with Austin, they used their business associations to exchange mineralogical knowledge. Additionally, to comprehend the nature and the distribution of lead ores in the Louisiana Territory, Jefferson needed Moses Austin to identify and catalog the minerals, which signals a cultural sea change for the improvement of a nation's economy.

By the late eighteenth century, as European states became increasingly interested in mining and metallurgical industries, in a number of locales of the young

United States, mineralogists also began to focus their attention on the study of mineralogy. Mining officials, merchants, miners, and national leaders all needed information about the location and properties of metallic ores such as lead, zinc, copper, and tin. In Europe, many trained physicians contributed to mineralogical knowledge, and in the United States, some physicians and laymen did the same. Additionally, many non-university-trained mining officials, merchants, and miners played a key role in the development of mineralogy. They tended to see improving mineralogy as a key to accessing government officials.[36]

Jefferson and Congress envisioned the Louisiana Territory's lead mines as a way to lessen America's dependency on foreign products made from lead and to establish the possibility of a dynamic domestic market for products such as musket balls and shot, sheet lead, white lead, and red lead. Moses Austin's *A Summary Description of the Lead Mines in Upper Louisiana* was the first American mineralogical survey of the Louisiana Territory and called for improving metallurgical methods at the Missouri lead mines. Austin discusses his confidence in English scientific advances and practices, which he believed had surpassed those of other civilizations. Notably, the mineralogical survey reveals the beginning of a transition away from Native American and French techniques as more miners adopted European experimenting, mining, and smelting practices. Like Europeans who believed their prospecting instruments to be superior to others, Austin also alleged that precise application of his apparatus would prove his techniques superior to local miners' methods. By the opening decades of the nineteenth century, both Europeans and Americans, especially coming from the East, considered their way of thinking and identifying the earth as profoundly different from the ideas and practices of the peoples they encountered in this new vast territory.

Sources

Information about explorations and eventual extracting, smelting, and trading practices at and beyond the lead mines fill French, Spanish, and American journals, memoirs, scraps of accounting sheets, census reports, scientific literature, mineralogical reports, medical reports, newspapers, and personal letters. As well, the reports of travelers, scientific observations, and physicians' reports communicated important facts about the changing conditions many faced in the lead mining environment. Understanding how people used a myriad of methods to produce lead means reading and rereading these sources.

The diaries, chronicles, bookkeeping records, population surveys, and technical writings left to us by European explorers are a set of rich sources revealing how mining lead was extracted, smelted, prepared for sale, and used in the Illinois Country and Upper Louisiana. Each of these sources also functioned as travel narratives revealing how people moved lead from the ground to the furnace and beyond. All reveal stories about the daily tasks associated with the social and culture aspects of lead mining. Desiring to find mineral wealth similar to what the Spanish had discovered in New Spain, Jesuit missionaries asked their Native American counterparts not only about fur trading, but also about the lead mines in the heart of the Illinois Country. Missionaries and ironmasters similarly described in their respective journals "discovering a very rich lead mine" only after Native Americans guided them to the location. The census report completed by French leaders at the request of Spanish officials after Spanish occupation of Upper Louisiana in the 1770s shows a continuing desire to see miners move away from using the Kaskaskia-French traditional mining and smelting amalgam and instead begin to adopt European mining practices in Ste. Geneviève's mining district.

Beginning in the 1770s, accounting sheets maintained by the French Creole François Vallé are used to highlight the significance of lead to the community at the center of Spanish Louisiana's administration—Ste. Geneviève. Vallé's and other Creole letter books narrate the movement of lead ore and recount early attempts at forming business associations at both Mine La Motte and a newly discovered mine at Castor Vein, while continuing to apply the Kaskaskia-French amalgam to extract and smelt lead ore. Their accounts consist of correspondence, bills, inventories, bills of lading, and transactions related to pig lead, which was regularly used as a form of payment and exchange to settle business accounts and worker contracts, and as a form of gifting to local Native American communities.

In 1804, as American leaders arrived in Upper Louisiana to begin to establish their government after the transfer of the territory from France to the United States, Thomas Jefferson wanted a mineralogical assessment of the new territory's lead mining district where the American settler and lead mining expert Moses Austin worked the ores since the late 1790s. Austin worked hastily to complete the first American mineralogical assessment of the mining district, which was then forwarded to the president. Austin's findings are noteworthy—he conducted experiments and told the president and Congress that "the mineral" was of superior quality and that "in the hands of skillful smelters" could produce "sixty and in some veins seventy percent" metal at two of the most significant mines, Mine à Breton and Mine La Motte. Since the late eighteenth and early nineteenth century was a time of burgeoning mercantile growth in the United States, Austin's report also directs attention to the role of science in westward expansion as American natural philosophers journeyed from their homes returning with new environmental knowledge.³⁷ These natural philosophers reformed science as they investigated nature with the emergence of scientific mineralogy in the United States.

Divergent sources contain some fascinating observations about connections between health and place from 1700 to 1840. As miners began to convert their mining and smelting practices from the Native American and French hybrid to the more European and American methods, material changes to the landscape and human bodies also began to occur. The western expansion into the mining district promoted a growth in the number of villages as American settlers arrived, which also called for doctors to ply their services among miners. By 1816, doctors began examining miners, diagnosing miners with lead poison, and documenting early accounts of the region's unhealthiness in connection with new lead production techniques. Dr. Hardage Lane of St. Louis, Missouri, published the first medical report of lead poisoning in the newly established *St. Louis Medical and Surgical Journal* after tending to a thirty-year-old male slave at the lead mines. While doctors' medical reports framed the work of mining and smelting as unhealthy and dangerous, new geological reports promoted the region as healthy. Medical and geological reports, which collectively made settlers and doctors recognize and define the environment as unhealthy for settlement, help us to ask new questions about land unhealthiness as lead production increased in scale.

Chapter Layout

This narrative moves chronologically from 1700 to 1840 and is divided into four chapters. Chapter 1 follows the arrival of the Kaskaskia Indians and French settlers to the middle Mississippi valley and the convergence of their prospecting, mining, and smelting practices. A detailed explanation of how longstanding Native American prospecting, extracting, and smelting practices converged with French methods on this lead mining frontier is presented. The amalgamation reveals the influence of Native American knowledge and practice, as opposed to only impositions of European mining applications. Amalgam is not only a metaphor for the melding of techniques but also for the process that united Native Americans and Europeans to create a mining middle ground.

Chapter 2 presents the next phase of the story, as miners continued to employ the syncretic methods designed by Native American and French miners to become successful in mining and smelting endeavors. The chapter highlights how Spanish officials made a number of efforts to promote changes in administration in accordance with the Bourbon Reforms in the late colonial period (circa 1760 to 1810). To increase economic productivity, especially in mining, Spanish officials arrive with new ideas to boost the lead mining district by introducing new innovations associated with digging deeper shafts, constructing stone furnaces, and allowing skilled American and British miners to immigrate into Upper Louisiana. This period may be considered as the start of a "revival" in mining in which Spanish officials are credited with the rise in production. However, the evidence shows that the mining industry continued to expand as ordinary miners played a role in increasing production while still using the long-established Native American and French mining methods.

Chapter 3 continues the mining narrative in the months following the Louisiana Purchase when Captain Amos Stoddard requests that Moses Austin, the natural philosopher and businessman, complete the first mineralogical and chemical assessment of the region's lead ore for national development. The miners' experiences with chemistry and mineralogy would encourage the flow of lead from Mine à Breton and Mine La Motte to local and urban spaces. Also developing near the two primary mining sites, Mine à Breton and Mine La Motte, were new factories to manufacture shot, sheet lead, and red lead. However, with the increase in smelting and production of products, local families became concerned and troubled by the sulfur and arsenic polluting their water. Over time, the buddles that were built to wash lead ores with water produced toxic liquid and toxic sludge polluted with arsenic, mercury, and sulphur.

Chapter 4 describes the years leading to Missouri statehood in the 1820s, as promoters of lead mining successfully encouraged American and European miners possessing European skills to relocate to the mining district. American geologists and entrepreneurs desired to see miners extract and smelt lead ores efficiently by employing what they considered to be sophisticated knowledge and complex machines. Most significantly, improvement became the motivating force as promoters and boosters hoped to impose power over nature and transform the land. These changes did not just affect how Americans viewed the landscape; their vision included industrializing the mining district. New technologies and practices enabled miners to dig more shafts, build more furnaces, and export more lead. Miners now combined their mining and farming operations near their settlements year-round instead of following a seasonal schedule, which was the practice of Native American miners. Additionally, as travelers, geologists, and miners made their way west to inspect the lead mines, they combined regional definition and regional promotion as a way to recognize regional resource potential. Most significantly, in the mining district, after Americans installed their structures, settlers began to assess the healthiness and unhealthiness of their farms and settlements.

CHAPTER 1

Early Mining and Smelting in North America

Introduction

In September 1700, about twelve hundred Algonquian-speaking Kaskaskia people and the French Jesuit naturalist Father Jacques Gravier began their migration into *le pays des Illinois* close to the Mississippi River. *Le pays des Illinois* (Illinois Country), which became the center of North America, was the region east of the Mississippi, and the region on the western edge was named *Louisiana* by René-Robert Cavelier, Sieur de La Salle. When the Kaskaskia and Father Gravier reached the eastern bank of the Mississippi River, they were directly across from the site of the present city of St. Louis, Missouri. Also assembled there was a community of some two thousand Tamaroa, Cahokia, other tribes of the Illinois Confederacy, and French settlers and traders.[1] Within a few weeks, on October 9, Father Gravier, a few French traders, and a small number of Tamaroa Indians traveled along a network of Native American trails, and after arriving at the Meramec River, they boarded five canoes. After traveling two miles, one of the Native American guides pointed out the location of a lead mine. The following day, Sunday, October 10, Father Jacques Gravier noted in his journal, "We discovered a very rich lead mine near the Meramec River, and the ore from this mine yields three-fourths metal."[2] Gravier no doubt heard about the Meramec mines from the tribes of the Illinois Confederacy. Although it was silver that the French were seeking, Gravier was pleased that the discovery of lead would contribute to national wealth, and he went about collecting, examining, and measuring the quantity of lead ore in the samples he extracted.[3]

The location of the mines visited by Father Gravier appear to be in the vicinity of what is now Old Mines in Washington County near the present-day Big River, which was considered part of the Meramec in the eighteenth century. Within a few years of the Tamaroa directing Father Gravier to their lead mines, French settlers established a colonial outpost in 1703 close to the Kaskaskia village in *le pays des Illinois* between the Missouri and Ohio Rivers in the vicinity of the present-day town of Kaskaskia on the eastern edge of the Mississippi River. It was here that the Tamaroas, Kaskaskians, French naturalists, fur traders, and settlers began to forge an alternative middle ground of mutual dependence.[4] In addition to exploiting the region's furs, after Native Americans guided the French to their lead mines, together they mined, smelted, and exchanged lead. By the time of Father Gravier's arrival at the lead mines, Europeans were manufacturing lead into musket balls, inflicting wounds and death to many.[5] Beginning with Father Gravier, all naturalists and subsequent French mining engineers sent lead ore samples and dispatches from Louisiana to Paris to highlight the vast mineral wealth of the Illinois Country. Eventually, the lead mines that Thomas Jefferson mentioned in *Notes on Virginia* would be discovered by Antoine de la Mothe Cadillac. With a small Tamaroa Indian party guiding them, Cadillac, his son, a few leading Kaskaskians, and a group of soldiers and laborers, including most likely some African slaves departed from the village of Kaskaskia followed small streams and well-marked Native American trails to the site of the Native American lead ore diggings, that Cadillac named Mine La Motte in 1715 in the state of Missouri.[6]

The lead mining regions in the Mississippi valley were known to the French as early as 1700, demonstrating the importance of lead ore to explorers, entrepreneurs, and settlers in *le pays des Illinois* throughout the eighteenth century. As French settlers continued to arrive, the Illinois Confederacy miners guided them across the Mississippi River revealing additional mining sites. Similar to the Gravier and Cadillac expeditions, they all traveled across overland trails and on local river systems until they arrived at the lead mines in *le pays des Illinois*. These early eighteenth-century mining expeditions are clearly representative of the importance of lead to French explorers, natural philosophers, and mining experts who established intercultural relations with Native Americans; however, they were not the first to navigate and explore this region in search of lead mines near the Mississippi River.

Native Americans had been extracting lead from these mines for centuries. Local Native Americans knew and worked lead ore, which they had treasured for its decorative qualities since the early days of the Mississippian mound builders.[7] The Illinois Confederacy miners who guided Father Jacques Gravier and Antoine de la Mothe Cadillac to their lead mines were not the first Native Americans to extract and smelt lead ores in the Mississippi valley. For centuries, Native Americans moved up and down the Illinois, Missouri, Ohio, and Mississippi Rivers traversing the interior landscapes of the continent creating formidable alliances with other

Native American tribes. Endeavoring not to erase North America's precontact history by falling into the trap of omitting indigenous people's cultural practices, it is important to remember that just as the Kaskaskia, French Jesuit missionaries, French fur traders, and French settlers explored and settled in *le pays des Illinois*, so too did early Native Americans enter this region from distant places at varying times. Both groups carried traditions into the continent's center bordering the Missouri, Ohio, and Illinois Rivers near the convergence of the Mississippi River.[8] The history of eighteenth century Native American and French settler lead mining in *le pays des Illinois* did not begin with French naturalists, traders, trappers, missionaries, or the Kaskaskia.

Lead mining history in the Middle Mississippi valley begins long before Europeans migrated to *le pays des Illinois*. Upon the arrival of the Kaskaskia and Father Gravier, they were greeted by Native Americans living in a number of villages along the east side of the Mississippi River. Since the Kaskaskia and Illinois both spoke Algonquian, they learned to exist in a relaxed confederacy that included the Cahokia, Peoria, Michigamea, Moingwene, and Tamaroa. With no official political structure, each tribe became part of the confederacy spreading into parts of what are now Iowa, Missouri, and Wisconsin, inhabiting over sixty towns. Their neighbors the Sac, Fox, and Shawnee lived to the north, south, and east of the Illinois tribes. Across the Mississippi River in what is now Missouri, the Siouan-speaking Iowa, Missouri, and Osage had an established presence.[9]

The initial attraction to this region for many of these Native Americans and their ancestors was an environment rich in wildlife. This is not to suggest that when Native Americans dominated the Mississippi valley it was a beautiful and natural Garden of Eden where precontact Native Americans did not inflict the same harm on their environment as later non-Indian settlers eventually would. In fact, they did alter the landscape around them in important ways.[10] For example, by 8000 BC, human hunter movements and climate change resulted in the extinction of a number of large mammals on which peoples had come to depend. Following the thinning of local game, some people moved, while others adapted to their environment by selectively burning forest to improve their yields from hunting and gathering. These prehistoric actions promoted the mosaic quality of forest that greeted early French settlers.[11] Similar to other indigenous communities, Native Americans practiced fire burning to extend the edge of a grass and forest border until it penetrated the drier forest. This pushed back the edge, thereby allowing grasslands to advance over hundreds of years.[12] Even nonagricultural Native Americans expanded what may have been pockets of natural grasslands at the expense of forest through annual burning.

Eventually, as the seasons turned cooler and wetter and with the introduction of seed crop cultivation around 300 BC to AD 300, ceremonialism developed among Mississippi valley peoples. By AD 900, after Native Americans experienced several centuries of wet and cool years, a more complex culture emerged. The change in climate supported maize cultivation that spread from Mesoamerica to North America and became the centerpiece of Mississippian culture. At the culture's peak, Mississippian sites stretched from the Atlantic Ocean and the Gulf of Mexico north to the Ohio River valley and west beyond the Mississippi into present-day Oklahoma. As the Mississippians cultivated maize, the population expanded and became increasingly dense.[13]

In another effort to alter the environment, by the tenth century the Mississippians started moving earth and engineering their landscape on a grand scale that would continue through much of New World prehistory. Groups began to construct large earthen works in the Mississippi and Ohio valleys and across what is now the southern United States. Large quantities of both earth and stone were transferred to create various raised and sunken features, such as agricultural landforms and settlement and ritual mounds. Mounds of various shapes and sizes were constructed throughout the Americas for temples, burials, settlement, and effigies. The most immense mounds were located on the west bank of the Mississippi, just south of the Missouri River, near the site of Cahokia north of Kaskaskia, where 104 earthworks would later be called "Mound City." These mounds were considerably smaller than the 120 dirt pyramids located across the river in Cahokia, which during its peak years became the largest Mississippian site in eastern North America and the largest urban center north of Mexico.[14] Sustaining a dense population required abundant food resources, and maize became the most important in Cahokia diets. Additionally, they depended on resources from neighboring regions.[15] The Mississippians in Cahokia had long depended on exchanges not only with over fifty nearby villages that stretched south near present-day Ste. Geneviève but also with people at considerable distances along the Mississippi, Missouri, and Illinois Rivers and along overland routes to the west. Cahokia essentially became the primary crossroads of long-distance exchange.

Along these same well-worn pathways, the Mississippians of Cahokia, Illinois, established trade networks with the Zuni and Pueblo and may have also assimilated their lead mining and lead ore smelting practices. By the year AD 300, in the middle Rio Grande Basin, the Zuni were already working with glaze paint, which may have been central to a lead-mining enterprise that continued through the glaze-ware period. The Pueblo began to produce glaze paints for decorative motifs on their pottery. By grinding the ore into a fine powder, miners were able to mix it with plants or other minerals in order to manufacture a particular color of glaze paint. The Zuni may have transferred their glaze-ware knowledge to the Pueblo, and then the Pueblo likewise exchanged their skills with Native Americans living along the Mississippi River. Following routes across the Rocky Mountains to the Great Plains, indigenous people's glaze-ware knowledge and mining practices would come to influence Native American communities in the center of North America. After the Mississippians discovered lead deposits in their region, they started mining ore to make glaze paint for their pottery. It is unfortunate that we cannot ask the original miners to describe their skills or how they accumulated and passed their knowledge about mining onto those who would follow in their footsteps. However, it is likely that the lead mining practices of Mississippian societies came out of indigenous practices elsewhere in North America.[16]

Known for their well-developed horticultural production of maize and beans, the Mississippians coordinated their seasonal cultivating schedule with their lead mining activities. Following the planting of seeds, miners traveled on foot and by streams and water arriving at the lead mines to begin prospecting and extracting the lead ore. In many cases, if the Mississippians located galena, they would begin to grind it into powder, or they would begin to melt the lead ore on-site near the mines. Mississippian archaeological sites show evidence of Native Americans grinding galena crystals into powder to manufacture a metallic blue-gray pigment. Mississippians also exchanged galena in the form of crystals, nodules, and shiny cubes, which they used to bury with the dead.[17] Researchers have discovered 500 pounds of lead in multiple forms from burial grounds.[18] The galena discovered at Mississippian worksites aids our understanding of how they and later Native Americans practiced crushing and grinding lead ore to manufacture paints or ceremonial powder. In some cases, the Mississippians also traded either the powdered or melted lead with other native peoples. A longstanding valued commodity for Native Americans, the galena extracted from the Mississippi valley mines entered interregional exchange, making it one of the original North American lead-mining centers centuries prior to European contact.[19]

Lead is a metal that has been of great practical use to civilizations both in death and in life, and lead has had a great usefulness as a constituent of pottery, shiny glass, and paint.[20] Similar to the North American Pueblo and Mississippian peoples, Assyrians and Egyptians used galena to glaze pottery. Moses mentioned it, and there is evidence that the Assyrians and Egyptians and nearly all ancient people knew and used lead.[21] Evidence exists of the use of paint as early as 2500 BC to decorate Egyptian tombs. In Egypt, early interior decorators received most, if not all, of their commissions for making gorgeous resting places for the dead.

As with the Mississippians, early civilizations used lead when burying the dead. Some may recall Portia and her three caskets—one of "gaudy gold"; one of silver, the "pale and common drudge between man and man"; and a third of "meagre lead, which rather threatens than dost promise aught."[22]

Easily cut with a pocketknife or dented with a fingernail, the extremely malleable lead ore could be shaped in catlinite molds; made it into pipe, rolled it into sheets, without breaking. In addition to manufacturing lead into musket balls, European merchants and fur traders molded lead into baling seals to identify goods with information such as the location, manufacture, size, and quality of the textiles and furs contained in the cloth bundle or wooden barrels shipped back to British or European cities.[23] The watertightness, flexibility, and bonding strength properties of lead also made it useful in Europe for building roofs to carry off the rainwater. In Europe, artisans had a longstanding practice of producing beautiful ornamental leadworks. Once in North America, French settlers most likely recalled the fine specimens of ornamental leadwork on the cathedrals of Amiens and Rheims and had heard about the leaded dome constructed in 1553 on Barnard's Inn Hall in London. After arriving in *le pays des Illinois* in 1700, the Kaskaskia and others of the Illinois Confederacy learned to extract, grind, melt and use lead prior to European contact in the Mississippi valley. After the Kaskaskia, Father Gravier, and French fur traders established their new Kaskaskia Village in 1703, lead still represented an important trade commodity.[24] The extremely malleable lead ore was polished and shaped into cubes and rectangles; fashioned into perforated beads, bird effigy pendants, and baling seals using catlinite molds; or made into musket balls using a musket ball casting device.[25]

While it is possible that the techniques used by the Tamaroas and Kaskaskians miners to extract and smelt lead ore originated with the Zuni Pueblo peoples and Mississippians, Father Gravier and later French natural philosophers interested in mineralogy, mining, and the production of metals all looked back in time to the German physician and natural historian Georgius Agricola. He was a mineralogist and metallurgist born in Saxony of the Holy Roman Empire. Agricola entered the University of Leipzig at the age of twenty, finally earning his degree in medicine. Primarily a physician, Agricola became a court historian in 1530 and a city physician in 1533. Agricola's medical training helped him to become a profound observer of the nature of mines he investigated in the sixteenth century, as he developed a keen interest in prospecting, mining, and smelting of ores. Agricola's magnum opus is a series of six works on mining, metallurgy, geology, and animals used in mining.[26] The well-known pioneering work, *De Re Metallica, On the Nature of Metals [Minerals]*, was published posthumously in 1556. Together, the twelve volumes are a comprehensive systematic study, classification, and methodical guide on all practical aspects of mining, the mining sciences, and metallurgy. Agricola's written analysis was based on direct observation and site investigations in the mining environment.[27] *De Re Metallica* along with *De Natura Fossilium*, also written by Agricola, are two of the earliest comprehensive "scientific" approaches to mineralogy, mining, and geological science.[28]

As a trained physician scrutinizes the human body, Agricola documented the mining activities he surveyed being practiced by miners, who were also contemporaries of the early Native American miners in the Middle Mississippi valley. Agricola recorded every aspect of the miner's day as he surveyed, read, or listened at multiple mining locations. Agricola identified lead ore as a solid that melted when heated and returned to its original form once cooled. Long before Agricola, in *Meteorologica III, Part 6,* Aristotle also described "metals as things mined" essentially in the same terms as Agricola.[29] Depending on the knowledge conveyed by Agricola and early natural philosophers who were interested in minerals and metals, the eighteenth-century mineralogist Father Gravier also examined metal ores and the occurrence of veins for their utilitarian value. This governed the attention of mineralogists. Like Aristotle and Agricola, the natural philosopher Father Gravier and those miners who would follow viewed mineralogy as a tool to study, understand, and ultimately exploit minerals to promote local and national wealth.[30]

Agricola's *On the Nature of Metals [Minerals]* would serve as the standard reference for natural philosophers, miners, and mineralogists for two hundred years. In essence, centuries before its time Agricola carefully followed principles of the modern scientific method. To describe the major mineral classes of various ores, he practiced the art of assaying ores to examine their many chemical substances.[31] According to Agricola, the purpose of miners assaying lead ore, or any mineral, was first of all to determine whether some metal was present in the ore. Secondly, the assayer needed to determine what proportion of one or more metals might be in the lead ores being examined. Recall when Father Gravier documented in his journal not only the importance of discovering a rich lead mine near the Meramec River but also the importance of assaying the lead ores, which he learned yielded three-fourths metal.[32] After taking a few samples from the lead mines west of Kaskaskia, he proceeded to test a small amount of lead ore for its quality and measurable content to determine its value. Father Gravier's practice shows a similar type of detail to Agricola's methods, highlighting the extreme care necessary in the assaying process. Natural philosophers, miners, and mineralogists understood that any error made in the measurement of the quality of ores would be multiplied many times over when the bulk of ores was smelted. An incorrect analysis would lead to mine owners losing valuable metal during the smelting process.[33]

Following Gravier's mineralogical assessment in *le pays des Illinois* lead mines in 1700, he sent his findings back to France, and subsequently, additional skilled miners and natural philosophers crossed the Atlantic, entered the Gulf of Mexico, and traveled up the Mississippi to the same lead mines. Like Agricola in Saxony, they began to document how a nonliterate people found and exploited minerals. It is unfortunate that early Tamaroas and Kaskaskians miners left no written record describing their mining skills or how they acquired their prospecting and smelting knowledge. However, similar to Agricola, who was the first to write down in detail the kinds of mining activities and technologies Saxon miners put into practice, eighteenth-century French natural philosophers also observed and documented in their mineralogical reports the longstanding prospecting, extracting, and smelting skills practiced by Native Americans.

The first French natural philosophers to include in their reports Native American mining techniques were Antonine de la Loire des Ursin, a gentleman "skilled in mineralogy," and Sieur de Lochon. Both arrived in Kaskaskia Village in 1719 and were guided by members of the Illinois Confederacy—Tamaroa who also accompanied Father Gravier in 1700 and Antoine de La Mothe Cadillac in 1714—to their lead mines where they described in detail Native American mining techniques and the mining landscape. Des Ursin's "Relation of the Journey to the Mines of Illinois" describes the route to the mining region and how the Tamaroas prospected and extracted lead ore. Additionally, in 1723 the French mining engineer Sieur Philippe de La Renaudière wrote the second detailed account of Native American mining in the region, noting its seasonal practice. La Renaudière also begins to outline in detail how the local environment adjacent to the mines offered the possibility for establishing the first French mining settlement and where European mining operations could be installed.

The final extant report comes from Antoine Valentin de Gruy who in 1741 just prior to getting married and buying a home in Kaskaskia, sought the assistance of Native American miners. He requested guidance to Mine La Motte. Like La Renaudière, De Gruy also describes aspects of the local environment in the mining region and offers the first descriptions of Native American lead smelting techniques. Additionally, De Gruy noted how French miners worked one season per year using "primitive methods" and expressed his concern about the changing forest landscape as the mining region's trees were being stripped for smelting lead ore. De Gruy's analysis highlights how miners used exploitative methods such as the digging of numerous trenches in the earth and the use of timber for smelting. The reports by

French mining engineers Des Ursin, La Renaudière, and De Gruy offer a window into Native American mining knowledge and practice in the region. Due to the lack of financial backing from French officials, French miners who settled near the Kaskaskians—in a small area of the vast territory between New Orleans and the Saint Lawrence, lying in the basin of the Mississippi River—learned and adopted Native American lead ore mining techniques, thereby creating an alternative middle ground.[34]

Long before citizens of the United States migrated to and settled in the Louisiana Territory in search of minerals, and before President Thomas Jefferson received the first mineralogical report about the lead mines from Moses Austin in 1804, Kaskaskian and French miners created a middle ground centered on mining. Together they engaged in extracting, smelting, and trading lead ore to sustain and defend their communities. Native American and French mining techniques continued to be applied even after the Spanish gained control of the Upper Mississippi valley in 1763. Successively, following the Louisiana Purchase citizens of the United States migrated to the Louisiana Territory to extract and smelt lead ore. A close reading of early eighteenth-century mineralogical reports left to us by Des Ursin and De Lochon, La Renaudière, and De Gruy underscore the significance of lead ore to Native Americans, and to French and American settlers. These French observations also note a longstanding tradition of Native American extracting, smelting, and molding lead ore in the Middle Mississippi valley. Traditional practices that French miners adopted made it possible for lead products to flow through Native American and French communities of New France where the village of Kaskaskia became the earliest site for the production and exchange of both Native American and European lead products following contact in the Mississippi valley.

Native American mining knowledge and practice also filtered into the cartographic catalogs of European geographers.[35] In addition to French miners and natural philosophers' mineralogical reports being used as tools to inform potential French settlers of mining opportunities in *le pays des Illinois*, cartographic reproductions also were used as marketing tools to alert settlers and merchants contemplating commercial ventures associated with lead ore in New France. By the early eighteenth century, Jesuit missionaries, explorers, travelers, and prospectors documented Native American mining traditions not only in mineralogical reports, but they also conveyed the locations of the lead mines in visual sources. The maps from the period of exploration passed through the hands of explorers to French cartographers in Paris who created and published maps during a period that represents the collection of Native American environmental knowledge and the potential for mining activity. The first map to reveal the exact locations of the mines appeared in 1703 and was authored by the French cartographer Guillaume Delisle.

Guillaume Delisle, who never visited the New World, accomplished his early eighteenth-century map depicting the Native American mines by interviewing explorers who returned from New France to Paris. As was the practice of early European mapmakers, Delisle collected firsthand accounts of *le pays des Illinois* geographical and environmental knowledge from soldiers, missionaries, fur traders, and explorers who worked closely with Native Americans. Or the environmental knowledge reached him through interviews with leading figures familiar with Father Gravier's exploration of the lead mines in the fall of 1700. When Gravier was introduced to the lead mines adjacent to the Meramec River by members of the Illinois Confederacy, he was on his way to meet two officers and explorers in the colony along the Gulf Coast, Jean-Baptiste Le Moyne de Bienville, the founder of New Orleans, and Pierre Le Moyne d'Iberville. Although early attempts by Iberville to establish a colony in the Mobile area had failed, unofficial attempts of missionaries, traders, and explorers already proclaimed a French presence in the Mississippi Basin by 1702. It appears that Iberville traveled to and from France multiple times during his sojourns in North America; and prior to his final departure from Louisiana in April 1702, he met Father Gravier, who remained in the colony in Mobile until February 1702 before returning to the Illinois Country. Prior to his return to *le pays des Illinois*, Gravier conveyed his knowledge of the Native American lead mines to Iberville before he departed for France. After arriving back home in France, Iberville maintained his interest in Louisiana. In fact, he was called on for counsel on several occasions by a number of ministers. He repeatedly relayed his theory for French expansion in America, which most likely included Father Gravier's written observations about the Native American lead mines.[36] Communication of the locations of Native American villages, rivers, and mining landscapes to French officials would also have been conveyed to Delisle who began his work producing a detailed map of the still-elusive Mississippi valley.

The Delisle maps, which circulated throughout France in order to guide newly arriving explorers, began the visual reproduction of the Native American lead mines' locations.[37] In his 1703 map and subsequent maps, Delisle clearly showed major rivers, tributaries, and streams leading to the lead mines. Delisle and Jacques-Nicolas Bellin, another French mapmaker, described the region as the *country full of mines*, a designation that appeared after hearing about Father Gravier's lead mining experience and reading the Des Ursin, La Renaudière, and De Gruy accounts. Beginning with the French cartographer Delisle Native American environmental knowledge and mining practices were transformed into visual form on his 1703 and 1718 maps. Later in 1744, Bellin continued this practice by representing Native

FIGURE 1. Thomas Kitchin *Louisiana Map, 1765*. It depicts the former French Louisiana Territory, including Tamaroas and Kaskaskians Villages, and "country full of mines" is labeled.

American mining knowledge and practices with his maps. Following the publication of both the Delisle and Bellin cartographic reproductions of the *country full of mines* in the Mississippi River valley, knowledge of these early Native American lead mines circulated throughout Europe. As French miners and entrepreneurs continued to extract, smelt, and trade lead ore in the years after Louisiana came under the control of the Spanish, the British mapmaker Thomas Kitchin also labeled the region as a *country full of mines* in 1765 (figure 1). Eventually in the late eighteenth century, around the time of Thomas Jefferson's *Notes on Virginia*, Spanish officials began to encourage American miners to settle near the mines and begin extracting and smelting lead ores in what became the state of Missouri.

Discovering Lead Sources

Many are familiar with the travels and writings of René-Robert Cavelier, Sieur de La Salle, Louis Jolliet, and Father Jacques Marquette, however, we have overlooked the mining accounts of Des Ursin, who was also was accompanied to the mines by De Lochon, "a gentleman skilled in mineralogy." The Des Ursin mining expedition also included two Frenchmen from Kaskaskia and their five slaves; an officer with six soldiers; and five Native Americans, two of whom were Tamaroas and three of whom were Kaskaskians. Additionally, the mining account of La Renaudière, is also considered here as both were official reports that offer a perfect picture of the mining country more than three hundred years ago. Another report written by De Gruy in 1743, provides a picture of how miners not only extracted lead ores, but also how miners smelted the ores. A few months before Native American miners guided La Renaudière to their lead mines, the French mining engineer Sieur Marc Antonine de la Loire des Ursin wrote an account based on his observations of the same mining landscape introduced to Father Gravier by members of the Illinois Confederacy. Both the Des Ursin and De Lochon account and the La Renaudière mining narrative revealed to explorers, entrepreneurs, and miners the region's mineral wealth, but each account also offered insight into longstanding Native American extracting and melting techniques that became culturally noteworthy for their assimilation by French miners. Together the Native American and French miners from the community of Kaskaskia carried prospecting and extracting knowledge with them to Mine La Motte, which highlights the value of lead ore to early eighteenth-century settler and indigenous communities. Most significantly, La Renaudière's account foreshadows how the Native American and French miners at Kaskaskia amalgamated their mining practices to jointly prospect, collect, assay, and smelt lead ore on the frontier.[38]

With the discovery of abundant lead mines, the French began to produce lead baling seals using catlinite molds. Prior to the early years of French settlement, Native Americans sold, traded, or offered lead and fur products as presents.[39] By the early eighteenth century, French settlers and native peoples were deeply committed to the business of fur trading. It was just one of the mainstays of the French colonial economy. Lead seals were the colonial equivalent of modern merchandise tags, which conveyed information concerning the circulation and taxation of goods by marking everything from packaged tobacco to bundles of various trade goods like furs. Lead, a soft metal, easy to mold through applied pressure, was an ideal substance for these objects that were fixed to bolts of cloth in order to testify to the quality of the material and to signify that it was ready for market.[40] Lead was also an important commodity for frontier settlements and homes, which required musket balls for protection and gifting. A type of small projectile fired from muskets, musket balls were usually made of lead by pouring molten lead into a musket ball mold and trimming off surplus lead once it had cooled. In fact, one villager was responsible for manufacturing musket balls.[41] As a native currency, little blue-gray pellets became increasingly significant in the form of musket balls for the protection of French and Native American frontier homes and for baling seals for the fur trader.

After hearing about the abundant lead mines located in the vast unsettled territory just west of the Mississippi River, and seeing one of Guillaume Delisle's maps on November 15, 1718, the former ironmaster Philippe de La Renaudière of Maubeuge and his wife, Perrine Pivert, boarded the ship *Comte de Toulouse* to begin a long strenuous journey across the Atlantic from France to New Orleans and ultimately up the Mississippi River to the Illinois Country.[42] The ship's attendance roll shows him as one of the employees of the Company of the West. The couple was listed as Renaudière—the "Clerk of the Company and Conductor of Mines and [also] his wife." Answering the request of the French regime, and given the title "mining engineer," La Renaudière went to the French colony to prospect for and exploit the lead ore where Native Americans had mined many years prior to European contact. While it is unclear why they remained in New Orleans for a lengthy stay, sometime in 1723, after La Renaudière became the "Commissioner of Mines" for the Provincial Council, he and his wife departed New Orleans for Kaskaskia. Sometime prior to August 1723, La Renaudière and a group of Native American miners began a forty-mile journey to Mine La Motte, and along the way he documented what the environment could offer potential settlers interested in mining and farming. Only weeks prior to the birth of his daughter, Marie, on August 23, 1723 La Renaudière noted in his description of Mine La Motte that "on the other side of the St. Francis river a large number of mountains may be seen the color of whose stones gives strong indication of mineral wealth." Carefully following his Native American guides, La Renaudière planned to collect and assay lead minerals and to survey the local landscape for the possibility of establishing a settlement near the mines. He conveys his findings to the French Company of the Indies about the mining landscape's economic viability in his mineralogical report.[43]

In his overlooked account, La Renaudière describes what his Native American guides wanted him to see and writes, "After we penetrated rock and found a little silver ... we found sure indications of the presence of copper, mingled with veins of lead one half foot in thickness." He continues, "As one begins to dig in some places, the mineral is only one foot below the surface, where pieces of lead weigh from 20 to 30 ounces." Using his mineralogist skills, La Renaudière notes, "In the location where the veins are well formed, the mineral is found to be good." As a mining engineer, after he collected the lead ores, he now proceeded to assay the minerals. After melting pieces of lead ore, and conducting a few calculations, La Renaudière was capable enough to determine whether the lead in the region would "produce as much as forty-five percent" after smelting. La Renaudière continues his report noting how much lead could be extracted from a bulk of lead ore. He also informs readers about the number of workers and slaves required—to "extract, melt, and refine about 10,000 pounds each month by the steady work of eight miners ... and [with] thirty Negroes placed under their management, [lead] can be delivered over land and by water."

Always aware of the need to capture the attention of French officials, La Renaudière describes the location of the mines as the ideal place to establish a settlement by including specifics about the landscape. He notes, "In the neighborhood of the mines there are small prairies [to] serve as retreats to numerous horned buffalo, bears, and deer. There are many persimmon, plum, and pecan trees and grapes; these fruits furnish nourishment to the Indians. . . . This is very fine country, it would be a good place for settlements; there are rivers in which fish abound and a large number of fowls." La Renaudière envisioned how a good living could be made, "and approximately three hundred million pounds of lead each year" could be extracted, melted, and delivered throughout the Illinois Country to Kaskaskia, Fort de Chartres, and New Orleans by land and water for economic development.[44]

La Renaudière not only observed how the Native Americans travelling with him extracted and melted their lead ores, he also documented how they molded and designed items in catlinite molds.[45] He writes, "About one and one-half feet down to procure mineral from these openings, they are like trenches, one must follow the veins which have quantities of lead mineral ... after which they [made] lead."[46] After Native peoples extracted and melted their lead, they manufactured and designed items using catlinite molds.[47] Catlinite, also called pipestone, is a type of metamorphosed mudstone, brownish-red in color, which is fine-grained and easily worked. Native Americans in the Mississippi valley used it to make ceremonial peace pipes or calumets. Additionally, native peoples made lead ornaments in these catlinite molds, shaping the clay into small triangles, circles, or earbobs. French settlers also came to refer to catlinite as pipe clay and admired its characteristics, being fine-grained and malleable for molding various shapes and sizes before it hardened. Eventually the French also integrated the use of catlinite molds to make important European religious symbols and trade products, crucifixes and baling seals, respectively.[48]

When evidence supporting prehistoric mining was initially discovered, researchers were puzzled about how early Native Americans discovered lead deposits buried and hidden from direct observation. Des Ursin and La Renaudière both suggest that the important and foremost step occupying the minds of miners was obviously locating lead ores. Significantly, Des Ursin and La Renaudière observed Native American prospecting techniques that revealed their knowledge of the environment.⁴⁹ Agricola believed that experienced miners knew how to find veins from surface indications, and therefore, it mattered when miners searched for lead ores and when those mines would be worked. It was June and August, respectively, when Des Ursin and La Renaudière traveled to the lead mines, and based on the condition of the vegetation, the Native Americans had developed an acute sense of how to determine the location of minerals. They looked for clues about the location of veins by carefully searching for discolored vegetation. La Renaudière documents how indigenous peoples followed the color of plants, stones, and soil to locate minerals. He describes how Native Americans guided him to "a large number of mountains where the color of plants and stones gives strong indication of mineral wealth."⁵⁰ After being told by his Native American guides to dig a trench in the location of the faded vegetation, La Renaudière confirmed Native American acumen when he discovered a vein of galena running underneath the unhealthy plants. This event reveals how Native Americans understood environmental indicators. Moreover, La Renaudière recognized that Native American miners applied the same methods that Europeans had to discover lead ore. Both parties' prospecting skills included following environmental markers in open spaces that lacked clusters of plants in rows, thereby discovering a vein of ore underneath the soil's surface.

When the ancestors of the Winnebago (or Ho-Chunk) first made their homelands in Wisconsin, Minnesota, Iowa, Missouri, and Illinois, they built thousands of effigy mounds during the Late Woodland period. The Winnebago also applied the same methods of prospecting–depending on vegetation color to guide them to lead ores. Early geologists attributed this method to them, a practice later adopted by Euro-Americans.⁵¹ The plant *Amorpha canescens*, known as lead weed or prairie shoestring, commonly grew in elongated patches over clay-filled lead crevices. When a traveler visited the region near the Winnebago homelands in the vicinity of the Galena, Illinois, lead mines, he noted, "We saw the mineral plant with its blue leaves and most beautiful [purple] flowers, growing in clusters, bunches and rows, indicating where beds and veins of lead ore existed beneath the surface."⁵² John Weaver, a botanist who specialized in prairie vegetation, wrote that this plant's roots often penetrated the clay to a depth as much as forty feet. He went on to suggest that it was "perhaps the most conspicuous and characteristic shrub of the upland. *Amorpha* may develop into bushes two-and-one-half to four-feet tall and when abundant," the landscape appeared to be a "leaden" color.⁵³ One Euro-American who adopted this Native American practice was the prospector, miner, and smelter Esau Johnson. He noted, "I then went out on the prairie searching for my oxen, and in a ravine I noticed a place about one rod wide and near twenty rods in length, that the grass and weeds looked very thrifty and blue, and all around it there was very little of anything growing; I looked it over and then commenced [to dig] on the south side of the range. As I spaded and shoveled off the fourth time, my spade touched mineral all the way across the range."[54] From his discovery, Johnson recorded that he extracted over 3000 pounds of lead ore before nightfall on the same day. Ultimately, he raised a total of 154,550 pounds of mineral from the same location. The Winnebago and their ancestors, and European and Euro-American miners all understood the importance of identifying and following environmental markers to veins of lead ore lying beneath the soil.

During Des Ursin's expedition to the lead mines near the Kaskaskia worksite, he also considered soil coloring among the layers of earth when identifying veins of lead ore, based on what he learned from his Native American guides. On the first day of the digging, Des Ursin wrote, "The earth is very black and heavy; then we found two feet of yellow earth also with lead . . . deeper down we met with a layer of ground, black and yellow, mixed with pieces of lead."[55] The Des Ursin account depicts how the Kaskaskia understood the importance of recognizing various shades of color when prospecting for galena.[56] During his expedition to the mines, La Renaudière also recognized the presence of other minerals by the natural appearance and shading of nearby stones. He remarked how "we found a quantity of stones of verdigris, which is a sure indication of the presence of copper, mingled with veins of lead one half foot in thickness . . . on the other side of the St. Francis River where a large number of [colorful] stones gives strong indications of mineral wealth."[57] Both French accounts note Native Americans using environmental knowledge, a practice familiar to European miners who applied it in their mining environments back home.[58]

The Native American guides accompanying La Renaudière also understood that spring waters were sometimes indicative of metal-bearing veins. Water action and distinctive taste enhanced the likelihood that lead ores could be found. Europeans adhering closely to Agricola's written observations when searching for lead ore also understood that there could be numerous places where water movement carried lead ore nodules to areas where the water had an acidic taste. After tasting the water in the mining area, La Renaudière ascertained that galena would be found in the area where his Native American guides directed him. La Renaudière related, "In the neighborhood [of the mines] there are many mountains . . . the waters flowing from them are bitter. There is no doubt that if these were excavated and dug into, very good mines would be found" where miners could extract "approximately three hundred million" pounds of lead "each year from many trenches."[59] Continuing his reliance on his Native American mining guides, after describing their prospecting techniques, La Renaudière begins to call his readers' attention to the methods they used to extract lead ores from mining trenches.

The Kaskaskia miners' prospecting and mineral extracting practices were similar to those of Europeans in a variety of ways. They watched for environmental markers while prospecting for minerals. Indigenous miners understood that plant, stone, and soil color often combined with acidic-tasting water suggested the possibility of mineral location. After observing the surface of the soil and finding lead ore, the Kaskaskia assembled their stone, wood, and animal bone tools and began digging six-to-nine-foot-deep trenches.[60] Since most Native Americans used lead ore either for paint or for ceremonial purposes, they extracted only small quantities. In many cases, due to the weakness of their tools and body strength, when they encountered rock they would simply abandon their trench and continue searching elsewhere for minerals. Finally, when their prospecting and extracting season was complete, they carried nodules of galena back to their villages to begin crushing ores, easily combining the smaller grains and powder with plants to make pigments for their pottery; or they melted and molded ornaments.

Extracting in the Neighborhood of the Mines

After arriving at the mining area and searching for lead ore, La Renaudière and Des Ursin produced accounts that offer a glimpse into past Native American extracting methods. In Des Ursin's account, he noted how "the [Tamaroas and Kaskaskians] have made an infinite number of holes from which they drew lead in this neighborhood where there is such an abundance of similar mines. At a depth of seven feet we found a layer of lead, which at its narrowest was eight inches wide [and] you find similar mines everywhere on the surface of the earth."[61] Likewise, La Renaudière explains how the Native American miners discovered lead ore veins, writing that the lead ore appears "as a small layer of lead four feet below the surface of the ground which is yellow, intermingled with black, green, gray and reddish; below it is a very hard rock, mixed with grains of lead, six inches thick; deeper down is another layer, three to four inches thick."[62] The Native Americans also instructed La Renaudière to dig trenches where they pointed, and he located galena: "In some places only one foot below the surface, as one begins to dig, going down to the rock, were pieces of lead weighing from 20 to 30 oz. The distance from the rock to the surface varies from seven to eight feet according to the spot that is opened." Instructed by the indigenous miners to move to yet another location, La Renaudière states, "In order to procure much mineral from theses openings, they are like trenches, one must follow the veins which are found on the rock, which are about one foot thick, some are less."[63] In addition, the practice of extracting lead ores from trenches along the surface also meant that the lead ores were free-formed. Comparing both La Renaudière's and Des Ursin's narratives with Agricola's observations of miners extracting ores in Saxony reveals that the Native Americans worked their mines along the surface in similar fashion to the European miners.[64]

In Europe also, miners often began with surface mining. However, their objective was to mine not only the minerals located within feet of the topsoil; they also endeavored to trace the mineral veins deeper beyond the surface by constructing vertical shafts with connected galleries, all lined with copious amounts of timber to safely access additional blossoms of lead ore.[65] According to European standards, their shafts and galleries were high, wide, and roomy enough for miners to pass each other while excavating and removing soil, rocks, and debris to access mineral loads. The mineshafts extended beyond a depth of four to eight feet; and some reached over thirty feet to seventy or even one hundred feet deep to access ores. However, Des Ursin and La Renaudière and those miners who would arrive after them lacked the European-style tools described by Agricola, which made digging shafts and constructing galleries possible. So Des Ursin and La Renaudière began to assimilate the Native American technique of digging "only four feet deep" or "six feet deep," in the trench style.[66] It should also be noted that in the Des Ursin narrative, he tends to interchange the terms for indigenous trench and European shaft, signaling to the reader the significant amount of lead ore in *le pays des Illinois*. Des Ursin uses *shaft* to describe the abundance of lead ore in the *country full of mines* in a way that Europeans could understand.[67]

The La Renaudière account also notes that little tunneling was to be done along with surfacing mining. He also uses the terms *shaft* and *trench* and *pit*, signifying that these mines were irregularly disposed to being shallow along most of the general surface area, and rarely did the "pits" exceed a depth of ten feet. By referencing both terms, La Renaudière is also recognizing the varied mining practices used by the Kaskaskia. There may have been an outcropping of lead ore over large areas, or he may be calling attention to how mining was systematically prosecuted along a known lead-bearing belt of rock. Additionally, La Renaudière may be embracing the Native American mining technique while simultaneously using the term *shaft* not only to help Europeans understand the mining region but also to promote the economic value of the *country full of mines* in the middle of *le pays des Illinois*. He writes how "six men can operate three shafts, two at each shaft. . . . In order to procure much mineral from these openings, they are like trenches, and one must follow the veins which are found on the rock." Moreover, by calling attention to the veins intersecting with the rocks, La Renaudière is informing officials that in order to dig shafts similar to those constructed in Europe, capital and European expertise were equally important to "extract in the neighborhood of these mines" where he "found much lead in several places," and to test the mineral.[68] Des Ursin and La Renaudière remained cognizant of their object, to encourage French officials to send the necessary capital to mine according to European standards, or settlers would be forced to assimilate the Native American trench-mining practice.

The Native American mining guides, Des Ursin, and La Renaudière all prized the many lead ore pellets albeit for different reasons. When it came to determining the value of the lead ores Europeans and Native Americans discovered, the goal was to locate pellets of pure lead shaped like grain or granular sulphuret of lead—galena or potter's lead ore. Shaped like moderate-sized cubes, the dark blue "fine steel grain" mineral at Mine La Motte was the most common of all the lead ores and known to possess a low percentage of sulfur, which diminished the required smelting time considerably and decreased the required amount of timber for smelting.[69] European miners understood that in the first stage of smelting the goal is to remove the sulphur from the lead ore by roasting the ore in air, converting the lead oxide to sulphur dioxide gas that is released.[70] Similar to other European miners who settled and focused on mineral extraction in the New World, so too the French brought with them to *le pays des Illinois* mining traditions rooted in their long mining experiences, valuing the viability of a mine's economic development.[71] For Native Americans, the small granular sulphuret of lead crystals could be easily transported back to their villages, ground into powder, and mixed with crushed plants to produce a metallic blue-gray pigment for ceremonial practices. The lead crystals could also be exchanged with other Native American peoples for foodstuffs. These nodules and galena cubes have often been recovered from indigenous burial sites by archaeologists.[72]

Des Ursin and La Renaudière observed and documented a myriad of long-standing Native American prospecting and extracting methods; however, Des Ursin and La Renaudière do not address whether the Kaskaskia miners assayed their lead ores after recovery. Since Native Americans used nodules and galena cubes to make pigments, for use in burials, or for ceremonial purposes, they tended to extract small quantities of lead ore and may not have practiced assays. Assaying was an important part of the total European mining experience. The early eighteenth-century mineralogist or mining engineer would have referred to the observations and writings of Agricola to correctly assay minerals. In addition to classifying and describing metals as—things mined—the assayers of Agricola's time had an intimate knowledge of many chemical substances and their reactions. According to Agricola, the purpose of assaying was, first, to determine if some metal was present in the ore, and second, to determine what proportion of one or more metals might be in the extracted ores. Basically, it was a sampling process in which small amounts of the ore were tested for both qualitative and quantitative content to determine the economic value of the metal—in this case, lead.[73] Father Jacques Gravier was the first French natural philosopher to extract and assay the lead ores from the "very rich lead mine." After assaying the ore, he recorded it yielding "three-fourths metal."[74] La Renaudière and Gravier no doubt believed, after collecting, examining the quality, and measuring the quantity of lead ore in their samples, that these mines would contribute to national wealth.[75]

Des Ursin and La Renaudière both understood the importance of assaying ores. However, if any engineer or miner did not know how to conduct assays, he would send them to a mineralogist for testing. Des Ursin is representative of the miner lacking experience, writing, "I sent to Mr. Bienville the samples which Mr. Lochon selected here: 20 gros of ore . . . and Barrel No. 1 is from the shaft of Mr. De la Mothe. Barrel No. 2 from the last large shaft. A little barrel of what we have melted here and of which I send you a sample is from our last shaft."[76] Des Ursin understood his limits and therefore required a mineralogist to report not only the quality but also the quantity of lead contained in the ore. However, one of the former ironmasters Philippe de La Renaudière of Maubeuge is representative of the mining engineer with assaying experience. He documents a test he conducted on the lead ores from Mine La Motte, noting, "In locations where the veins are well-formed, the mineral is found to be good, and produces as much as 40 to 45 per cent. . . . These last mines are very rich in lead. They produce as much as 80 percent."[77] La Renaudière assayed lead ore to determine not only its quality but also to determine the quantity of lead extracted after the smelting phase. Although the Native Americans who guided the French to their mines took away small quantities of granular sulphuret lead crystals to their villages to produce a pigment for ceremonial practices, La Renaudière tested and weighed the lead ore to determine its economic value.

In the early eighteenth century, as an employee of the Company of the West, Philippe de La Renaudière crossed the Atlantic with environmental knowledge to prospect for lead ores and with the skills, tools, and techniques to test the quality and quantity of the lead ore he extracted. La Renaudière must have been pleased with the discovery of quality lead ores, because two years after the Kaskaskia miners guided him more than forty miles to Mine La Motte, there was a change in the administration at the mines. The pursuit of furs and metals that first lead French settlers to the region continued. After La Renaudière and Des Ursin completed their surveys of the mines, more French settlers began to arrive and make their living by trading furs, cultivating wheat, and mining lead on a seasonal schedule.

A Country Very Fine to Establish a Settlement

The vision of La Renaudière and Des Ursin to create a settlement solely focused on lead production started to become evident in 1725, although in its infancy, after Philippe Renault arrived in the *country full of mines* to extract and smelt lead ore seasonally. In 1721, after the French Regime Commissioners decided to send skilled "mining engineers" from Maubeuge to the Illinois country, in addition to La Renaudière, Philippe Renault was also sent. Renault, also a "mining engineer" and referred to as the Director General of Mines and the former Maubeuge ironmaster, carried with him across the Atlantic environmental knowledge and mining skills to apply in *le pays des Illinois*. Both La Renaudière and Renault became the Director of Mines in the *le pays des Illinois*. Philippe Renault would have agreed with Des Ursin and La Renaudière, describing the lead mines as appropriate for establishing a nearby settlement with mills and forges. In 1719, Des Ursin envisioned a settlement where "it is easy to construct mills ..., such and as many as you would wish." He continued, "The country is very fine to establish a settlement and the land is as good as might be desired." For the next two decades, Philippe Renault remained active in *le pays des Illinois*—and appears to have become wealthy from his seasonal mining enterprise. Most important, not far from Kaskaskia, he established the new village of St. Philippe that became the logistical base for Renault's lead mining operations on the east side of the Mississippi, thereby moving mining operations further away from Kaskaskia. It was from this location that significant quantities of lead and furs entered the interregional trade between the French and Native Americans north to the Great Lakes and south to New Orleans.[78]

Extant Illinois records not only chronicle Renault's activities in this period but also note the importance of lead to the community as a form of payment for services rendered. For example, on June 9, 1726, Philippe Renault made an agreement with Joseph Adam "to build for him, at [Adam's] own cost and expense, for [Adam's] labor and for [Adam's] maintenance ... the most beautiful [barn] in Kaskaskia ... fenced. ... The frame shall be made of walnut, of sassafras or of mulberry, of wood well squared and cured and the loft of the barn shall be at thirteen feet above the ground and the posts four feet in and the loft fourteen feet high."[79] The construction of the barn indicates Renault's increased wealth, but it also highlights what role commodities played as a form of payment for Joseph

Adam's service. The price for the construction of the barn and fence was a filly two years old, 200 pounds of flour, a year-old pig, 15 bushels of maize, and 200 livres of lead, which Renault was expected to send sometime during the winter to Adam at Kaskaskia.[80]

By the 1740s, the French had established the Illinois Country of the Province of Louisiana. The territory included all the French-claimed lands from the mouth of the Ohio River north to the Great Lakes, as well as the valleys of the Mississippi, Missouri, and Ohio Rivers. Originally settled by French Jesuits and the Kaskaskia, the village of Kaskaskia was by far the largest community, and Fort de Chartres was the military outpost and seat of government for the Illinois Country—all located along the eastern bank of the Mississippi about fifty miles south of present-day St. Louis. In 1741, Antoine Valentin de Gruy arrived in the Illinois Country and met Philippe Renault, and since Renault was recalled to France, he purchased Renault's livestock at St. Philippe.[81] A year later, De Gruy also purchased a number of African slaves, and the next year he purchased a house in Kaskaskia.[82] Just as Renault had utilized African slaves at his mining operations, Antoine Valentin de Gruy's slaves helped him to continue where Renault left his mining operations.

Sometime after Renault established the village of St. Philippe, he began purchasing African slaves directly from New Orleans to work at the mines. When the French began to establish colonies in the New World, chattel slavery had vanished from Western Europe, after the French jurist Antoine Loisel declared in 1608 that all Frenchmen were free.[83] Yet at the same time, most Europeans recognized the enslavement of non-Europeans as perfectly normal in civilized Christian society. Since French colonies were short of labor, Native American slaves were used in Canada as early as the 1670s, and there were some efforts made to purchase African slaves and bring them to the St. Lawrence valley.[84] Further south, during the initial days of the colony's formation, French colonists in Louisiana attempted to purchase a supply of African slaves as well. When the first French Canadian missionaries and fur traders settled near the mouth of the Kaskaskia River in 1703, they did not bring slaves with them from Canada. Instead, the first African slaves to reach Illinois Country came from Lower Louisiana via the Mississippi River. Following the founding of New Orleans in 1719, the Company of the West and the Jesuits with their financial resources purchased African slaves for shipment to the Illinois Country. For the most part, the Jesuits used their slaves as agricultural laborers and as domestic servants, but there were early plans to utilize slaves in mining.[85] During the Des Ursin expedition to the rich lead mines, he made it known that "the sooner we shall get negroes the better it will be" to work the lead mines and to obtain a "much higher profit from the mines as will permit."[86] Many French colonists like Des Ursin who were interested in mining operations went as far as to state that "you can imagine that the soldiers do not work at the mines ... and Frenchmen are unfit for this kind of work."[87] Des Ursin and others agreed that African slaves were necessary in order to effectively work the mines, for the work was both backbreaking and dangerous for French settlers.

It was thought that Philippe Renault brought hundreds of black slaves to work his mines when he arrived in *le pays des Illinois*. However, given the small number of slaves in Louisiana in the 1720s, Renault would have had to purchase every slave in the entire colony to match the number he was credited with owning. Moreover, in another extant document from 1726, there is a description noting that Renault owned twenty "Negro slaves." When a census of the Illinois Country was conducted in 1732, it showed that the Illinois Country had grown significantly. The southern *le pays des Illinois* communities now included Kaskaskia, Fort de Chartres, the Cahokia Mission, and St. Philippe (figure 2). The combined population of all four settlements included "159 white men, 39 white women, 17 white children, and 33 Negroes" belonging to Renault.[88] With that many slaves, the lead mining operations conducted by Renault and subsequently by De Gruy continued along the seasonal schedule. The African slaves did not remain at the mines on the west side of the Mississippi year-round; instead they alternated their labor between the mines and working the wheat fields.

When They Go to Make Lead

By the mid-1740s, French settlers continued to describe lead mining contacts with Native Americans. In spring 1743, Antoine Valentin de Gruy approached the Kaskaskia to learn if there were lead mines in closer proximity to the Village Fort de Chartres than Mine La Motte. De Gruy was in search of a mine with better access to the rivers to facilitate quicker transport. Again, the Native American miners were called upon to be guides to another mine. Most importantly, when they joined De Gruy, he observed and documented another mutual intersection between Kaskaskians and French settlers who smelted lead ore using Native American techniques. After confirming with his Kaskaskia guides that mines did exist closer to Fort de Chartres, De Gruy "accompanied by two Indians and two Frenchmen carried their implements necessary for making excavations" across the Mississippi River. Along the way he noted, "On the other side of the river we took a well-beaten trail ... [with] numerous rocks that would be useful for the construction of forges."[89]

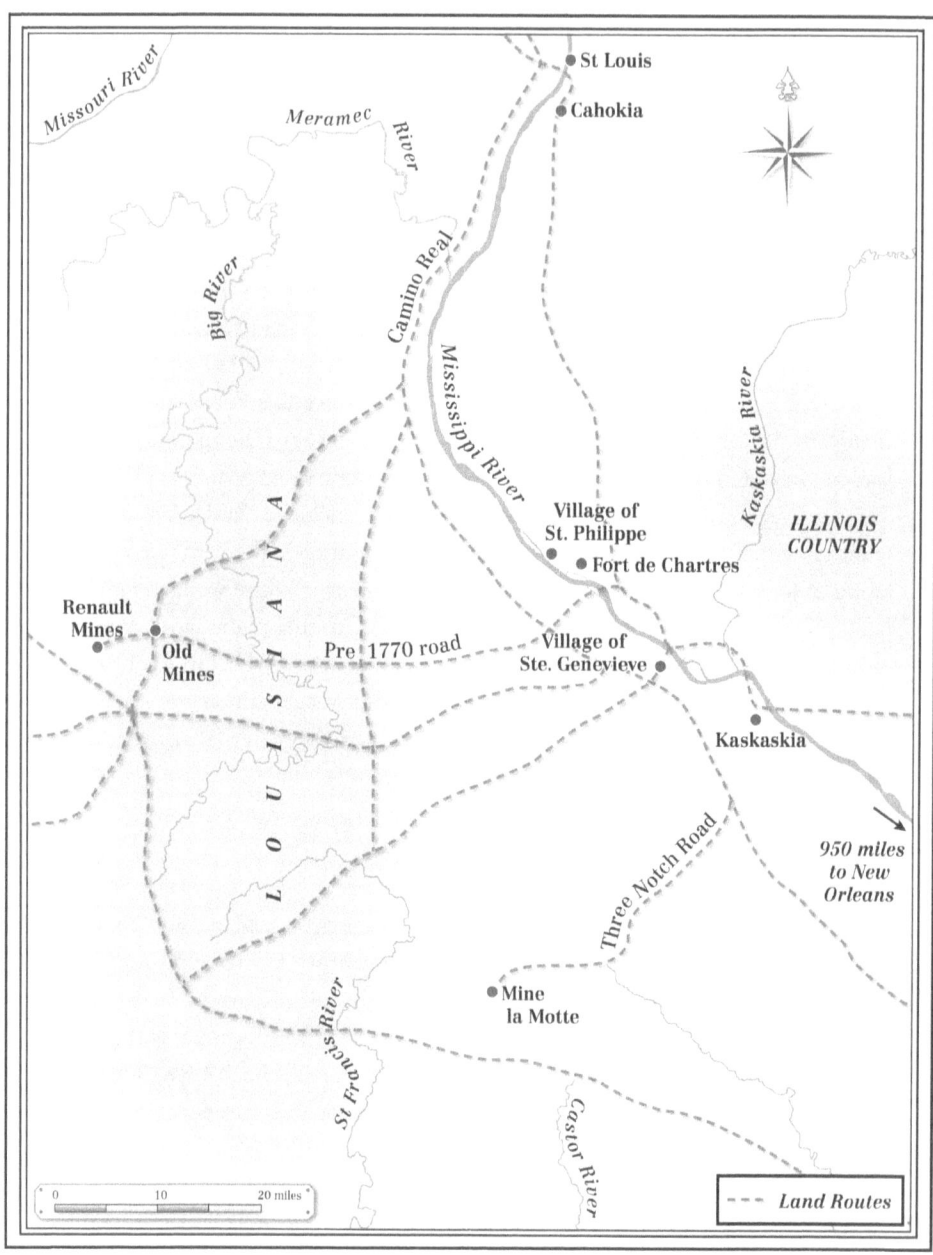

FIGURE 2. French and Spanish period villages and land routes to and from the seasonal mines, early 1700s.

In many parts of the world, prehistoric peoples invented metal smelting. Melting or smelting technology may have been accidentally discovered during pottery-making as a tangential effect of firing painted glazed pottery in kilns and learning the ability to manage oxygen flow. In these communities, learned practices enabled people to heat ores to molten temperatures high enough to remove impurities, extract and mix metals, and produce alloys with new physical properties. Researchers suggest that virtually everywhere in the world that much metal melting happened, smelting soon followed. In North America, such heating technologies were practiced in areas of present-day Mexico, near the southern borderlands of the United States, and in the Mississippi valley.[90]

The simplicity of the process of smelting lead ores can be illustrated by the account of Missouri hunters and settlers. They had learned to procure the lead necessary for making their shot and bullets by building a fire, finding a hollow tree, smelting the pieces of lead ore, and finally, picking up the metal from the ashes. There was also one more requirement—a simple resmelting before casting.[91] According to Paul Craddock, a process this easy is unlikely to leave much of a lasting record, and therefore we lack direct knowledge of the early history of lead production. While it is difficult to locate the surviving remains of early Native American lead smelting, returning to La Renaudière's account and a fresh examining of De Gruy's narrative, which records his expedition and historical and ethnographic accounts, together offer new insight into early lead smelting practices at North American lead mines.

The intercultural dialogue between French and Native American miners as outlined in the Des Ursin and La Renaudière mineralogical reports offers very little information about how, when, and where the Native Americans melted their lead before pouring it into molds to shape it into various ceremonial artifacts. However, in 1723 while walking alongside his Native American guides, La Renaudière recorded, "When they go to make lead, they carry with them only their weapons and ammunition."[92] La Renaudière documented in detail most aspects of Kaskaskia Tamaroa mining activities; they were likely following the techniques of earlier Native Americans who were the contemporaries of the miners Agricola had observed. The phrase "make lead" hints at a possible surviving record of Native American lead ore melting knowledge prior to contact. Although La Renaudière fails to mention how the Kaskaskia, after extracting lead from trenches, dispersed into the forest to gather timber to melt their lead ores, they apparently did make lead. Or it is possible that "making lead" meant Native Americans were grinding galena crystals into powder to manufacture a metallic blue-gray pigment as their Mississippian ancestors had done. Or did Native Americans in fact develop a method of melting lead ores, which is what De Gruy's narrative describes. Similar to the Des Ursin and La Renaudière mineralogical reports, De Gruy offers another eighteenth-century assessment of Native American and French miners working together and deciding to mutually assimilate lead-smelting practices.

Lacking financial support from French officials to construct masonry furnaces, European miners were forced to amalgamate the Native American smelting practice, similar to the mid-eighteenth century hunters and settlers. All they needed was plenty of timber. Amalgamation represents the process of combining or uniting multiple mining practices into one form to create a hybrid of smelting techniques, which was a requirement to make, trade, and mold musket balls along the frontier.[93] In turn, the genesis of Native American smelting traditions combined with French practices reveal a type of technological middle ground where Native American and French miners' smelting abilities converged.[94] Uncovering these associations reveals a Native American presence, but also a European smelting practice in the *country full of mines*.

The smelting amalgam reveals a convergence of indigenous and European environmental knowledge and skills, as opposed to merely the imposition of European techniques. To understand the lead ore melting phenomenon fully, first, we need to consider early Native American smelting skills. Then, we must examine French practices to explain what exactly was involved in the convergence between European and Kaskaskian methods. Though lead ore does not occur as a native metal except in very rare circumstances, the ores are easy to smelt. In many cases, since it may be almost a pure mineral, it can be reduced to metal under quite moderate temperatures.[95] Galena, the type of lead ore that Kaskaskian and French miners encountered in abundance, contains about eighty percent lead, and the ore was usually smelted directly without the necessity of long roasting to remove sulphur.

In Europe, smelting and molding lead was an important part of the mining process. The smelting techniques that Philippe de La Renaudière and Philippe Renault imported to the *country full of mines* in North America had been used by European miners since ancient times. Western Europe is scattered with mounds of slag from primitive smelting foundry furnaces of the Bronze Age. The mounds also reveal that copper was treated in shallow pits with charcoal. Over time, great skill and experience was acquired to obtain the maximum amount of metal from ores so that the least amount of metal might succumb to either the slag or fire. Following the extraction of galena from the earth, to separate the ores from the surrounding impurities miners were required to conduct a smelting. The smelting process described by Agricola was the earliest description in the metallurgy literature.[96] According to Agricola, the smelting of gold and silver ores, which were considered to be rich in metal, required a considerably shorter smelting period. Similarly, galena also required a shorter smelting period. However, the smelting of lead and copper usually required longer periods of heat. In Europe, lead smelting times extended over three days and three nights. Agricola describes the ancient furnace used to smelt lead or copper as a simple shallow pit in the ground lined with charcoal and covered with ore. Using these pits, smelters were able to produce crudely shaped round cakes, which settled at the bottom of the pits, measuring eight to ten inches in diameter.[97] Because of a number of similarities, it appears that French miners easily blended their knowledge and skills with the Native American miners. The amalgamation begins with exchanging methods and ideas about how to extract lead ore, but from the European perspective, also how to smelt lead ores in greater quantities, which was necessary to produce an abundance of musket balls for local and distant trade and protection.[98]

De Gruy fails to name a particular Native American miner or a particular French miner who extracted and smelted lead ore for frontier communities in 1741.[99] However, he may have been observing a well-known French miner, François Vallé, who conducted seasonal mining operations at Mine La Motte around the same time. Born on the outskirts of Quebec City, the French Canadian Vallé migrated to *le pays des Illinois* and quickly became involved in lead mining at Mine La Motte. Endeavoring to make lead mining his primary enterprise, Vallé initially established his operations at Kaskaskia, traveling across the Mississippi River to "work the diggings, to extract minerals, to cut wood and build furnaces to smelt lead, and transport pig lead by horse" back to Kaskaskia.[100] After accumulating enough pig lead using the Native American and French-style furnace, Vallé departed Kaskaskia on a number of trading expeditions. He often traveled to outposts in Upper Canada and Detroit by way of the Wabash River valley, and along the way he supplied pig lead to frontier villagers. In all probability, Vallé used a type of furnace that De Gruy observed Native American and French miners using. However, he was not the first.

In North America, early Pueblo miners roasted their ores "in an open fire," and "then baked or partially smelted" the ores in "a kind of subterranean funnel-shaped oven-furnace."[101] Therefore, at contact, the French were probably introduced to this longstanding Native American furnace and smelting practice, which they would have recognized. The furnaces European miners constructed in Europe appear to be similar to the type of furnaces used by Native Americans. The difference in the construction of early European furnaces was the stone walls, and they were not circular in shape. For example, European furnaces consisted of a "main wall against which a series of furnaces were built." At large metal works, the wall built of "brick or stone" was approximately fifteen feet high and had "six furnaces spaced six feet apart." Agricola suggested that stone was preferred to brick because it resisted fire longer.[102]

In the late nineteenth century, a Wisconsin archaeologist, W. Y. Woods, visited the Mississippi valley lead mines and discovered a hole with stones laid in circles about three feet in circumference. After excavating one of these subterranean holes, Woods believed he had discovered an object that was at one time used as a furnace. Woods described the shape of the furnace with the stones laid in circles underground as like an inverted cone, with sides carefully "walled up with flat stones, and around the edge was left bare, which ran around the mouth of the hole."[103] After Woods removed the earth resting in the furnace, at the bottom he recovered seventeen pounds of lead ore. In total, Woods recovered thirty of these small furnaces he suspected were used by Native Americans. There were similarities between how Agricola's contemporaries constructed their masonry furnaces and how Native Americans built their subterranean conical stone furnaces. Europeans dug a hole in the earth and "then built square walls along the sides and back to hold the heat of the fire more effectively." They also left the front of the furnace open for easy cleaning. After the hole was dug and the walls were constructed, "wood about twelve feet long" was "laid in the area in four layers," in alternating directions. In addition, Europeans built this particular type of furnace on the slope of a hill. Finally, just as Native Americans placed their lead ore under the small logs used for fuel, the European smelters also placed "pieces of ore" under the wood before igniting the fire.[104] What developed after contact was an amalgamation of indigenous and European smelting practices in similar furnaces, aiding early miners like Vallé to continue their mining enterprises.

Although De Gruy fails to attribute the smelting knowledge and technology he observed to the local Native Americans, what he describes was the hybrid of Native American and French furnace construction techniques. At the mining site, De Gruy observed that "When an individual has extracted enough of the ore to supply him with a living for the rest of the year, he stops digging and sets about the smelting process" by first cutting down at least three large trees and then chopping each into five-foot sections.[105] De Gruy's account methodically depicts how French miners smelted lead ore using a combination of features from the Native American subterranean oven and the aboveground French-style log furnaces. De Gruy observed miners collecting large amounts of timber to construct "log furnaces" over a type of catch basin.[106] De Gruy considered the method time consuming, calling it a "primitive method" to smelt lead ores. De Gruy's observations of smelting clearly depict another significant occurrence in the cross-cultural exchange, which continued long after the arrival of American and British miners. The De Gruy report opens a window into the manner in which "eighteen or twenty" miners designed a smelting amalgam to melt lead ore in log furnaces on this early North American frontier.[107] Most significantly, the De Gruy report, as well as archaeological evidence, not only shows how Native Americans invented a method to melt ores but also how French miners further amalgamated their practices by incorporating indigenous techniques.

De Gruy's depictions of assembling furnaces show how Native American and French miners blended more than their prospecting and extracting practices. Most significant was the furnace design. Native American and French miners began to construct their furnaces after "chopping down three big trees, and then cutting them all into three-foot lengths" and then into even smaller pieces of timber. Next, before placing the logs or timber above the hole, around the edges of the excavation a miner proceeded to "dig a small hole like a basin in the ground." The miner continued to build and shape the log furnace by positioning "three additional shorter logs." Thereafter, miners proceeded to add two additional "logs of the same length crosswise." Similar to Woods's description of the furnace shaped with stones underground, according to De Gruy, the entire furnace resembled a "funnel-like box." Next, the miner filled the inside of the box with wood and began to layer ore on top of the timber. Then, miners positioned three logs, shorter than the first three, in the same manner.[108] Finally, they filled the entire box with lead ore and wood, after which the furnace heap was set afire "from below."[109] During the 1740s, for many metallurgical processes including smelting lead ore, the timber required was in adequate supply and nearby, and twigs could also be used with brushwood and thinner wood for additional kindling.

Miners told De Gruy these early furnaces only smelted a part of the lead, and it was often necessary to repeat this process as many as three times, or additional firings, "to extract all of the material [lead], which collects in the bottom of the hole." The repetition of this process must have required three or four supplementary furnaces to procure a larger yield of lead. After the ores were assayed and it was discovered that the mineral contained a high percentage of sulfur, the ore was reheated three times: Miners placed the slag in another log furnace for the second roasting, to yield more lead. These first two processes, called roasting, were necessary to drive off the sulfur, and the third heating was considered the real smelting process when using these log furnaces.[110]

Following the Native American practice of designing a basin to collect the melted lead, miners excavated a hollow earthen basin below the burning heap; after the third firing, as the molten lead flowed into the conical earthen basins, the pig lead was shaped into "small flat disks each weighing" between "sixty and eighty pounds." De Gruy also noted that the disks resembled "a rough oval" because of the earthen basins used to create them.[111] Most disks were two feet long, six to eight inches wide, two to four inches thick in the middle, and became thinner toward the edge. Each oval-shaped leaden mass generally weighed between thirty and forty pounds.[112] Molding the lead into disks facilitated easy transport between the mines and villages. Following contact, the French adopted the same technique to transfer their lead to Kaskaskia or Ste. Geneviève efficiently. The French used horses and carts to carry four or five of these bars each trip, a mode of transport that may have been adopted by the Native Americans. The log furnaces that produced these pig lead disks yielded about forty percent lead; the rest of the ore was reduced to lead ash and lost.[113] The French had a name for lead ash, *scorie* or slag.[114] The slag consisted of various sizes of lumps of lead that fell through the logs before being touched by the fire.

Although French miners were concerned over the inefficiency of the furnaces, they continued the practice throughout the eighteenth century. To acquire additional amounts of lead, Kaskaskian and French miners would occasionally search through the ashes for ore that had not succumbed to the fire. Miners understood the possibility of procuring a further yield from the lead ashes by remelting the slag. The larger pieces, consisting of ore partly desulphonated were picked out from the ashes and added to the next smelting.[115] It was realized that a considerable amount of lead remained in the ash heaps, which were near each log furnace site throughout the mining landscape where De Gruy envisioned establishing a settlement. In addition to early French miners voicing concerns over the inefficiency of the furnaces, local villagers also understood that the continual cleaning, smelting, and remelting of lead ores exposed the surrounding landscape to toxins where La Renaudière and De Gruy envisioned building a settlement.

Unhealthiness Where Innumerable Creeks and Streams Flow

The progression of mining between 1725 and 1743 centered on lead ores being exploited only at Mine La Motte and Philippe Renault's mine. Renault's mine was being worked in a haphazard fashion, and although he built a masonry furnace, which was listed among his assets, the common form of smelting continued to be done in the log-heap-style furnace. Nevertheless, lead was being dug in some quantity according to the production figures for Mine La Motte, which was roughly a daily output of 1500 pounds of lead worth about $3,730.26.[116] In this embryonic stage of the mining industry, the village of Kaskaskia was the lead depot where sturdy little horses carried the smelted oval-shaped lead disks.

In the face of these circumstances, another village was established a few miles from Renault's mine and Mine La Motte prior to 1748. The establishment of this village so close to the lead mines highlights how the original inhabitants clearly comprehended that "insalubrity" or "impure" water could in principle promote illness. The earliest record of this village, which is now called Old Mines, is a baptismal record from the Fort Chartres Church of St. Anne. There, the infant of Pierre Wivarenne of Picardy, France, and his wife, Marie Anne Rondeau, "habitants du Village des Mines," was baptized on September 28, 1748.[117] According to oral tradition, the village was situated on a branch of the Mineral Fork, now called Old Mines Creek. Since the immediate vicinity of the village was rich in lead ore, it may also be assumed that surface mining was done in the region between Renault's pits and the village.[118] It is speculated that the village was placed above the mines on this little creek in order to have safe drinking water, because lead washing, a required step prior to smelting lead ores, made the water close to the furnaces unwholesome, especially for the animals. When the villagers articulated their concerns before deciding where to situate their settlement, they had a clear understanding of the industrial effects of frontier lead production.

Eighteenth century settlers understood the principle of pollution when it came to lead washing, which made the water close to the furnaces unhealthy, especially for watering animals. The French colonists who settled the village of Old Mines appreciated the value of drinking good water as opposed to drinking bad water polluted with lead particles, which was a threat to human health and was to be avoided. Conevery Bolton Valenčius notes that nineteenth-century settlers understood that illness or good health depended on the water that people came in contact with and its effects on the body. If water in contaminated condition was continually used by villagers, they knew they had to take great care for their own health as well as the health of their animals. In the inhabitants' minds, waters saturated with lead "might promote cholic in many persons" and their animals. The consumption of noxious water, like noxious air, would lead to a weakening and slowing of both human and animal labor.[119]

Georgius Agricola's *De Re Metallica* (*On the Nature of Metals* [*Minerals*]) not only became the primary textbook and guide for miners and metallurgists interested in prospecting, extracting, and smelting ores in the sixteenth century but also was an important reference for those who settled near active mining sites. Agricola's documentation of technical details enabled miners and mine owners to perform multiple operations at the mining settlements. As a trained physician, Agricola not only devoted his time to his patients working the mines and furnaces, but he also dedicated his time to reading Greek and Latin authors who referenced mining and discussed mining work with "the most learned among the mining folk." Agricola believed that his education in the arts and sciences contributed to his mining and medical success in two primary areas, philosophy and medicine. According to Agricola, philosophy helped the miner to "discern the origin, cause, and nature of subterranean things; for then he will be able to dig out the veins easily and advantageously, and to obtain more abundant results from his mining." Secondly, Agricola understood well the many diseases to which miners were susceptible; so he also became a fluent caregiver for "diggers and other workmen" in the mining fields and relayed his knowledge to miners who also needed to understand medicine to be able to assist in the healing of sick and injured miners.[120]

Agricola's research and documentation called on physicians and miners to devote their attention to the sometimes waterless and dry environment of subterranean mines where the stirring of dust particles could penetrate into the worker's "windpipe and lungs to produce difficulty in breathing, and the disease which the Greeks call [asthma]." He also noted that the dusty and "corrosive qualities [eat] away the lungs, and [implant] consumption in the body ... [leading] to a premature death." Of chief concern for Agricola was miners' exposure to "bad air," which was the result of "sulphurous and aluminous" vapors emitting from the mines; he described noxious vapors and stale air as harmful both to the miners and to those living in mining settlements. The theory of using respiratory protective devices to reduce or eliminate hazardous exposures to airborne contaminants did not begin with Agricola. The idea was first proposed by Pliny the Elder (AD 23–79), a Roman philosopher and naturalist, who discussed the importance of wearing loose-fitting animal bladders in Roman mines to protect workers from the inhalation of the red oxide of lead. In addition, many centuries later, Leonardo da Vinci (1452–1519) endorsed the use of wet cloths over the mouth and nose to protect against inhaling harmful airborne agents.[121] Like the Roman miners before them, Agricola also recommended that sixteenth-century miners should "fasten loose veils over their faces" so that "the dust will then neither be drawn through these into their windpipes and lungs, nor will it fly into their eyes." According to Agricola's observations, the mines spontaneously produced "poison and pestilential vapor ... containing more noxious fumes ... emitting pungent vapors which kill the miners if they linger too long in them." Agricola not only dedicated a section of his written work to the "maladies of miners," he also cautioned villagers living in the surrounding landscape, which he labeled as "metalliferous localities" due to the noxious vapors and stale air that harmed those living in nearby mining settlements.[122]

By the eighteenth century, European natural philosophers also began to make their own contributions to the reservoir of writings by Pliny the Elder, Leonardo da Vinci, and Georgius Agricola on incidents of lead poisoning that was the result of toxic vapors. For example, in the early 1700s, the Italian physician Bernardino Ramazzini, known as the father of modern occupational medicine, described industrial lead poisoning or "metallic plagues" among tradesmen like potters and portrait painters in *De Morbis Artificum Diatriba* [*Diseases of Workers*]. In *Diseases of Workers*, he describes potters glazing the pottery with calcined lead whereby they "received by mouth and nostrils and all the pores of the body, all the virulent parts of the lead ... and were thereupon seized with heavy disorders."[123] It is difficult to gauge to what degree miners and villagers were acquainted with the literature on lead poisoning in the early eighteenth century, however, the accumulation of treatises and pamphlets by this time was extensive.

After Agricola's and Ramazzini's early attention to the occupational hazards miners and workers faced when working in lead-mining environments, in 1656, the physician Samuel Stockhausen (like the inhabitants of Old Mines) observed the dangers to animals from lead exposure. Stockhausen documented how animals near the Goslar lead mines in Northern Germany were susceptible to colic. Similar to Agricola, Stockhausen also practiced in the middle of a "country of lead mines" and he demonstrated how the emanations from smelting lead ore produced colic in local animals as they came in contact with lead particles. Most significant are his observations on the effects of vapors on "cows, horses, and sheep pastured in the neighborhood of lead mines in Germany." He noted that when animals "graze in such pastures washed by streams from the mines there was a tendency to leave a slight deposit of lead salts" which caused the animals to contract lead colic.[124] Europeans and Mississippi valley miners and settlers alike understood that encountering dust or vapors associated with mining had environmental impacts on spaces where water flowed, where plants grew, and where settlers came in contact with both. As more and more physicians began to see themselves as the beneficiaries of an enlightened age, in the period of emerging science, they were also determined to extend the benefits of their acquired knowledge to others.

As European natural philosophers observed the effects of vapors on farm animals in the neighborhood of lead mines and shared their knowledge about lead poisoning, others were writing about maintaining good health. For example, the Swiss physician Samuel Tissot published *Advice to the People in General, in Regard to Their Health* in 1761, and the work was translated into numerous languages and reappeared in frequent editions. According to Tissot, good health depended on the availability of good water. Tissot noted, "Out of twenty sick Persons, who are lost in the country, more than two thirds might often have been cured, if being only lodged in a place defended from the injuries of the air, they were supplied with abundance of good Water. The bad quality of water is another common cause of country disease."[125] He noted that when both air and water were unwholesome, wherever they are found or wherever they flow was of deep concern. This also became a concern for the Upper Louisiana settlers who decided where to locate their villages. Tissot surmised, "Bad water, like bad air, is one of the most general causes of diseases." He would go on to recommend that "government officials should forbid, under the most severe penalties, all such adulterations, as tend to introduce the most painful cholic, obstruction, and a long train of evils, which it sometimes proves difficult to trace to this peculiar cause; while they shorten the lives of, or cruelly torment people."[126] Both urban and rural village dwellers became concerned with activities that may cause bodily harm, and started to judge for themselves the "salubrity" of air and "the purity of water" as they kept an eye to any future consequences, which they tried to control.[127] Ever since the arrival of French miners, the smelting of large quantities of lead began to change the taste of the local water flowing from the creeks and streams as witnessed by settlers in the village of Old Mines.

Flowing streams of water in the mining district not only offered La Renaudière and De Gruy an alternative route to deliver their pig lead "to the Illinois country by water," but miners also had access to nearby streams and creeks to wash ores before beginning the smelting process. In Europe, after lead ore was extracted from surface mines, the ores were then taken to a sorting table where women, boys, and some men sorted the ores by hand. In effect, they put the good ores into wooden tubs that were later carried to nearby devices for washing ore in a variety of troughs, sluices, and ditches—mixing the ore with water to separate the fine grains from the lumps. At Mine La Motte and Old Mines, Native American and French miners did not construct sorting tables, so slave women, boys, and old men sorted all the ores by hand before beginning the handwashing process in ditches, creeks, and streams; washing separated the fine grains from the lumps before the ore was placed in furnaces. It is not clear if weirs were available to catch and hold the heavier grains of lead, or if other catchment mechanisms were used. On the other hand, if swirling bowls were made by French and Native American miners to mix the ores with water, the local streams and creeks would have become polluted. Whether mining operations were small or large, miners depended on watercourses, and villagers depended on creeks and streams in close proximity to water mules, oxen, and themselves during the mining season. As colonial mining began to scale up in the 1750s, mine owners and their slaves and *engagés* (French indentured servants) continued to tap the same water sources using their well-established amalgam.

A Very Fine Hunting and Mining Country

The middle ground where Native Americans and French colonists exchanged their mining and smelting practices also became a place where French miners learned to combine their planting and harvesting seasons with hunting and lead-mining, as Native Americans had done for centuries. While the French looked forward to establishing mining settlements, Native American communities understood the importance of exploiting the seasonal diversity of their environment and practiced planting, hunting, and mining according to a particular seasonal cycle.[128] La Renaudière described the location of the mines as the perfect place to establish a settlement after witnessing small prairies where buffalo, bears, and deer gathered, and because of the many rivers and streams—"fish abound and a large number of fowls." When Timothy Flint traveled to the area, he too was impressed with the "place with an abundance of game, and a summer forest where a variety of flowers, such as red lilies" blanketed the landscape.[129] When fish were spawning, Native American families gathered there; when it was hunting season in the summer or fall, the same families scattered over many square miles of land. Similarly, settlers learned and adopted these longstanding Native American traditions—hunting the numerous turkeys running around the countryside, passenger pigeons darkening the skies, and ducks covering numerous small river systems—to find the greatest natural food supplies. Like the New England coastal peoples, the indigenous groups living along Mississippi valley river systems found that between March and October they were free from all anxiety regarding food supply.[130]

Similar to New England subsistence schedules, late winter was a time of plenty in the central Mississippi valley when Native Americans planted crops during the month of March. Between April and May, the arrival of migratory birds alerted the communities to the coming of healthy ducks laying large eggs. While the men fished and hunted, women and children gathered bird eggs. Beginning in June, indigenous communities began their summer hunt, living along the plains while hunting bison, turkey, and antelope. In July, the women harvested their crops and gathered nuts, berries, and other wild plants as they became available.[131] The late summer harvest began in August, and from late September through November, the fall hunt would commence. They hunted bear, rabbits, and migratory birds. From December to February, the Kaskaskia lived in small longhouses and clustered in smaller groups to hunt for bears. According to the mining expeditions calendar in the accounts of Des Ursin, La Renaudière, and De Gruy, French miners learned to integrate extracting and smelting lead ore according to the Kaskaskian seasonal timetable that coincided with their hunting and farming practices.

In addition to the mining and smelting amalgam the French, the Tamaroas, and the Kaskaskians practiced in the middle ground, they also blended their mining operations in the same cyclic fashion. Miners worked at the mines during a period sometimes called *la campagne* (the campaign).[132] French settlers used this expression to describe the nature of the lead mining enterprise during the early eighteenth century as the expression is used in military parlance in a wide-open field. Groups of men leaving their villages, homes, and families to penetrate into the countryside fighting the natural elements to discover, extract, and smelt lead ore. Following their Native American neighbors' calendar, the mining schedule began on April 1 and ran for several months permitting the miners to return to their villages in time to assist with the first summer harvest.

Miners traveled to the lead mines in April or during the early summer harvest. They crossed the Mississippi River by bateaux, and then traveled on horseback along wooded trails for approximately four days. By the mid-eighteenth century, miners began to take advantage of a second mining season. It ran from August to December, which was from the end of the second harvest until the first frost.[133] As the French and Native American miners continued to practice their set of amalgamated techniques near Old Mines and Mine La Motte, the French miners desired to establish a settlement to work in closer proximity to the lead mines. No longer did they want to depart for the mines during the spring and fall hunting seasons after planting their crops.[134]

Since the Tamaroas and the Kaskaskians guided both Des Ursin and La Renaudière to their lead mines in the early 1700s, both miners had written about settlement possibilities. Immediately after arriving at the mines, Des Ursin noted, "It is easy to construct mills here, such and as many as you would wish. The country is very fine to establish a settlement and the land is a good as might be desired." A few years later La Renaudière wrote, "By forming a settlement at the mines … the lead might be shipped on a small river, which passes the mine at a distance … and falls into the Mississippi." Both Des Ursin and La Renaudière envisioned the prospect of shipping lead from the mines in sufficient quantities to supply all French posts from the Great Lakes to the Gulf Coast, as De Gruy noted in 1743. In fact, De Gruy wrote about the need to create a settlement with the supreme objective of extracting and smelting lead ore year-round.

De Gruy's report is also replete with geographical descriptions outlining a number of landscape features surrounding the mines, calling attention to the express purpose of planning a European-style settlement. De Gruy reported that accessibility to the mining region by land or water made the location acceptable for a portage. He believed that lead ore could be cleaned in the flowing water before smelting, or better yet, the fast-moving water could propel machines for cleaning ore or for pumping the mines clean of underground water or debris. As he moved about the landscape similar to those French miners who came before, De Gruy also stated, "Miners required prairies filled with excellent timber" for smelting ore and building homes and mills. Finally, he documented the abundant meadows with "a great quantity of fruit trees," which would supply the miners with nourishment.[135] Like the Spanish, the French also carried with them old European mining traditions to the New World. Living and working in closer proximity to the lead mines meant the ability to convert lead ore into financial gain, but transitioning from seasonal mining to year-round mining would be a slow process.[136] Until then, the French adopted the Native American practice of combining hunting and farming with mining.

The accounts of Des Ursin, La Renaudière, and De Gruy describing the samples of lead ore they extracted from *le pays des Illinois* made their way into the hands of French officials back in Europe. De Gruy wanted "very much to find something worthy of attracting the attention" of a French official to encourage the creation of a mining settlement as opposed to following the Native American seasonal mining calendar.[137] At some point De Gruy began to correspond with Jean-Frédéric Phélypeaux, Comte de Maurepas, who was hopeful that the *country full of mines* might prove a benefit to the colony. He was a powerful figure in France during the reign of Louis XV, and his responsibilities included overseeing the administration of both the royal court and the French navy. Born at Versailles, in a family of administrative nobility, Phélypeaux requested that De Gruy send him a precise report to assist him in reaching a decision about the possibilities for constructing a mining settlement.[138] Maurepas requested the total production of the lead mines during the previous decades. More specifically, he desired to learn the location of new mines, the method of extraction, the number of settler homes, and the number of furnaces used for smelting. Furthermore, Maurepas requested an accounting concerning the skill level of those currently engaged in mining activities.[139] Although no report matching Maurepas's request is available, an account of the mineral exploration by De Gruy is useful. One of De Gruy's reports offers a sense that Maurepas was made aware that one of the significant challenges to extracting and processing lead ore year-round was French miners' continuing adherence to the Native American seasonal mining schedule, prospecting, extracting and smelting lead ore during the early spring, summer, and fall months.[140]

Like those explorers who came before, De Gruy also was familiar with English mining literature such as the engineer Thomas Houghton's instructions for establishing a European-style mining settlement. Houghton recommended that a miner focus on three environmental factors when considering a settlement: the ability to access the mining region, the availability of water, and the overall circumstances the landscape had to offer.[141] Closely adhering to the Houghton guidelines, when De Gruy set out from Fort de Chartres to appraise Mine La Motte more thoroughly, he notes, "I was thinking that it would be advantageous for the colony of Louisiana to locate some of these mines close to the rivers, which would very much facilitate their exploitation." De Gruy also mentions numerous small prairies in the area of the mines, which "served as a retreat to numerous buffalo, bears, deer, skunks, and other animals." Like La Renaudière, De Gruy also portrays the area as a "very fine hunting country, where the rivers contain plenty of fish and a large number of water fowl."[142] His account also mentions a great quantity of fruit trees, such as persimmons, plums, and pecans, and grape vines. He notes the fruits and nuts as having "supplied their diets" while Native Americans extracted and smelted lead ore. Like earlier accounts, De Gruy's narrative became a vehicle for French officials to determine how much to invest in developing a mining settlement.

Pleased with the surroundings, De Gruy turned to his Native American guides to learn more about the location of these rich lead deposits. They informed him that the mines were approximately twenty miles from the Village of Kaskaskia. The Kaskaskia guides told De Gruy "numerous mines could easily be found along many of the riverine systems." In an effort to promote the Mississippi valley lead mines to Maurepas, who might be inclined to send additional French miners to immigrate to the region to exploit the mineral resources, De Gruy included the guide's information in his report. Like La Renaudière, De Gruy also expressed an interest in the method of transporting smelted lead. The Kaskaskia informed him that horses should have no difficulty covering the distance in less than a day. De Gruy noted in his report, "The route to this deposit appeared practical for pack horses."[143] Based on their information, De Gruy acknowledged that the "passable road" would also make it easy to negotiate "pack horses" for carting tools and smelted lead. In addition to land passage, De Gruy noted that "the mine being in an area with innumerable streams all flowing into the St. Francis River," was ideally located only two miles from a navigable river.[144] The location of the mines, just twenty-eight miles from Kaskaskia in the "well-watered area would make it possible to deliver the smelted lead to the village, and then transfer the lead throughout the Illinois country." All three miners described how "carters" could travel twelve miles by water and sixteen miles by land to reach Kaskaskia.

As an economical and efficient means of transport, La Renaudière proposed shipping the lead on one of the smaller rivers, the St. Francis, which flowed into the Mississippi near the Arkansas River. La Renaudière thought that since the "St. Francis was navigable during the rainy seasons, if miners placed their lead on pirogues near the mines, it would reach the Mississippi River in eight days." He estimated that between five and six thousand pounds of lead "may be carried in a pirogue from the mines to New Orleans by way of the Arkansas River." La Renaudière anticipated the route to be equal to the distance between the mines and Kaskaskia.[145]

De Gruy also depicted the mining site as well watered, making it possible to deliver the smelted lead to either Kaskaskia or Fort Chartres using the navigable rivers. Recall that according to European standards, running water also suggested that miners could easily construct mills to grind large chunks of galena into smaller, more manageable pieces for the smelting. De Gruy used his report to sway French officials into believing that once a mining settlement was established, workers could then conduct the business of selling or trading their lead to the local community, or even shipping lead bars to New Orleans.[146] The numerous streams of water flowing through the mining district also suggested local river systems, mountains ranges, and other natural resources to support the development of a settlement. Like mining engineers who came before, De Gruy also reported that "numerous smaller rivers made it easy to construct mills here."[147] He also understood that water mills could supply water to clean lead ore before smelting. Instead of advocating travel between Fort de Chartres and the mines, each of their reports emphasized the possibility of Europeanizing the lead mines by establishing a permanent mining settlement. In essence, reporting the mining environment's general conditions to Maurepas was a promotional technique that De Gruy and others employed to entice their readers to support a settlement at the mines. La Renaudière clearly documents how "in the neighborhood of the mines there are many mountains of no mean height . . . if a settlement could be formed here . . . a good living could easily be made." After describing the mining landscape, La Renaudière lists multiple examples of European machines that could be constructed at the mines.

For example, because of numerous instances when miners encountered water obstacles, Des Ursin noted that installing pumps to extract the water was the best solution, thereby making it possible to reach the veins of lead easily. Des Ursin understood the necessity for pumps to drain the water from mines in order to make them operable at increasing depth. Beginning with Agricola, mining manuals illustrated mine-pumping equipment. Agricola used illustrations to show the mechanical art of mining, from the simplest European tools to the more complex water pumps. Europeans carried mining troubleshooting ideas and solutions to the North American mining sites to extract the most lead ores efficiently.

As another encouraging factor to support the establishment of a mining settlement, De Gruy notes a sufficient amount of timber. Late seventeenth-century and early eighteenth-century travelers describe the Mississippi valley as a place "filled with excellent timber."[148] In addition to lead ore surface mining, the next operations at the worksite were to test the lead ore's quality and to begin gathering large quantities of timber to construct numerous open-air log furnaces for the many firings; De Gruy considered this a crude, wasteful smelting practice that limited the possibility of having greater success at Mine La Motte. He also saw the short mining season of only four or five months, the absence of a permanent mining settlement, and the shortage of skilled workers as hindrances to further development. Although De Gruy considers what by now had developed into the Kaskaskia-French amalgam to be primitive, he recorded a significant amount of lead produced. De Gruy provided court officials with the production amounts of Mine La Motte. In 1741, he estimated that miners produced 2300 *saumon* bars weighing about 161,000 pounds at 70 pounds per bar.[149] In addition, the total amount that miners produced in 1742 was 2228 *saumon* bars weighing approximately 155,960 pounds.[150] Therefore, even though De Gruy clearly was critical of the present amalgam, the production amounts remained relatively satisfying to local and court officials, and the Native American and French techniques continued to be practiced. However, to counter the above impediments, he proposed the establishment of a mining community to grow the enterprise. He also suggested that the government consider sending prisoners to extract lead and cut timber for three years. He considered that such a commitment would support the year-round extraction and production of lead.

As Maurepas may have been considering how to best expand the small colony near the lead mines, he possibly thought the best course of action to make the colony profitable was to motivate the miners toward agriculture rather than toward the extraction and smelting of lead ore. Maurepas believed that they should be persuaded to concentrate on pursuing agriculture, as it would provide security for their families.[151] By the 1740s, life in and around the mines resembled the older towns on the other side of the Mississippi. Ste. Geneviève and parallel communities in the mid-eighteenth century looked more like another colonial region, the Chesapeake—and more like the Chesapeake than New England because the settlements near the mines directed much more of their agricultural production to distant markets. Unlike farming, fur trading and mining provided settlers with a degree of freedom, and most settlers rebelled when presented with the prospect of farming for a living. Local villagers, merchants, and miners were pleased with the production of the lead mines because they did not have to worry about the leakage of grain lead on the trans-Atlantic voyage and could avoid the difficult upstream transportation of lead from New Orleans to colonial posts located along the Mississippi River.[152] Another advantage was that a new commercial field would be established in the colony. As time elapsed, miners and farmers continued to extract, smelt, and manufacture musket balls, which became increasingly valuable to frontier fur traders and farmers needing to provide security for their homes. Settlers manufactured musket balls using a burnishing mill, arsenic, ladles, skimmers, scoops, testers, iron molds, scrapers, big covered boilers, and a gun barrel used to make the shot. Miners also suggested that manufacturing lead products in the colony would save on transporting similar items across the Atlantic. However, it was incumbent on the colonists to have a skillful lead manufacturer to make musket balls for the local and distant villages.[153]

During the French and Fox Wars in the 1730s, fighting between the French and Fox caused an increase in the demand for musket balls. However, because the lead trade was disrupted in the Mississippi valley so severely, by 1733 the Governor of Illinois reported that he was only able to safely procure enough lead for use by his own men and had none to export to New Orleans.[154] Although the Fox and French continued their clashes, the Native Americans near the lead mines still welcomed the French traders who came to live among them. They repeated their practice of forming trade alliances with the French who desired to exchange European manufactured goods for furs, hides, and lead.[155]

In light of these developments, miners continued to ignore the French officials. One colonist even sent his lead to the French market, as is disclosed in a letter to Maurepas, who approved the shipment of over thirty thousand pounds of lead belonging to Mr. Desclozeaux on the ship *La Charenete*.[156] Maurepas had arranged for the remission of the freight charges on his shipment. He also suggested that others might send their lead as ship ballast but made it clear that lead could not be permitted to displace other colonial products like wheat flour.

In 1754, smelter and merchant Karpen de La Gautrain, working near Mine La Motte, wanted to manufacture enough musket balls to supply a consistent flow to the royal warehouses and the entire colony. La Gautrain described himself as "already engaged in the lead industry" and requested expert workers be sent to the colony.[157] He required one or two workers capable of making "grain lead and rolled lead." He asked officials to try to locate a skilled individual and send him with the necessary equipment. La Gautrain even offered to pay all the expenses for the laborers' travel, their tools, and wages. La Gautrain's proposal to pay a workforce suggests his willingness to "further risk at" his "expense the working of mines" and the manufacture of a necessary commodity. Nothing is known of the response to his requests for skilled workers. It seems safe to assume that the court was slow to act on such colonial matters. The problem of securing knowledgeable workers and manufacturers continued to hinder the growth of the French lead enterprise.[158] Although it remained a viable colonial industry for local consumption, after Spain came to possess the region in 1763, new knowledge and technology began to influence mining and smelting on the frontier.

In conclusion, De Gruy's expedition and reporting of Native American and French prospecting, mining, and smelting represents a continuum of the amalgam of cultural contact and the creation of new alliances. His narrative uncovers a number of changing patterns in the way Native Americans and French miners applied their environmental knowledge and tools to prospect for and extract galena. Most significantly, De Gruy's travels through the *country full of mines* provide a window into a longstanding indigenous smelting technology, which French miners adopted in order to manufacture lead bars for local and distant consumption. The De Gruy narrative is also useful for estimating the mining progress for the years between 1719 and 1743. Only Mine La Motte had been exploited. Even though miners worked only one season per year using "primitive methods," by 1743 the landscape had changed. De Gruy's analysis documents the effects of the exploitative methods of miners—digging numerous trenches in the surface of the earth and stripping the mining district of timber for smelting.

According to the Mine La Motte production records, miners continued to extract a large quantity of lead. Miners using Native American methods began to be viewed as mining in what De Gruy described as "a haphazard fashion."[159] Kaskaskia continued to be the lead depot. To transport the lead from the mines to Kaskaskia, Vallé directed his sturdy horses to carry the lead bars. To ease the movement of lead, local miners, after the lead was sufficiently melted, placed a large stick into the molten lead before it hardened. When the lead became solid metal, the stick was removed, leaving behind a hole through which carters could thread rope for ease of carriage. These details of production note the hard but somewhat profitable early stages of a developing mining industry.

During the years of French domination of the Mississippi valley, the region witnessed the continuation of a thoroughly native lead enterprise. Assisted by Native

Americans, De Gruy and French miners located additional lead mining sites. Independent of France and depending only upon the "strength of their arms," the settlers and the Native Americans carried the enterprise forward.[160] As a native currency, lead became increasingly significant in the form of baling seals for the fur trader, and in the form of musket balls for the protection of French and Native American frontier homes.

The prospecting, extracting, and smelting amalgam created by the convergence of Native American and French miners' knowledge and practice allowed them to establish a small enterprise. As the enterprise grew, new villages were established. One of the villages, Old Mines Creek, situated on a branch of the Mineral Fork, became significant following the arrival of Moses Austin in 1797. Austin came to Spanish Louisiana from Virginia with a number of knowledgeable British miners and smelters. These men would import their European extracting and smelting technique, as well as establish a permanent mining settlement. New technology would not erase the presence of the Native American and French practices. Instead, both European and Native American methods continued to develop alongside each other at two separate settlements, and lead trading continued within the Mississippi valley.

CHAPTER 2

Tracing Eighty Years of Early Mining Associations

Introduction

The battles were over, and the British along with their North American colonists could begin to enjoy new land acquisitions. For French colonists, the signing of the Treaty of Paris in 1763 proved to be harsh. All French territory, including *le pays des Illinois,* in North America was lost, and the British received Quebec and the Ohio valley. The port of New Orleans and the Louisiana Territory west of the Mississippi were ceded to Spain for their efforts as a British ally, and the Mississippi River became an international boundary. The region where early Native Americans and recent French settlers amalgamated mining and metallurgical practices was now in the hands of Spain. Forty-five years earlier, recognizing the possibilities for shipping at the Mississippi delta, where the Mississippi River meets the Gulf of Mexico, early French settlers founded the city of New Orleans. The colonial settlement quickly grew into a rich port city, shipping timber, agricultural products, and perhaps most notably, high-quality furs and lead ore transported downriver from the *country full of mines* in *le pays des Illinois* for quick delivery to Europe. Although the treaty was dated 1763, it would take until around 1770 before Spanish representatives departed the Mississippi delta toward what the Spanish administration called Upper Louisiana and Illinois Country. Soon after their arrival, French, Creoles, and Africans, both slaves and free, relocated from their villages on the eastern border of the Mississippi, controlled by the British, to the western border of the Mississippi, settling in either Ste. Geneviève or St. Louis now controlled by Catholic Spain. The Spanish administration organized Ste. Geneviève into a district that included much of the lead mines. Referred to by the Spanish as the Ste. Geneviève District, the village also became the administrative center where French and Creole miners maintained early French and Native American syncretic mining techniques.[1] For the most part, Spanish officials, instead of interfering with local affairs, focused their attention on defense, the fur trade, and controlling the immigration of foreigners, which ultimately preserved and encouraged the expansion of the amalgamated mining business.

At both Old Mines and Mine La Motte, miners sustained the lead mining and smelting traditions as the mining enterprise further developed. Lacking the necessary capital to dig deep shafts to extract greater quantities of galena, and still unable to construct masonry furnaces to smelt lead ores more efficiently, French Creole miners and their slaves remained successful by applying the amalgamated practices even though there were environmental consequences. By the 1740s, De Gruy already observed and documented miners repeatedly cutting and collecting copious amounts of timber to construct their "log furnaces" three times over. Surviving accounts from the François Vallé mining enterprise also reveal that the log furnaces required large amounts of timber for multiple firings during the three smelting phases, which resulted in a landscape with large areas of tree stumps.[2] In addition, after Moses Austin emigrated to Upper Louisiana in the late 1790s to settle near Mine à Breton, he expressed that "A continuation of smelting in this manner will exhaust the timber."[3] Smelting in log furnaces resulted in deforestation, which impacted the regional environment.

During the Spanish period, Old Mines and Mine La Motte continued to flourish, and eventually Mine à Breton became another primary source of lead mining and smelting. The lead business in the Ste. Geneviève district was connected to the village of Ste. Geneviève and remained in the hands of local French mine owners who sent engagés and slaves to do the difficult work of extracting and smelting lead ore. It is difficult to determine the percentage of pick-and-shovel labor at the mines, how many times slaves were rented out by their owners, or what percentage of engagés worked at the mines who were paid in lead.[4] Shortly after Spanish officials arrived in Ste. Geneviève, mine owners began to build small houses for engagés and slaves to live adjacent to the mines during their seasonal mining expeditions. Additionally, this period is tangentially connected to the beginning of Spanish reorganization of government, promoting changes in administration known as the Bourbon Reforms. The Spanish administration's organization of Ste. Geneviève into a district was one outcome of a Bourbon reform—the creation of large districts headed by an official directly responsible to the crown in Spain. Reforms also became one of the driving forces to shape silver mining and metallurgy during the Mexican late colonial period.

In Mexico, a mining bonanza followed the innovations associated with the Bourbon Reforms, as a royal commission began to inventory some 453 sixteenth- and early seventeenth-century mines in the Mexican mining districts.[5] The Bourbon Reforms during the late colonial period (circa 1760 to 1810) included administrative projects designed to centralize authority and increase economic productivity. In Mexico, from 1732 to 1770, a downturn in mining was related to problems with production, resources, and financing. However, beginning in 1767, with an increase in silver production, economic activity and population growth increased until 1809. David Brading described this period as a "revival" and has credited the economic boom to a number of Spanish officials who received assistance from the crown resulting in reinvestment in old mines to revitalize the mining industry. Brading also suggests that the mining recovery was dependent upon new capital to support the infrastructure and technology needed to refurbish many of the mines.[6] Although there is little doubt that silver production resurgence was due to the efforts of colonial officials and crown incentives, ordinary miners also played a role in Zacatecas's silver production and population revival.[7]

When Spanish officials finally made their way to Upper Louisiana in 1770, a second wave of Bourbon Reforms, in 1760 to 1810, would affect French Creole mine owners, miners, engagés, and slave mining labor who were now considered part of the Spanish empire. However, evidence of administrative projects designed to centralize authority and increase economic productivity began two years prior to the formal arrival of the Spanish governor of Louisiana, Don Antonio Ulloa. In 1768, after learning about Mine La Motte and how a Native American "had discovered a new lead mine," Ulloa commissioned a survey of the lead district. Ulloa believed "that if this mine is as good as is said, it would be to his majesty's benefit [and] could supply all of the Illinois where a lot of lead is used in private houses."[8] Ulloa also hoped to build a permanent year-round settlement at the mines. However, his hopes for such a settlement would not come to fruition until the last decade of the eighteenth century after Pierre-Charles de Hault de Lassus de Luzières arrived in Upper Louisiana's Ste. Geneviève District in 1793 and changed the Spanish immigration policy, allowing non-Catholic immigrants to settle.[9] In line with the Bourbon Reforms that shaped Mexican mining and metallurgy, Ulloa and De Luzières both desired to centralize authority and begin to see miners curtail the use of the Native American and French mining and smelting amalgam, and instead transfer European scientific practices to the Ste. Geneviève mining district.

In this age of emerging science, the old immigration policy practiced under Ulloa allowed only a small number of immigrants from Catholic Europe to settle in Upper Louisiana. However, beginning in the 1790s after De Luzières became in charge of the district, a new immigration policy was ordered to encourage population growth as foreign entrepreneurs possessing new scientific and technological acumen, as well as capital to invest in the business of mining and smelting, arrived in Upper Louisiana from the United States and Great Britain. Many of these entrepreneurial natural philosophers brought with them more advanced European mining practices that would eventually begin to supplant the amalgamated practices currently in place. In addition, De Luzières hoped that foreign settlers would bring the necessary capital to finance digging and constructing deeper shafts and erecting masonry furnaces to smelt lead ores using more economically and ecologically beneficial methods.

Before 1763, during the closing years of French control, smelted lead ore became more significant, and the lead enterprise continued to make progress under Spanish authority. Therefore, it is difficult to gauge the exact influence the Bourbon Reforms had on the direction of mining and metallurgy in Upper Louisiana. At Mine La Motte and Old Mines, the increasing power of lead and the negotiating and enterprising skills of local French Creoles made lead ore the center of Upper Louisiana's local currency. For example, as the population increased, the musket ball proved beneficial for the trader, for the wilderness home, and for nearby Native American villages, demonstrating that all occupants in Upper Louisiana were utterly dependent upon the security that smelted lead provided. The village of Ste. Geneviève grew from twenty-three persons in 1752 to six hundred in 1769, with most of the increase coming after the Spanish named it the administrative central district. In the Spanish censuses conducted between 1769 and 1785, African slaves accounted for approximately forty percent of the population.[10] In this middle-ground community that also included fur traders, merchants, and Native Americans, the people continued to exchange goods for pig lead and turned it into musket balls to defend their homes and villages.[11] In the years following French domination of the Louisiana Territory, with little guidance from Spanish officials, Creoles, new settlers, and slaves continued to prospect, extract, and smelt lead ore. As De Gruy had observed earlier, French Creole miners depending only on the "strength of their arms" added to the well-established amalgam the formation of local business associations, which carried the early lead mining enterprise forward with additional local and distant exchanges.[12]

On the eve of Spain gaining control of Upper Louisiana, French Creole families continued to arrive in the Mississippi valley. Many new settlers became connected to lead mining, formed mining associations, and expanded the lead mining business. Accounts and ledgers of new arrivals such as François Vallé, Nicolas Noel dit

La Rose, Pierre and Ann Gadobert, and Jean Baptiste Datchurut together highlight a significant growth in the business of lead mining and smelting activities in Spanish Louisiana. The accounts of mine owners consist of correspondence, bills, inventories, bills of lading, and transactions revealing the importance of pig lead as a form of payment to settle business accounts and to seal labor contracts and pay engagés. Pig lead, worth the equivalent of five cents a pound in the 1770s, was also used by settlers to purchase clothing, household items, foodstuff, musket balls, and to purchase miscellaneous items from New Orleans. Pig lead was practically at the center of every transaction connected to sustaining life in the colonial Ste. Geneviève district and beyond. In addition to settlers, Native Americans traded pig lead or offered it as part of their gifting practices within the confines of the Mississippi valley. Similar to the Mississippians, local eighteenth-century Native Americans worked to turn commodities into gifts and ordinary utilitarian tools, and in many cases, like the French the Spanish, also practiced gifting. From 1790 to 1803, Native American tribes, like the Osage, received hundreds of guns, blankets, needles, mirrors, and hatchets; thousands of knives, beads, and bracelets; and lead, as well as gun powder, rope, silk, shoes, and wool.[13] For these multiple uses of pig lead, Vallé, La Rose, Gadobert, and Datchurut attempted to form business associations at both Mine La Motte and a newly discovered mine at Castor Vein, where mine owners, engagés, and slaves applied various Native American and French mining methods to fulfill a number of cultural practices.

This forward-looking colonial enterprise was focused on developing the local and national economy through mining business associations, extracting, smelting, and trading pig lead as this embryonic industry passed under the dominion of Spain. Lacking support and capital from officials in control, French settlers learned to depend on one another and used their associations to exchange mineralogical knowledge as well as pig lead. Additionally, with the lack of financial support and possibly technological acumen to construct stone furnaces, the French and Native American practice of gathering profuse amounts of timber to smelt lead ore in multiple firings in the simplest way, by piling it onto log furnaces, also continued.[14] Since a single or double firing of lead ore recovered only forty to fifty percent, engagés and African slaves were forced to cut timber to build smelting furnaces anywhere they could until the introduction of the reverberatory furnace in 1798, which allowed for the recovery of a larger percentage of lead. Until then, as smelting increased so too did the consumption of fuelwood, which had a significant environmental impact on the eighteenth-century landscape.

Another chronic kind of environmental change came from the production of mining waste. Smelting lead ore in either a stone furnace or a log furnace left behind large piles of tailings, mine waste, and scoria (pebble-sized twists of furnace discards). The scoria or slag was the scum that formed by oxidation at the surface of molten metals during smelting; and the scoria raked out from the smelters was laced with lead, zinc, manganese, and mercury.[15] Additionally, mine waste came in the form of discarded rocks piled up near mine trenches and pits and finer tailings generated after processing the lead ores. Leftover tailings were also deposited around the various mines and furnaces, creating high embankments stocked with lead, zinc, cadmium, copper, and arsenic that spread into the environment as waterborne sediments or wind-borne dust. By the early 1790s, the environmental effects of mining and smelting became a function of the quantity and concentration of pig lead produced, and mine owners and Spanish officials decided to reprocess tailings and slag. To do so they called on Pedro Vial, a Frenchmen from Santa Fe, New Mexico, who had experience in smelting lead ashes in Mexico. Spanish officials expected Vial to transfer his technological acumen to begin a similar reclamation project in Upper Louisiana.[16] As in Mexico, Upper Louisiana officials desired to promote a boom to mining by bringing in miners familiar with new technologies to reclaim higher yields of lead from lead ore, scoria, and tailings. Eventually, after De Luzières wrote, printed, and distributed a recruitment pamphlet in the spring of 1797, the American entrepreneur Moses Austin immigrated to Spanish Louisiana with the new scientific and technological expertise to improve lead mining and to decrease the environmental impacts on the landscape.

Lacking financial support from the Spanish government, colonists employed social structures and relations around lead ore, which, in a sense, explains the emergence of rural capitalism in Upper Louisiana with lead at the center—a set of social relations in which laborers mined lead, manufactured pig lead, and used pig lead as a form of wages for their work. These social relationships in the countryside existed not only in agricultural and fur trading circles but also were developed by mine owners and miners in early American mining landscapes.[17] Therefore, as this early mining enterprise developed, miners and slaves unable to dig deep shafts to extract ore and unable to construct masonry furnaces to smelt their ores more efficiently reconciled themselves to the amalgamated practices. They collected lead ore from shallow trenches, and they cut copious amounts of timber to melt the ore.

Early Enterprises and Business Amalgamations

In Upper Louisiana just thirty-five miles inland from the Mississippi River in the Ste. Geneviève district, miners used either roads or the St. Francis River to move their lead ore and pig lead to nearby towns in this riverine colonial landscape.

Following the transfer of French Louisiana to the Spanish empire, a small number of French and Creole families dominated the lead mining business.[18] Notably François Vallé, Nicolas Noel dit La Rose, Pierre and Ann Gadobert, Jean Baptiste Datchurut, engagés and African slaves worked at both Mine La Motte and the newly discovered mine near Castor Vein. The Vallé, La Rose, Gadobert and Datchurut accounting records highlight ore-cleaning practices, new lead ore discoveries, and the seasonal nature of lead mining. Other fragments of the accounts open a window into the continuing application of the traditional extraction methods and the log furnace smelting amalgams; and they offer a picture of how the physical environment changed as mine owners, engagés on labor contracts, and slaves extracted lead ore by scratching the surface of the earth and chopping down trees in this heavily wooded area to melt the ore at Mine La Motte. The Vallé, La Rose, Gadobert, and Datchurut ledgers also demonstrate how commercial connections in Upper Louisiana more resembled the United States rather than European colonial commercial and cultural regimes as local business associations supported the frontier pig-lead economy.

Among the new settlers to trickle into Kaskaskia and eventually Ste. Geneviève from Canada in the 1740s was François Vallé. Vallé was born on January 2, 1716, in a town in French Canada in what is now Quebec. He was the fifth child of Charles Vallé and Geneviève Marcou La Vallé's twelve children, and due to limited economic opportunity, François moved in 1739 to the town of Kaskaskia in the Illinois Country. Prior to getting married, Vallé conducted trading expeditions as far away as the trading posts in present-day Indiana and Upper Canada, selling lead. In 1748, François Vallé married Marianne Billeron, the daughter of Léonard Billeron, a local notary, in Kaskaskia. Since Marianne could read, and Vallé was illiterate, which was common among frontier people, Marianne continued to play an active role in François's business affairs during her lifetime. She also helped to teach their six children, four boys and two girls, who all became literate. As early as 1753, Vallé owned property in Ste. Geneviève; and in 1754 he sold his home in Kaskaskia, and the family moved to Ste. Geneviève in the summer of that same year.[19] Despite the growing tensions between Great Britain and France and the war that finally erupted in 1756, the Illinois Country was relatively peaceful in the 1750s.

Another prominent lead miner and merchant was Vallé's son, François Vallé Jr., who was born in Ste. Geneviève in 1758. François Vallé Jr. married Marie Carpentier in Ste. Geneviève in 1777, and they had thirteen children. After the Spanish government gained control of the Louisiana Territory, Vallé would eventually become the first commandant of Ste. Geneviève and a member of the American government after the formation of the Upper Louisiana Territory in 1804. The bills and receipts from the François Vallé family business letter books reveal the importance of early mining activity at Mine La Motte—as De Gruy described, "digging to extract mineral, cutting wood to build log furnaces to smelt the lead, and transporting the lead by horseback" to Ste. Geneviève.[20] Multiple documents show that the Vallés joined other colonial miners at Mine La Motte and continued using the traditional methods to extract lead ore from shallow trenches and melt their lead on log furnaces seasonally. When not at the mines, they remained in Ste. Geneviève working their fields, growing wheat and produce. The Vallés also desired to erect a stone smelting furnace, but either lacking the capital or the technological skills, they along with other miners applied the longstanding traditional method of mining and smelting lead ore.[21] Although Vallé desired to establish a settlement closer to the mines thereby expanding the business, he and other business associates were content with their engagés and slaves following the traditional planting, harvesting, hunting, mining, and smelting seasonal schedule, or living near the mines on a part-time schedule.

Prior to the region being transferred from the French into the hands of Spain, Vallé continued to invest in mining operations, but he stopped mining himself by the late 1750s. He used the profits from his early mining and trading ventures to buy land that was farmed by slaves, and he became a part owner of Mine La Motte with La Rose sometime thereafter. Vallé owned sixty-three slaves by 1766 and eighty-four slaves by 1781, making him the largest slaveowner in Upper Louisiana and creating a significant slave population in the Ste. Geneviève district. Since Vallé was no longer digging and smelting lead ore, he sent the slaves who were not working his estate and fields to Mine La Motte; he remained in Ste. Geneviève tending to his family and businesses. The fifty-two-year-old La Rose was known as the "miner resident of Mine La Motte," but La Rose was not digging for lead ore alone; Jean Duval engaged himself to La Rose, and the contract was to run one calendar year beginning in November.[22] At this early date during the mining and smelting season, La Rose instructed Duval and Vallé's slaves to build small cabins where they would live for the three-month mining season before returning to Ste. Geneviève to join the other slaves in Vallé's fields to bring in the harvest.

The business association arranged between La Rose and Vallé was formed after Vallé's slaves and La Rose made a new lead discovery at the Castor Vein. The terms of agreement stated that "every fourth bar of lead was to be paid to Vallé for six years," which is the period that the company operated.[23] Within a few years of forming their business association, Vallé and La Rose also reached an agreement with Jean Baptiste Datchurut. Datchurut was a Ste. Geneviève trader who acquired land near Mine La Motte after the previous owner started to establish a seasonal mining settlement, as La Rose had instructed Duval and Vallé's slaves to do. When Datchurut took possession of the area near Mine La Motte, "the property included a house [and] three slave cabins where five slaves stayed while mining lead ore and cultivating the land for a little under four months at a time." There were ten horses, some household equipment, and some tools for exploitation, which workers used to extract and smelt lead ore.[24] A number of extant documents show that at the mine were "two hundred and four lead bars weighing approximately fifteen thousand pounds of lead in total." There were also two heaps of lead ore, ready for smelting, lying on the ground. Datchurut's account books also show how engagés and slaves most likely worked closely together mining, smelting, and farming. It is difficult to determine what percentage of the pick-and-shovel labor at the mines was completed by either Vallé's slaves or slaves rented out to La Rose by other slave owners in Ste. Geneviève. The lead mining accounts also note that various engagés labored at the mines and were due wages. Although Duval's contract to La Rose was for one year, most labor engagements at the lead mines were for shorter terms than the usual engagements, running for a month at a time instead of a year, which would work well with the three-month mining season. On the other hand, slaves not only worked at the mines but also cultivated the fields located near the mines for foodstuffs. Thereafter, mine owners began to construct permanent slave cabins, which indicates not only the seasonal nature of mining but also the possibility of moving toward yearlong work.

Mining accounts also provide additional insights into the role of engagés, slaves, and pig lead in the mining district. Engagés prized pig lead as a form of payment for their services. The agreements between Vallé, La Rose, Datchurut, and Gadobert, who later joined the association, include similar contracts with engagés, good for a calendar year from the signing date. While Jean Duval engaged himself to La Rose "to do everything he was ordered that was honest and legal" for that period, other accounting contracts record payments to workers and slave owners. Some contracts with slave owners who rented out their slaves to prospect, mine, or smelt lead ore during the mining seasons show owners, and sometimes engagés and possibly rented slaves, requesting pig lead as the form of payment.[25] Since the days of early French settlements, cash was hard to come by on the frontier. With no banks in the region, business deals were often conducted using credit in the form of furs and pig lead. Paying for services performed in pig lead, worth about five cents a pound, allowed settlers to purchase miscellaneous items from New Orleans, and it became the center of multiple types of transactions to help settlers sustain their frontier lives.

The accounts from the couple Pierre and Anne Gadobert not only reveal the position of women in the colonial Ste. Geneviève frontier community but also the continuing importance of slaves and engagés in their roles as miners. The prolonged absence of her husband gave Anne Normand Gadobert additional power and ample opportunity to act as a deputy to her husband's affairs. Women like Anne Gadobert undertook additional responsibilities while their husbands were away from the family to protect their own interests and that of their families.[26] The Gadobert marriage and multiple surviving business contracts make it clear that wives were entitled to at least one-half of the family estate. Moreover, when conducting important family business, husbands and wives were joint owners and both signed the contract of exchanges regarding goods. In 1774, when Pierre Gadobert left Ste. Geneviève on a trading expedition with Vallé, he expressly stated in an affidavit "he turned his mining affairs over to the charge of my wife," Anne Normand Gadobert. Madame Gadobert had the ability not only to handle the family's mining business affairs but also to invest in new projects she deemed important to their family enterprises related to mining lead ore.[27] For example, while her husband was away, Madame Gadobert operated her husband's share of Mine La Motte, as evidenced in the balance sheet for the mining season in the following chart. Lead was cheap at Ste. Geneviève in 1773, bringing in just over five cents (sous) per pound, which was the equivalent of the Spanish piastre or peso.

Value of lead delivered to Ste. Geneviève	780.94 piastres
Wages for *engagés*, smelting, tools, and foodstuffs	639.74 piastres
Blacksmith's wages for the season	18.00 piastres
Transport for lead	69.85 piastres
Total expenses	727.59 piastres
Net profit	53.35 piastres[28]

From an investment standpoint, Madame Gadobert's balance sheet shows that wages for engagés could consume a large amount of profits from lead mining. However, if the investors were slaveowners who chose to rent slaves out to the mines, using slaves instead of engagés would have substantially increased their profit margins. Additionally, the wages paid to local engagés highlight the importance of slaves to the lead mining enterprise. The Spanish accepted the longtime French practice of using African slaves in Upper Louisiana; however, a number of restrictions may suggest why the Gadoberts chose not to use slaves for mining. In the district of Ste. Geneviève, blacks were required to carry passports to cross the Mississippi River, which may have impeded the transporting of pig lead to small villages with limited access to the Mississippi River. The movement of slaves was restricted; for example, they were forbidden to congregate in their cabins, and they could not leave their cabins after a noted time in the evening. Recall that when the mine owner Datchurut took possession of Mine La Motte, on the property there were a number of houses. One house was for the engagés, and three separate slave cabins housed five slaves for the period they were at the mines. The need for slave cabins even for a short period would have resulted in additional expenses on the Gadoberts' balance sheet, thereby influencing their choice not to rent slaves. Finally, forty percent of all households in Ste. Geneviève owned at least one slave, and the Gadoberts may have just made the choice to use their slaves for fieldwork, clearing land, cutting wood, or domestic work on their Ste. Geneviève property, which was how the majority of slaves were used.[29]

Evidence that the miners at Mine La Motte still used the Native American trench extracting method appears throughout local account records. Recall how De Gruy noted "a vein of lead mineral a foot wide, and two feet down [he] found the same vein and extracted two hundred [pounds] of mineral" from below the surface of the earth. Without the required capital to dig deep shafts, slaves and engagés were most likely unacquainted with the utility of mining machines and could only procure lead ore from the surface.[30] In Europe, in order for miners to obtain minerals beyond twelve feet, they installed a windlass above their shafts to lower tools and to raise ore, water, or debris from their mines, and they followed veins by constructing galleries.[31]

The records of Vallé and La Rose further support how the French and Native American amalgam continued and allow us to gain a somewhat sketchy visualization of how French Creole engagés and enslaved miners exploited minerals. Prior to the report received by the Spanish governor of Louisiana, Don Antonio Ulloa, on May 26, 1768, regarding the Native American who "discovered a new lead mine," François Vallé lent two of his black slaves, one male and one female, to Nicolas Noel dit La Rose in the spring of 1757. The three individuals set out from Ste. Geneviève to discover lead mines on the "condition that Vallé would receive one-half ownership of any lead veins they discovered." Acting as prospectors, Vallé's two black slaves and La Rose did in fact make a discovery, which they called the Castor Vein near the Castor River Road, making Vallé a half owner in the new mine.[32] Local accounts also noted how the long, low heaps of mineral were being extracted from "gaping pits rimmed with red clay," indicative of trench methods.[33] It is difficult to really know what methods La Rose and Vallé's slaves used to locate the Castor Vein; however, since most "miners [were] contented with [digging pits] to three or four feet in depth," the French and native prospecting and mining amalgam remained intact. Although the Mine La Motte workers applied the mining amalgam by extracting deposits from the surface, they probably hoisted the lead ore using ropes and Native American buckskin baskets called *mococks* in lieu of the European-style windlass.[34]

Lacking capital investment from the Spanish centralized administration, the French Creole miners and their slaves continued to extract lead ore using the syncretic techniques, after which they cleaned their lead in preparation for the smelting process. It was important to scrape away as much of the spar and soil as possible using a small, sharp pick. The desirable size for the "lump of ore was the bigness of a man's two fists and weighed about fifteen pounds." Without the use of a stamping mill, miners crushed large lumps into smaller chunks by hammering or stamping on them before beginning the smelting process. The miners at Mine La Motte continued to manually beat lead ores until Americans arrived with the financial investments to construct European stamping mills to mechanically crush large pieces of ore.[35] Still following their seasonal mining and smelting schedule, miners dispersed into the forest to cut enough timber to build log furnaces by rolling large oak logs into place.

Helping us to visualize French Creole engagés and enslaved miners applying the traditional smelting amalgam to lead ore, sections in the Vallé letter book reveal workers and slaves transporting large amounts of timber across the mining district. Descriptions of miners building log furnaces and cutting large amounts of wood meant that the log heap furnace was the smelting method still employed at Mine La Motte. During the second mining season after miners returned from the first harvest, according to Vallé's records, miners alternated [changed their work patterns in alternate mining seasons] by conducting additional mineral extraction and then began "chopping . . . wood for the furnaces—estimated for forty furnaces."[36] In the late eighteenth century after Moses Austin sold his lead mine in Virginia and settled in Upper Louisiana, he noted miners building French Creole–Native American–style log furnaces to smelt their lead, stating "miners extracted their lead ore [and then] they dispersed into the forest to cut enough timber" for their furnaces. Austin states that miners smelted their lead by "depositing mineral in a pile of logs" and that the piles were then set on fire, consuming the entire log furnace.[37]

One did not have to be a miner or mine owner to understand that local farmers and their wives could easily extract galena using the Native American trench method. The mines offered farmers an alternative means to supplement their yearly income. Already, many locals could be found "working the mines between August and December after the harvest" as they "depended on the mines to furnish them with lead to purchase all imported articles."[38] The local business associates' accounts reveal how French Creole engagés and enslaved miners continued the longstanding seasonal planting and harvesting, hunting, and mining cycles but also reveal a change in the work habits of a number of individuals. The accounts of La Rose show an additional smelting season, which increased the wealth of some merchants.

In August 1769, La Rose entered into a contract with Pierre Gadobert for a partnership in the enterprise for 15,000 pounds of lead. The amount of lead in the yard amounted to four heaps, which he had mined, and which totaled 22,000 pounds of ore. Another miner's pile contained 18,000 pounds; and yet another miner's pile was appraised at 2,000 pounds of ore.[39] Terms of the agreement show what La Rose retained and mined when he entered into the partnership with Gadobert. Both La Rose and Gadobert established a large house, slave cabins, some cultivated land, and a yard for the lead bars in Ste. Geneviève. After the mid-eighteenth century, following the early intercultural mining practices established between the Native Americans and early French colonists, we learn that French Creoles and African slaves assimilated the traditional cyclical pattern of seasonal planting and harvesting activities with hunting and lead mining to exploit the seasonal diversity of the environment. Following the Native American calendar, the mining schedule still began around the beginning of April and continued for several months. During this cycle, miners returned to the village in time to assist with the first summer harvest.

During the mid-eighteenth century, miners began to take advantage of the second mining season, which ran from August to December, nearing the end of the second harvest until the first frost. With two mining seasons, mine owners directed engagés and slaves to build houses and slave cabins. As the location started to resemble a mining settlement, in addition to surface mining, and throwing chunks of lead ore atop their log furnaces, miners, engagés, and slaves also began to cultivate fields, thereby taking full advantage of the mining country's rich alluvial soils. With the addition of another smelting season, the quantity of the product, pig lead, increased. The accounts that have survived highlight the years between 1769 and 1775. They are incomplete; therefore, it is difficult to summarize any production costs and profits. Nonetheless, they do offer a basis for estimating the difficulties attending a business in the days when most accounts were settled in lead. The accounts also suggest how the forest near the pits must have been scarred with areas of stumps and how lead ashes would have been piled next to the smelting sites.[40]

These accounts require imagining the effect that French Creoles and enslaved miners' activities at Mine La Motte had on the ecosystem after they exploited and smelted the minerals. The accounts reveal a physical environment where the presence of the Kaskaskia-French amalgam still existed, and where those engaged in mining at Mine La Motte had produced a considerable quantity of mineral. The ore heaps totaled 122,000 pounds, which after smelting would have yielded about 61,000 pounds of bar lead. With the 105 bars in La Rose's possession probably weighing 8,000 pounds, the exploitation of this section of Mine La Motte for a year was about 69,000 pounds of lead.[41] During the eighteenth century, after Ste. Geneviève became the main depot for shipping pig lead to New Orleans, many visitors to the mines stated, "The lands . . . are beautiful and fertile . . . and already some inhabitants are thinking of settling there."[42] Their comments suggest the importance of seeing farmers and miners of Ste. Geneviève construct a self-sufficient mining community, similar to ironworks in the eastern United States, except where lead was the ore of choice.[43]

Since Mine La Motte was the central mine before the arrival of Moses Austin, we can offer an estimate of the relation between the scattered Mine La Motte production figures and the lead shipments to New Orleans. In 1770, the portion of Mine La Motte under the control of La Rose probably yielded 69,000 pounds of lead. The Vallé exploitation may have added about 9,000 pounds to this figure after 1772. Another part owner's section of the same mine produced about 9,000 pounds. This would indicate that in a good year Mine La Motte might produce 78,000 pounds of lead. In 1772, of the ten Ste. Geneviève citizens who sent goods to New Orleans, seven sent their lead shipments totaling 60,000 pounds.[44] This amount was probably a large part of the lead production of that year. The shippers were Antoine Bernard, Henri Carpentier, Benito Vasquez, Esteban Barre, Pablo Segon, Diego Fortier, and Daniel Fagot Lagarciniere. These individuals may have conducted exploitations at Mine La Motte, or they may have procured the lead in trade. Lead was the one commodity that the pioneers could keep for a period, so it may be that part of this shipment had been mined at some previous time; however accounts also show that most often shippers and conveyers of lead wanted to be paid in the mineral.

When her husband was away, Madame Gadobert became the deputy who controlled their mining business. In this role, she often had difficulty conveying lead ore or pig-lead disks. For example, in January of 1775, Madame Gadobert was determined to transport lead from the mines to Ste. Geneviève even though it was during the winter, explaining that the "weather was favorable and any delay might cause the lead to be lost." She turned to the family attorney, Henri Carpentier, who in turn enlisted the help of Vallé (who he worked closely with while serving in the Ste. Geneviève militia) with setting up a bidding process. Carpentier, a

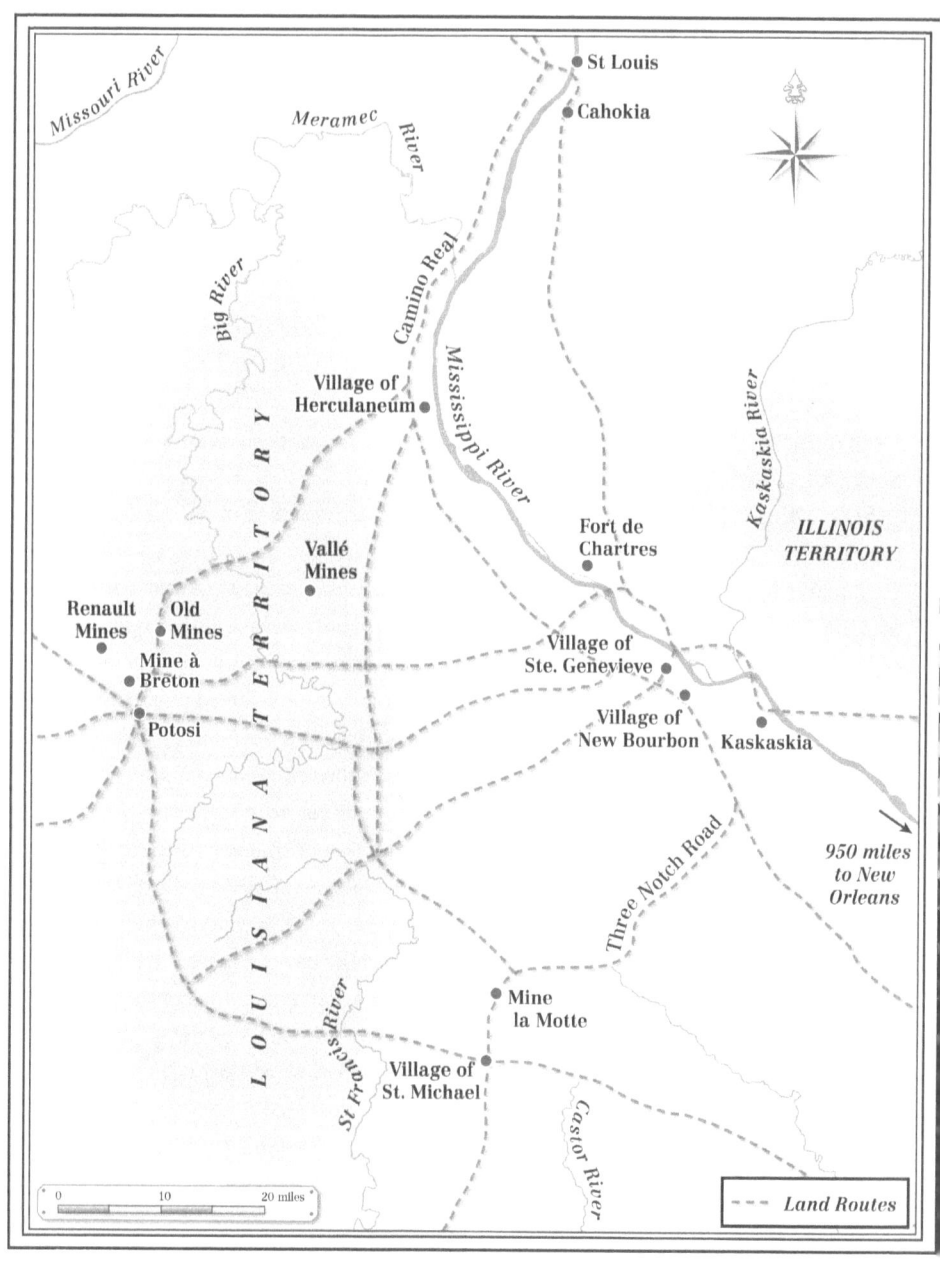

FIGURE 3. French and Spanish period villages and land routes to and from the seasonal mines, late 1700s.

person of means who owned five slaves, understood the urgency of the matter, so he and Vallé decided to make an announcement the following day at the church door. As the people left the church "in great numbers" after mass, he shouted "that he would immediately take bids on the transportation of the pig lead from the mine to Ste. Geneviève. He also insisted that the bidder must complete the delivery by March 10th and take precautions to ensure that the metal would be safely transported." The cost to convey the lead was approximately ten cents a pound. Madame Gadobert also promised that whoever secured the bid would have access to all of her horses, which were ready and available to use as a "pack train." Once the bidding started, Carpentier joined in by voicing his willingness to convey the mineral in exchange for 800 pounds of lead; another resident bid 644 pounds of lead; and it appears that Vallé bid 643 pounds of lead, which lead to the former resident bidding 642 pounds of lead at high noon, and no one else entered the competition. In the end, Madame Gadobert paid 1,284 pounds of lead to have the ore transported back to Ste. Geneviève.[45]

The Vallé and Gadobert experiences not only offer images of a changing physical environment, but also show that Ste. Geneviève, named the center of the district by Spanish officials in 1770, was a place where it was common to conduct transactions, using pig lead as a form of payment to settle business agreements beyond the purchasing of miscellaneous items. Pig lead was practically at the center of every transaction connected to sustaining life in the colonial Ste. Geneviève district. This fact highlights how in the midst of this community's social and spiritual lives, they were constantly focused on the business of lead ore or pig lead. Villagers employed bidding on the church doorstep to convey lead ore from one place to another as rural capitalism emerged in Upper Louisiana. Mine owners and miners developed these business relationships in the mining frontier to coincide with farming, trading, and even religious practices.

Tracing Eighty Years of Mining Business

Well into the late eighteenth century, with the influence of the Bourbon Reforms in Mexico, a successful mining bonanza developed. Similarly, in Upper Louisiana Spanish officials also desired to remake Mine La Motte into a successful permanent mining settlement where miners would begin to apply European practices.[46] When Vallé and La Rose agreed to expand their business associations to include Jean Baptiste Datchurut of Ste. Geneviève, Mine La Motte had already become a seasonal mining settlement (figure 3). The landscape about the mines became dotted with tools, horses, and household equipment. A house for engagés and three slave cabins were constructed. When not working with lead ore, slaves were required to spend their time clearing trees, tilling the soil to improve the land, and growing crops.[47] Judging by the amount of lead extracted, smelted, and conveyed from Mine La Motte to Ste. Geneviève, this once seasonal mining location was transitioning into a European-style settlement with mining at the center of operations. Around this time, Upper Louisiana miners would have been very familiar with Gabriel Plattes's *A Discovery of Subterraneal Treasures*. First printed in the late seventeenth century and again in the 1780's, Plattes's outline of well-established European mining practices would have been a guide for Ste. Geneviève miners for how to reproduce English and German mining communities near the lead mines.[48] Plattes's manual also outlined the importance of forming a mining association, which Vallé and others used as a guide to organize their mining association. Now their business association turned to Plattes again to recreate Mine La Motte into a mining settlement closer to the source of their lead ore.[49]

To ensure that miners were content and successful at the mines for an extended period of time, Plattes recommended the following be considered when deciding where to establish a mining settlement. The construction of a settlement needed capital backing; therefore, the mining association played a key role. The business association must include wealthy merchants who possessed the capital needed to support exploration and digging. In addition, finances were needed to build roads, houses for miners and members of their families planning to relocate to Spanish Louisiana, and smelting furnaces.[50] With a permanent resident mining labor force, fields needed to be cultivated to grow foodstuffs to sustain engagés and slaves. In addition, the association needed to consider carefully the location of the settlement and its proximity to the lead ore, copious amounts of timber, and water. Acres of forest would be needed not only to build houses and slave cabins for workers and their families but also to provide fuel to smelt lead ore. Plattes recommended that mine owners acquire access to quarries to exploit the natural stones to construct masonry furnaces. Desiring to establish European mining practices, miners also needed to have access to flowing streams of water to power waterwheels for pumping water out of shafts and for cleaning and grinding lead ore before smelting.[51] Mine La Motte and additional mining sites in the vicinity were located near navigable rivers so that pig lead could be transported to Ste. Geneviève and to markets east of the Mississippi at the lowest possible cost.

As Vallé and those engaged in the business of mining and smelting lead ore continued to strengthen their associations, the possibility of establishing a settlement at Mine La Motte became a reality when Spanish officials offered Vallé and his associates the opportunity to obtain a land grant. At this point in time, the association consisted of Vallé, part-time miners, and farmers including Paul De Guire, Joseph La Chance, Gabriel Nicollet, Jerome Matis, Peter Chevalier, and Pierre Variat who all joined Vallé's petition to apply for land grants just south of Mine La Motte.[52] With a new village, miners could spend more time committed to their mining chores, as opposed to spending two days each week traveling between the mines and Ste. Geneviève. However, until they were granted the rights to the desired land, they continued mining seasonally as the Native Americans and French had done earlier. Their sprouting settlement in the far interior would be called St. Michel, and later became known as "St. Michael's," after more Americans than French moved into the village following the 1803 Louisiana Purchase. In 1799, when St. Michel's was initially established in close proximity to the lead mines it was a French Creole space where engagés and slaves extracted and smelted lead ore according to the traditional methods established almost seventy years earlier.[53]

The story of St. Michel begins when the Tamaroa guided Antoine de la Mothe Cadillac to their mines (which later became Mine La Motte) soon after he arrived in Kaskaskia in 1714. One of the early French explorations to Mine La Motte took place in the hills west of the Mississippi near the location where possibly the same Tamaroa miners guided De Ursin and La Renaudière. Both of their early eighteen-century mineralogical reports described why they believed the region was favorable for economic development, where miners could settle and enjoy a good living by extracting and smelting "approximately three hundred million pounds of lead each year" and delivering it throughout the Illinois Country by land and water. In 1723, La Renaudière described Mine La Motte as the ideal place to establish a settlement. He notes, "In the neighborhood of the mines there are small prairies [to] serve as retreats to numerous horned buffalos, bears, and deer. There are persimmon and pecan trees; these fruits furnish nourishment to the Indians.... This is very fine country, it would be a good place for settlements; there are rivers in which fish abound and a large number of fowls."[54] The location of Mine La Motte was also where Philippe Renault, his slaves, and other French settlers labored to get the lead out well into the 1740s. Then in 1757, François Vallé and his slaves found ore along the Castor River; and by 1773, other Kaskaskia and Ste. Geneviève residents became involved in mining and formed the first known partnership with Vallé. Although these mining activities were rooted in traditional practices, gradually a number of mine owners began constructing cabins scattered across the landscape to access the lead mines seasonally and then year-round. One engagé's contract stipulated that he work at Mine La Motte but also provided him with fifteen days to return to Ste. Geneviève to harvest wheat.[55] While a number of extant documents highlight the formation of a seasonal settlement, other François Vallé letter books show the emergence of a year-round settlement at Mine La Motte and the steady penetration of European labor techniques that were amalgamated with the longstanding traditional practices.

During the first half of the eighteenth century, there is no evidence that miners worked according to the division of labor common in Europe. However, by the second half of the eighteenth century, the names and work assignments of De Guire, La Chance, Nicollet, Matis, Chevalier, and Variat appearing throughout Vallé's letter book uncover the gradual influence of the Bourbon Reforms as European division of labor practices appeared at Mine La Motte.[56] The division of labor is the separation of tasks in any system so that participants may specialize at a particular work assignment such as surface mining, cleaning ores, separating ores, or smelting ores. In Europe, Adam Smith envisioned the essence of industrialism by determining that division of labor represented a substantial increase in productivity.[57] By 1785, Immanuel Kant noted how many crafts and trades profited from the division of labor as workers focused on one particular kind of work and did a better job than when one person did everything.[58] Vallé or his wife, Marianne Billeron, who was literate and active in François' business affairs, listed specific work tasks next to the names of De Guire, La Chance, Nicollet, Matis, Chevalier, Variat, and engagés; or in the case of a slave, the owner's name was listed. The letter books offer an image of a tightly knit community of Creole miners and slaves. It is not clear whether Vallé adopted these new labor practices because he hoped to extract greater profits; however, what it does signal is the disappearance of the amalgam where miners combined their strength to extract, smelt, and shape lead.

One of the first divisions of labor Vallé incorporated at Mine La Motte was the assignment of Nicolas La Chance as the overseer of all mining activities, a practice also outlined in Gabriel Plattes's manual. To fulfill his duties as the overseer, La Chance, under Vallé's instruction, would have supervised the mining activities, a typical European and American practice.[59] On American plantations, the overseer earned his wages by trying to get the most work out of the slaves. The overseer on a plantation was consider the "middleman" who was given the responsibility to manage the enslaved laborers as the master ordered. Overseers lived in a separate house away from the main house but in close proximity to the slave cabins. In the case of Vallé, assigning La Chance as the overseer allowed Vallé the freedom to be away from daily mining management at Mine La Motte. Similar to the plantation overseer, La Chance was also responsible for the implements and tools required for all the steps in the mining process. La Chance was a likely choice. Similar to Vallé, his family had roots in Canada and migrated to Kaskaskia. Additionally, the variety of work done by La Chance at the mines called for him to be familiar with various supplies. These activities also benefited the development of other economic activities in the village of Ste. Geneviève, such as agriculture and commerce.

In addition to fuel, food, and fodder for miners and the animals, water and wood for construction were also necessary. Food, for example, was increasingly grown locally. As previously noted, while away at Mine La Motte, slaves also cultivated crops nearby. Growing food near the mining settlement lowered the cost of conveying food thirty-five miles inland from Ste. Geneviève and most likely awakened the ingenuity of the overseer and mine owner. Miners began to till the soil in the valleys and on the slopes of the surrounding hills, establishing farms in the vicinity of mines to remedy the shortage of food and the considerable price to cart agricultural products into the mining district. Another new labor practice was to assign one individual to do all the hunting. The name La Malice appears in the Vallé letter book, and his responsibilities included hunting and gathering food from the surrounding area. He supplied meat, biscuits, meal, salt, and oil for the miners who were too busy prospecting, extracting, smelting, and molding lead ores into movable bars of pig lead. Other than these two specified work assignments, the Vallé account fails to show any additional divisions of labor among the other miners; therefore, one can assume that De Guire, Nicollet, Matis, Chevalier, and Variat continued to jointly prospect, mine, and smelt lead ores.[60]

The Vallé account books also note an additional implement La Chance would have become responsible for in the mining process. In addition to the tools described by De Gruy in 1743—the iron probe, the shovel, and the poll, which served as both a pick and a hammer—the pointed shovel, called a spade, became the ideal tool to penetrate the coarse fragments in the trench.[61] Miners used their shovels when they encountered "soil intermixed with stones." Since miners encountered limestone intermixed with lead ore, La Chance, similar to European supervisors, was now responsible for conducting nightly tool inspections to ensure their sharpness for the following day's work. Dull tools would have been passed on to the most mobile of the workers, the hunter, La Malice, who combined his hunting trips with stops at the blacksmith in Ste. Geneviève, where he could have the tools repaired or sharpened before returning to the mining settlement. Later in the 1790s, after the Virginia miner Moses Austin organized a settlement at Mine à Breton, he outfitted the location with a blacksmith shop. Thereafter, La Malice could cut his travel time by one-half day (as opposed to making the trip to Ste. Geneviève) and have the blacksmith at Mine à Breton fix or sharpen the tools.

To the prospecting and extracting amalgam, Vallé, his engagés, and slaves combined the European fire-setting technique to extract ores while surface mining.

The Vallé accounts list an expensive powder, steel, and other items translated as flame or light. When miners combined these articles, they successfully produced a fuse for fire setting to help with the extraction of ore from in between limestone. While the fire setting did not allow miners to dig deep shafts, it did help miners to crack limestone and recover lead ore, which was a significant change. Often miners would dig two feet down and find a vein of lead only to find a barrier in the form of a slab of limestone or some other obstacle. Now, fire setting along these rock formations offered miners greater access to more substantial amounts of lead ore.[62] The earliest record of miners using the blasting method at Mine La Motte was in the late eighteenth century.[63] Depending on whether the ground was wet or dry miners employed two methods of blasting. Miners placed the gunpowder into a powder burn or tin cartridges to protect the fuse—and then rubbed the paper with gunpowder. If miners encountered a wet area, they filled the space with clay to act as a drying agent.[64] The Vallé letter book and Plattes's mining manual both reinforce the gradual transfer of European mining acumen and technologies to Mine La Motte.

The Habitants Are Contented with Scratching the Earth

When the French Revolution refugee Pierre-Charles de Hault de Lassus de Luzières returned to New Orleans in the spring of 1793, after his initial visit to Ste. Geneviève, De Luzières began to organize a plan to establish a new community three miles south of Ste. Geneviève. De Luzières hoped to attract other French émigré settlers from other places across the United States to the hills overlooking the Ste. Geneviève fields where Vallé, De Guire, La Chance, Nicollet, Matis, Chevalier, Variat, engagés, and slaves farmed wheat. Although the Bourbon king Charles IV still ruled in Spain, the staunch royalist De Luzières determined that the new settlement would be named to honor the memory of the recently beheaded Bourbon king of France, Louis XVI. De Luzières noted, "I have decided that [the settlement] be called Nueva Bourbon. Not only in order to place the new settlement under the special protection of the sovereigns [governing] España, but also that the descendants of the new colonists may imitate the loyalty of their fathers for the king." According to De Luzières, the new village would "contribute to the promotion of industry to derive greater profits for the public." During his yearlong stay in Upper Louisiana, he compared the area to France—"the country of [Illinois] resembled his home country in climate, soil, productions, mines, and manufactures." He continued to describe the mining region as follows: "[the] country abounds with [mines] yet to be opened, as the habitants are contented with scratching the earth three or four feet in depth. These mines, if worked with intelligence, would come to be the source of great prosperity for individuals, and a national aim of the greatest importance."[65] De Luzières clearly had agricultural pursuits and lead mining on his mind when writing his descriptions of the Ste. Geneviève mining district. Similar to Ulloa's vision to apply the Bourbon Reforms to the mining area almost three decades earlier, De Luzières also desired to see miners cease using the seasonal mining and smelting amalgam and instead transfer European mining practices to the Ste. Geneviève mining district.[66]

By the end of 1793, De Luzières succeeded in establishing New Bourbon three miles south of Ste. Geneviève. It would become the district's second largest settlement behind Ste. Geneviève in 1800, with a total of 630 souls. During the 1790s, New Bourbon attracted farmers who would venture out seasonally to mine for lead ore, and more young miners from the village of Kaskaskia and other parts of the young United States also settled in New Bourbon before relocating closer to the mines. In fact, Vallé's mining supervisor, Nicolas La Chance, and his large family, after staying in New Bourbon, decided in 1799 to become one of the first to establish the village of St. Michel near Mine La Motte. This relationship between New Bourbon villagers and St. Michel villagers continued well into the 1810s, as close family members and residents of the two traditional French villages exchanged lead and property among themselves. De Luzières, who also became involved in lead mining, remained focused on improving and promoting lead mining, smelting, and lead manufacturing. He strongly believed it was necessary to apply the Bourbon Reforms at the Spanish Louisiana lead mines in order for the empire to see considerable economic growth.[67] To further this plan he set his sights on finding an experienced miner.

Moses Austin, the Virginia lead miner and manufacturer of lead products, arrived in 1797 to examine the lead mines. He recorded in his journal, "Nature has undoubtedly intended this Country to be not only the most agreeable and pleasing in the World, but the Richest also."[68] After Austin arrived, he too embraced the vision of improving lead production, and he initially turned his attention to bringing an end to the traditional wasteful smelting practices. Austin states, "The French inhabitants of this country have followed the mining business upwards of eighty years, yet they have not advanced in the art of smelting a step beyond their ancestors."[69] During Austin's initial visit, he toured the mining district with François Vallé. Austin also observed miners spending a majority of their time cutting numerous trees to obtain the significant amounts of timber required to build log furnaces and begin smelting and resmelting their lead ore to obtain higher yields of lead.

De Luzières also wanted to see miners discontinue firing the log furnaces especially after Austin expressed his plan to construct the first reverberatory furnace in the district at Mine à Breton. There was another furnace variation in use at Mine La Motte in 1797 that was different from earlier log furnaces. It resembled a furnace used in Europe. The construction date of these furnaces is uncertain, but it must have been in use during the period of Austin's correspondence with De Luzières. According to Austin, there were twenty such "French furnaces," which were similar to European furnaces and more complex than the log furnaces. These furnaces had three log walls lined with split logs standing upright. Miners placed about five thousand pounds of ore in the furnace and covered the ore with additional logs.[70]

Later, Austin and other Americans commonly called these peculiar furnaces "log hearth furnaces," which further notes the penetration of European technologies and their existence alongside of the amalgam.[71] The descriptions resemble the Scotch hearth furnace, which originated in England, and miners placed these furnaces on summits or western slopes of the highest hills. After Austin settled in Upper Louisiana, one of the miners who joined him, Josiah Bell of Derbyshire, England, would have recognized these furnaces. A similar furnace had been mentioned by Agricola.[72] Agricola describes a method used in the Saxon mines that was also used in Derbyshire in northern England to smelt lead ores in "very rude furnaces, or boles, urged by the natural force of the wind."[73] The log hearth furnace was constructed on a hillside with an inclined hearth surrounded by walls on three sides. The wall at the base of the incline was constructed with an arch for the admission of air.[74] Usually, two furnaces were built together, making the common center wall stronger. In addition, the design was more economical, since one crew could manage and regulate the heat of two furnaces.[75] When this furnace was constructed in the Mississippi valley is uncertain; it may have been in use since the last decade of the eighteenth century after miners in the Upper Mississippi valley corresponded with miners at Mine La Motte.

Most likely, the furnace was a hybrid of earlier Native American and European furnaces.[76] Following contact near Galena, Illinois, European and Native American miners designed a hillside trench furnace to yield more metal. One early description of such a furnace dates back to 1804. In that same year, Sergeant Nathaniel Pryor of the Lewis and Clark expedition purchased a supply of lead near Ste. Geneviève to make musket balls for his trip into the West. Pryor described the shape of what he calls a "Native American furnace as being similar to a mill hopper," which is also similar to the conical shaped furnaces that Native Americans constructed.[77] Further north, in the Upper Mississippi basin, the Fox also developed a very interesting method to smelt lead ore, which may account for the exchange of smelting technology between Mine La Motte and the Upper Mississippi lead mines. The New York glassmaker Henry Rowe Schoolcraft, who toured the mines with Moses Austin in 1819, also gives a very clear description of the furnace at Mine La Motte as well as at the Upper Mississippi lead mines. While touring through both mining regions, Schoolcraft used the expression "like the roof of a house inverted" to describe the same furnace.[78]

An Italian traveler, Giacomo Beltrami, who visited the mining frontier also described the furnace shape as similar to a mill hopper.[79] Beltrami was a passenger on the *Virginia*, which was the first steamboat to ascend the Mississippi River. Stopping at the Dubuque mines, he observed how the Fox "smelt the lead in holes which they dig in the rock to reduce it into pigs." A little reflection suggests that for Native Americans smelting lead ore in the Mississippi valley, this may have been a practical way to prolong the life of furnaces. Most significantly, these furnaces were very similar to another type of furnace Austin described when he witnessed miners depositing mineral into a "furnace of stone somewhat similar to lime-kilns." These furnaces consisted of a square area dug out of the earth and lined on three sides with stone or brick to make a three-section furnace. Similar to early European furnaces, the fourth side was left open so that miners could pile in timber to cover the ore, which kept burning for hours. The masonry walls retained the heat more efficiently. Miners then layered firewood in alternating patterns, and larger pieces of ore were placed on top, followed by smaller pieces of ore placed around the sides.[80]

Another variation of the same type of furnace used by Native Americans was a horizontal trench dug in the side of a hill, with smaller trenches running downhill to distribute the melted lead into separate holes to cool the metal. Miners filled the primary trench with ore and logs. It was then ignited, and the molten metal ran out the small trenches into the various holes. After the lead cooled, it resembled pig bars weighing sixty to eighty pounds each. Sometimes a hard stick or other object was placed in the center of the mold so that the lead would cool leaving a perforation to slip rawhide rope through for easy transport. Although it is not clear whether this type of furnace was used at Mine La Motte, American miners and Native Americans used the smelting method at the Galena lead mines. These early furnaces only yielded about forty percent lead; the rest of the ore was reduced to lead ash or was lost. When miners began to use this type of furnace is unknown; however, since Schoolcraft's illustrations were similar to sketches drawn by Agricola, the furnace design must have originated in Europe.[81] The use of this particular smelting furnace at Galena represents the region's transition as new alternative mining techniques converged with preexisting mining practices.

A Continuation of Smelting in This Manner Will Exhaust the Timber

In addition to Agricola calling attention to humans working in subterranean mines, he noted that particles of lead dust were penetrating the human body; he also noted that the fields were devastated by mining operations. In some European regions, miners were warned by law not to dig for minerals "so to injure their very fertile fields, their vineyards, and their olive groves."[82] Europeans also argued that the woods and groves were cut down in excess for the endless amount of wood for constructing timber machines and the smelting of metals. For example, Italians expressed that "when the woods and groves are felled, then are exterminated the beasts and birds, very many of which furnish a pleasant and agreeable food for man." Agricola also wrote of villager complaints about miners "washing ores and the poisoning of brooks and streams, . . .[which] either destroys the fish or drives them away." Similarly, the Upper Louisiana residents were also concerned about the devastation of their fields, woods, groves, brooks, and rivers, which hindered their ability to procure the necessaries of life.[83]

Lead mining in Upper Louisiana affected the local environment in a number of ways. First, communities established near the lead mines were deemed harmful to animal health. Visitors observed how horses, cattle, dogs, cats, and chickens dropped dead after licking lead slag or breathing fumes of arsenic and sulphur.[84] The second issue concerned fuelwood consumption. From an ecological perspective, the elimination of forests began the destruction of the subsistence base of the miners, their slaves, and their animals. It also compromised the subsistence base of the indigenous peoples, which hastened their relocation and the demise of their autonomous living culture. The surrounding landscape that the early eighteenth-century ironmaster Philippe de La Renaudière described would have been unrecognizable if he returned in the late eighteenth century when Moses Austin arrived and established Mine à Breton. Austin depicted the landscape as completely denuded of any tree or sizable shrub. Austin wrote that "a continuation of smelting in this manner will exhaust the timber." He attributed this to miners' employing the amalgamated smelting methods.[85] Due to the construction of numerous log furnaces to meet miners' smelting needs, deforestation began to have important consequences on the landscape. Tree growth was stunted, and observers interpreted this simple spatial association to mean that lead ore caused diminutive trees, although stunted post oaks were known to have occurred on cherty surfaces not known to contain lead.[86] The third and most detrimental effect was the occupational exposure of engagés, slaves, and animals to lead smelting when they performed their work assignments and came in direct contact with high concentrations of the melting mineral and breathed in poisonous gases.[87] People and animals had to stay away from the "poisonous effluvia" produced by smelting; perhaps that is why black slaves did most of the smelting.[88]

The consolidation of business associations affected the regional environment as colonial mining increased in scale after François Vallé, Nicolas Noel dit La Rose, Pierre and Ann Gadobert, and Jean Baptiste Datchurut joined their operations, tapping the once abundant natural resources. Vallé's accounts show that smelting in the log furnaces required large quantities of timber, so much so that the forest must have been scarred with large areas of tree stumps. The Vallé accounts highlight paying workers to cut copious amounts of timber but also describe workers carting timber from greater distances back to the mines as local nearby resources were becoming increasingly depleted.[89] The concern with the exhaustion of timber appears to have started around 1798 when Moses Austin warned that wood stocks were in danger of disappearing. Austin's concerns were similar to those of Alexander von Humboldt. During a visit to a Mexican mine and smelter in the late eighteenth century, Humboldt wondered how mining survived in a country "which wants combustibles, and where the miners are on table lands destitute of forests."[90] In 1804, after four years of traveling in South America and Mexico, Humboldt visited the United States. During those six weeks, if after spending time with President Thomas Jefferson, he had traveled to Upper Louisiana's lead mine district and inquired about the "destitute forests," the local miners would have answered by explaining how they just tap wood reserves at greater distances away from the mines.

In 1800, after forming a new association, François Vallé, St. Gemme Beauvais, and other owners of Mine La Motte sent a petition to the son of De Luzières, Charles (Don Carlos) DeHault Delassus. After his arrival in New Orleans in 1794, he was appointed to be the civil and military commander of the post of New Madrid, which allowed him to be near his parents in New Bourbon. By 1799, under orders from Spain, Delassus was appointed lieutenant governor and commander in chief of Upper Louisiana, and he was stationed at St. Louis until 1804. He presided over the transfer of Upper Louisiana to the United States in February 1804. In effect, their petition explained that because of the past exploitations of lead ore and the number of furnaces in operation "they had exhausted the nearby timber to such an extent that the continuation of their enterprise was handicapped, and to continue they needed to purchase about 3200 acres [five square miles] of additional territory."[91] What had concerned Moses Austin after he arrived was now realized. Delassus replied to the petition for additional forest lands noting that he did not have the power to sell such a grant of land, but in turn, he promised to recommend that the administrator look favorably on their petition because of their loyalty to Spain.

Two years passed before one of the Mine La Motte owners, Father Maxwell, the St. Michel parish priest and miner, presented their request to the Intendant Juan Bonaventura Morales in April of 1802. Spain appointed Juan Morales the intendant general of the Province of Louisiana on October 22, 1798, and was given the power to grant lands, and in April 1803, the United States acquired Louisiana. Attempting to lend aid to the Mine La Motte owners, Delassus endeavored to secure additional property; however, the first board of commissioners refused to confirm the grant, which was now 24,010 acres and also contained a rich lead mine. The local miners' request, Vallé's letter book, and Austin's concerns together suggest that prior to the Louisiana Purchase, and two decades before miners fully implemented European smelting technologies, the forest in the surrounding area was already exhausted of timber, forcing miners to request additional land specifically to supply their furnaces with fuel.[92]

Lastly, the production of mining waste also led to a more chronic form of environmental degradation. The mining waste materials came in various forms: unused rocks piled next to each mine's surface openings and finer tailings generated after multiple processes. Tailings consist of ground rock and process wastes generated in extraction of the desired product from the mine ore. In addition, the scoria raked out from the smelters was usually just left in the form of waste laced with lead and sulfur. Given the accounts of François Vallé and the early business associations he formed, it is difficult to know how much waste accumulated, or how much exhaust dispersed from the log furnaces into the atmosphere and blanketed the landscape following long periods of smelting. The multiple examples of establishment of new technologies in the Upper Louisiana mining frontier suggest the beginning of a regional redefinition of mining settlements, which had produced a chronic environmental effect on the landscape and those living in the vicinity of the mines.[93]

The Mining District's Reclamation Project

As administrator of the Upper Louisiana lead mines during the Bourbon Reforms, De Luzières continued his attention to developing the lead industry to promote the local and national economy. De Luzières and local miners hoped to obtain an additional yield of pig lead to sell or mold into musket balls. One of Luzières's goals was to import new technologies to extract more lead from the scoria that remained piled on the ground in the form of ashes near the smelters. Mounds of lead ashes were the result of smelting with log furnaces. In an effort to procure an additional amount of lead from the ashes, miners built their own log furnaces and resmelted the lead ashes. The accounts of various business associates describe a significant amount of ashes remaining near the furnaces and how miners often searched through these ash heaps for lead ore that had not succumbed to the initial smelting process. To accomplish this goal, De Luzières decided to look for a skilled individual from either Europe, the United States, or Mexico to set up a reclamation project and begin to resmelt the lead ashes.[94]

In 1797, De Luzières met Pedro Vial in St. Louis. Vial was a Frenchmen from Santa Fe, New Mexico, and he expressed to De Luzières his interest in smelting the lead ashes near the mines.[95] Vial may have visited the mines in 1787 for he appears on the St. Louis Spanish census lists as living at the home of a worker named Huche.[96] He later became a wilderness pathfinder, and on an expedition in 1792 that opened communications between Santa Fe, New Mexico, and St. Louis, he eventually arrived in St. Louis on October 3, 1792. Apparently, Vial did return to Santa Fe in 1797 when he had become interested in the lead mines of Upper Louisiana. Although Vial told De Luzières that he understood the technology to construct a special furnace to resmelt lead ashes and increase the yield of lead when they first met in St. Louis, that was not true.[97]

Santa Fe was an important trading hub where people might have had particular mining and smelting expertise that they had gained in the Southwest or Mexican mining zones. Before Vial returned to Santa Fe, he learned of a man named Joseph Miner who knew about minerals, the construction of furnaces, and smelting. However, before Vial reached an agreement with Miner, Francisco Luna of Santa Fe came to him and boasted about his familiarity with mining and smelting. Luna was a mineralogist, and it was incumbent on him to impress Vial with his skills by showing him his cabinet of minerals.[98] Vial offered him the opportunity to return to Upper Louisiana and begin to construct a furnace to smelt lead ore and ashes. As a skilled mineralogist, Luna would have been able to extract additional quantities of lead to yield another thirty-five percent of lead metal. Luna's skills reveal the beginning of a more efficient process for transforming lead ash waste into useful pig lead. This event also subtly emphasizes the transfer of European knowledge from the metropolis to the frontier.[99]

Although Vial noted that he was interested in the lead ashes, his immediate focus was on what Upper Louisiana mines could yield besides the mineral lead. He was of the opinion that the lead ore contained a large amount of silver, and he planned to bring an experienced miner from New Mexico to extract silver from the lead ore. The Spanish officials, who created a revolution in government, also became the driving forces to shape silver mining and metallurgy during the late colonial period of Mexico. The mining bonanza at the Mexican silver mines was associated with innovations. One popular improvement, the patio process, was applied to extract silver from ore. This innovation had existed since 1554 when Bartolomé de Medina invented it in Mexico. It was the first process to use mercury amalgamation to recover silver from ore by smelting at Spanish colonies in the Americas. The communications between Vial, De Luzières, and François Vallé are silent on whether Vial wanted to experiment using the patio process to extract silver from lead ores in Upper Louisiana. But it was his original intention to bring an experienced natural philosopher from Santa Fe and involve Luna with extracting silver from the galena.[100]

The agreement and association formed between Luna and Vial further notes the importance of forming business partnerships and the willingness of late eighteenth-century settlers to conduct experiments. Later that year, Vial and Luna traveled to St. Louis where they agreed to form a partnership before traveling to the Ste. Geneviève mining district. It was Vial's hope that Luna carried to the mining district knowledge of the patio process to begin extracting silver from lead ore and lead ashes. As part of their agreement, Luna committed to build a furnace to smelt lead ashes "in a satisfactory manner."[101] Although there are no additional documents regarding Luna's background, he may have become familiar with these new techniques through Freiberg Mining Academy in Germany or at the Royal Mining School in Mexico City in the eighteenth century. Freiberg is a German university of technology in the city of Freiberg, Saxony. It was established in 1765 and is the oldest university of mining and metallurgy in the world. The Royal Mining School in Mexico City dates back to 1793. As the Mexican mining boom continued to influence the economic activity of New Spain, it became a source of economic strength for the Spanish empire. It also represents the Enlightenment ideal to attain knowledge through scientific exploitation of mineral resources. Both schools were well-known during this era of emerging science and mineralogy.[102]

Luna and Vial did continue on their journey to the lead district where Luna constructed a furnace. What type of furnace he built is not clear; however, Vial did sign contracts with a number of Creole miners to buy their lead ashes for resmelting after the furnaces were constructed. Luna said that he had moved two hundred loads of lead ashes from the section of Mine La Motte owned by Jean Baptiste Vallé, François Vallé's brother. While Vial made little mention of the lead ashes to Luna while in Santa Fe, Vial became disappointed when Luna made his initial examination of the lead ore. Luna conveyed to Vial that it was doubtful that silver could be extracted from the ore, but Vial urged Luna to experiment further, because he still thought that the silver content was large enough to validate more experiments.[103]

Although the experiments related to smelting lead ashes failed to make a significant contribution to smelting practices, these attempts alerted François Vallé's brother, Jean Baptiste Vallé, and Charles (Don Carlos) DeHault Delassus to future possibilities.[104] Interwoven with these events regarding lead ashes is a picture of the future extension of the lead industry. Luna appears to be an industrious individual who personified the convergence of culture, natural history, scientific knowledge, and technology in Upper Louisiana's mines. Luna may be considered a member of the learned culture understanding the environment through the science and art of mining and metallurgy.[105] While working at the mines, Luna united the two sciences.[106] Luna also possessed a knowledge of chemistry, which guided him to experiment with the local galena and realize that the area's lead ore would yield very little silver. In 1711, the early French explorer Antoine de La Mothe Cadillac, who became governor of Louisiana, initially traveled the Mississippi to Illinois Country in search of silver mines. After resting for a time in the village of Kaskaskia, he found no precious metals, but instead Cadillac followed the instructions of his Native American guides, began exploratory diggings for lead, and found Upper Louisiana to be a "country full of mines." As a knowledgeable and skilled artisan, Luna appears to be what De Luzières really needed. Luna carried with him from Santa Fe, New Spain, innovative smelting techniques to increase lead production by reworking the lead ashes spread around the mines. However, local Creole miners and officials still required a furnace capable of doing away with wasteful smelting methods that produced an abundance of lead ashes.

In addition to attempts to reclaim lead from existing ashes, the construction of mining communities at Mine La Motte and Mine à Breton was in line with what was occurring along the United States' Atlantic coastline and in Europe. Beginning in the late eighteenth century, a good deal of what is now called technology transfer began after Europeans immigrated to the East Coast.[107] Moses Austin was representative of an American migrating west. He planned to encourage English lead miners to immigrate to the Louisiana Territory to help expand his lead plantation at Mine à Breton as they had at Chisel Mines in Wythe County, Virginia. However, a similar technology transfer occurred from the East Coast to the Mississippi valley. A few years prior to the Louisiana Purchase, Spanish Louisiana, like the United States, had become a place where immigrants carried their technological skills and mining management style from the newly formed United States and Great Britain to Upper Louisiana's lead mines.

Inventive Genius and Inclination to the Arts

To continue focusing on developing the local and national economy through lead production and in the interest of the emerging sciences, De Luzières turned his attention to bringing in new settlers with specific technologies to efficiently increase lead ore extraction and smelting. He set his sights on increasing the population in the depot for the lead trade, Ste. Geneviève.[108] As a necessary step to developing additional mining settlements in the district, he planned to reach out to Americans and Europeans with "inventive genius" and "inclination to the arts."[109] De Luzières believed that if enticed with land "a great number of families could be found who would" come from the United States.[110] The plan was announced in the United States through the distribution of handbills.[111] The timing of the policy coincided with the Panic of 1796–1797 when a number of downturns in the credit markets of Great Britain and the newly established United States preceded commercial downturns in American port cities that continued until after 1800. In the United States, the problems first surfaced in 1796 after the failure of a series of land speculation schemes that issued commercial paper backed by claims to Western lands. Investors in land schemes were not alone in their misery. Shopkeepers, artisans, and wage laborers, each of whom depended on overseas commerce, felt the impact as businesses continued to collapse between 1796 and 1799.[112]

Notwithstanding, almost immediately after the new immigration policy went into effect in 1796, Lieutenant Governor Zenon Trudeau reported that a wave of Americans was flowing into Upper Louisiana. In a letter of October 10, 1796, the governor described the numerous requests by Americans to come down the Ohio River to enter Spanish Territory.[113] Americans began to travel from Pennsylvania, Kentucky, Virginia, and North Carolina to the western edges of the Mississippi. De Luzières's plan to populate Spanish Louisiana demonstrates how an economic downturn in one country could promote settlement in another empire's colonial outpost near the lead mines. The distribution of flyers throughout the United States and Europe was aimed to attract artisans with the technological acumen to extract and smelt lead ore and to expand the industry by manufacturing large quantities of not only shot but also sheet lead and zane.[114]

First of all, many settlers either purchased pig lead or they extracted and smelted their own lead ore to make musket balls for their security. Settlers had in their homes all the tools and chemicals required to make musket balls. However, after the English plumber and inventor William Watts converted his home into the world's first shot tower to make shot, De Luzières believed erecting a shot tower near the mines would enable skilled persons to efficiently manufacture a considerable portion of pig lead into shot.[115] Additionally, sheet lead was an important product for roofing homes, sheathing ships, and also for rolling into pipe. Finally, although making zane was a tedious process, De Luzières understood that the effort would pay off as zane was coveted by glass manufacturers for bottle making. Following the American Revolution and the economic Panic of 1796—1797, De Luzières would have been aware that glassmaking was America's first industry and was continuing to grow in places like New Orleans, Pittsburgh, and Baltimore.[116] Each one of these important lead-based products was part of De Luzières's vision to promote economic growth both locally and nationally, but to expand, De Luzières also needed immigrants who understood how to apply the science and technology to manufacture each of these products.[117]

As fame of the vast lead ore deposits in Upper Louisiana, and the ease of mining it, quickly spread over the eastern seaboard, the news reached the ears of Moses Austin. Born in Durham, Connecticut, in 1761, Austin was raised in Middletown near the lead mines that supplied the Continental Army's musket balls and where he became familiar with mining. As the American Revolution was coming to an end in 1783, Austin developed an interest in a dry goods importing company in Philadelphia, and the following year the company sent him to Richmond, Virginia, with a plan to expand. One of the items the company sold was pewter buttons, which at the time were all imported from England. However, Austin later learned the art of making the buttons by using lead and zinc.[118] Eventually, to obtain all the lead necessary to make their own buttons, and other products, Austin recommended purchasing the nearby Chisel Mines in Wythe County, Virginia, to the Philadelphia firm.[119] In time, after acquiring the mines, Austin found it necessary to hire skilled English miners to help with the expansion of the settlement and persuaded his brother Stephen Austin to travel to England to hire experienced workers. While there, Stephen met Josiah Bell who helped him locate skillful miners familiar with smelting and manufacturing lead products. Stephen Austin and Bell returned to Virginia in early 1796, only to find Moses Austin preparing to travel to Spanish Louisiana to explore the lead mines advertised by De Luzières.

On December 8, 1797, Josiah Bell and Moses Austin mounted their horses for the long and arduous journey from Virginia to Spanish Louisiana. Following a short stay in St. Louis, Austin and Bell continued on to Ste. Geneviève where they arrived on January 20, 1797. Austin made the acquaintance of François Vallé Jr., who provided him with a two-horse wagon to ease his travels to the famed lead mines, thirty-eight miles inland from the village. After conducting a survey of the lead diggings, Austin commented, "a country with everything to make its settlers rich and happy could hardly remain unnoticed by the American people." Almost immediately after returning to Ste. Geneviève, he submitted his land grant application, near to Mine à Breton, to De Luzières, simply stating, "he desired to settle in the country."[120] In addition, prior to returning to Virginia, Austin and Vallé discussed plans about how to grow the mining business.

Skilled Persons to Manage the Extraction, Smelting, and Manufacturing of Lead Products

Transforming the mining frontier into a place where manufacturing lead-based products would promote economic growth on both a local and national scale required the continuation of mining associations similar to those arranged by François Vallé, Nicolas Noel dit La Rose, Pierre and Ann Gadobert, and Jean Baptiste Datchurut.[121] In addition, at the lead mines, Vallé, La Rose, the Gadoberts, and Datchurut considered it essential to begin mining and smelting year-round once they established more efficient settlements, marking the further supplantation of the amalgam by more Euro-American and European mining practices. In Europe, before starting iron plantations, a group of entrepreneurs pooled their resources to finance land purchases; then houses, furnaces, and waterwheels were constructed. Miners who lacked the capital to prospect, extract, and smelt minerals then joined with entrepreneurs and began mining the earth for its minerals. In the United States, merchants and miners formed similar associations by jointly signing contracts with American and European miners. For example, in 1793, Jacob Mark, General Philip A. Schuyler, and Nicholas J. Roosevelt formed the New Jersey Copper Mines Association with the intention of reviving the business on a larger scale. Their need for new engines and other machinery prompted members of the association and their European miners to set up a smelting plant, machine shop, foundry, and other necessary equipment. In January 1797, the New York merchant Jacob Mark joined with Nicholas Roosevelt and agreed to hire German and English miners skilled in metallurgy before mining and smelting production began. Mark and Roosevelt's plan was "to explore for metals in the Northwestern and Southwestern regions" of the United States.[122] In most cases, after the association was formed, members decided to employ "skilled persons" in metallurgy to conduct the smelting of lead ore before manufacturing lead products.

Similar to Mark and Roosevelt, Austin would convey to Vallé the importance of hiring knowledgeable miners skilled in metallurgy to increase the flow of pig lead and lead by-products down the Mississippi River to New Orleans and other East Coast ports. While Austin, Vallé, and Josiah Bell made plans to hire Europeans who could efficiently establish new techniques and construct European machines to extract and smelt lead ore, mine owners, engagés, and slaves continued to employ longstanding Native American procedures. Thus, although the formation of a mining association was organized to begin European mining practice, miners at La Motte continued to extract lead ore from trenches and smelt their lead in hybrid furnaces uninterrupted. Both mining and smelting methods appear to have existed alongside of one another, and both had similar outcomes—the extraction of significant amounts of lead ore for the production of lead articles for society.[123]

The association formed between De Luzières, Vallé, and Austin was an arrangement to encourage Austin to dig the region's first shaft; erect a reverberatory furnace; hire experienced miners from Derbyshire and Cornwall, England; and manufacture shot, sheet lead, and a key component for making glass bottles, zane.[124] By 1800, Great Britain was "annually produc[ing] the greatest quantity of lead, 12,500 tons," one more reason Austin looked to British miners and artisans from Derbyshire and Cornwall; and within a few years of Austin's arrival, experienced English miners began to immigrate to Upper Louisiana searching for veins of lead.

The invasion of new European scientific practices and technologies reveals the lead mines of the Mississippi valley as a zone of intercultural penetration.[125] In this age of emerging science, Moses Austin and the British and American miners who accompanied him to Mine à Breton began to put into practice European mining methods. They were the first to dig a mining shaft and to heap roast ores before smelting the ores in the region's only reverberatory furnace. The willingness of miners to adapt to European methods also represented the locals' appreciation of new technological exchanges. As this chapter highlights, although the amalgam survived until now, its coexistence alongside new settlers' mining practices was in question, and the old Kaskaskia-French amalgam practices began to disappear. In 1799, Austin reported, "The mines on the waters of the St. Francis are capable of furnishing vast quantities of lead," and he estimated that with the current 150 workers employed year round miners extracted over 200,000 pounds and 366,666 pounds of "lead ore" at Mine La Motte and at Mine à Breton, respectively. Austin also began to imagine that after "admitting one thousand men to be employed the year round, at the different mines now known" the manufacturing and exporting of shot, sheet lead, and red and white lead would considerably increase with prospecting for, extracting, and smelting quality lead."[126]

Take Lands and Lead Mines and Make Such an Establishment

When Moses Austin and Josiah Bell relocated from the United States and Great Britain, respectively, to the Upper Louisiana lead district, they carried with them new ideas, practices, and apparatus to mine and smelt lead ore. Austin planned to follow the explicit directions of Gabriel Plattes and firsthand experiences of British miners to help him recreate a late eighteenth-century mining community where miners extracted, smelted, and refined lead in close proximity to where lead could be found. A similar transfer occurred across the Atlantic between England and cities like Philadelphia, New York, and Boston. Samuel Slater, an English immigrant, is one example. As early as the 1790s, textile merchants in New England began experimenting with machines to replace the putting-out (subcontracting) system. To effect this transition, merchants and factory owners relied on the theft of British technological knowledge to build the machines they needed.[127]

In 1789, for instance, a textile mill in Pawtucket, Rhode Island, contracted the British immigrant Samuel Slater to build a yarn-spinning machine and a carding machine. Slater had apprenticed in an English mill and succeeded in mimicking the English machinery. Slater learned that some American states were paying British citizens who knew how to build cotton-spinning machinery to relocate and build the same machines in their states. Slater immigrated to New York carrying textile machine designs he committed to memory. Eventually, Slater would establish a number of mills in Rhode Island, where most of American industrialization remained for many years. In 1813, American industrial espionage peaked when Francis Cabot Lowell and Paul Moody re-created the powered loom used in the mills of Manchester, England. Lowell had spent two years in Britain observing and touring mills in England. Like Slater, he also committed the design of the powered loom to memory so that, no matter how many times British customs officials searched his luggage, he could smuggle England's industrial know-how into New England.

A few decades after Slater arrived in Pawtucket, Rhode Island, Josiah Bell learned about new prospects in lead mining and immigrated to Austin's Chisel Mines in Wythe County, Virginia.[128] Like Slater, the skilled English miner Josiah Bell transferred his knowledge about creating a mining settlement from England to the United States and eventually Spanish Upper Louisiana.[129] Bell was from the lead district of England, Cornwall and Derbyshire, and was partly responsible for launching and cultivating European mining techniques at Mine à Breton.[130] Bell came to the United States to assist Austin with his Virginia mines, before they both moved to Spanish Louisiana. Austin decided that Mine à Breton would be a good place to "take up such lands and lead mines and there make such establishment of the lead mine business and introduce the many improvements in mining with which he was familiar."[131] Like Mine La Motte, Mine à Breton exhibited certain geographical features conducive to a mining community.[132] The location of Mine à Breton afforded plenty of flowing water for washing ores and for constructing water-propelled machines. There were also plenty of prairies with "fruit trees and ... place[s] filled with excellent timber."[133] Austin wanted to discourage the local miners from using their "wasteful mining practices." He called them "the unskilled workers in the neighborhood of Mine à Breton."[134] Within a few months of Bell's arrival, Bell and Austin made their way back to the lead mines along the Mississippi River, carrying with them the knowledge and plans for the introduction of a European-style settlement in the mining region.

Before Austin departed Virginia for Mine à Breton, he sent a team of employees from his lead plantation to begin work on the Mine à Breton settlement. The initial team included his nephew, Elias Bates, as well as workers from the Chisel Mines—Kendall, Shouse, and Nall—and Matthew and Timothy Mullins, brothers from Great Britain.[135] Later, Josiah Bell, John Storts, Drury Green, and John S. Brickey traveled with Austin and his family to Mine à Breton. Before the first group of miners departed Virginia, Austin gave them an illustration of how the settlement should be laid out. The settlement map described to the team the location of Austin's home, living quarters for the miners, the furnace house, and sawmill providing "a good idea of the situation of the lead mines and mark[ing] out the place for the furnace to stand." Austin suggested that the furnace house be positioned "at the lower end of the Village," and he also outlined the specific dimensions not only "for the furnace house, but also for the saw mill," which he wanted "to be built in a good place."[136] Austin told the Mullins brothers and Storts to erect the sawmill first so that workers could begin cutting lumber to build living quarters, and that the construction of a shot tower, and sheet lead manufactory should be next on the list of projects.[137] By the time of Austin's arrival, the workers had accomplished a great deal. They interacted with the French Creole miners and slaves and also began to conduct mineralogical surveys to understand the widespread occurrence of lead ore scattered across the vast landscape.[138] Foremost, the sawmill was finished, and the workers started to construct the reverberatory furnace house.[139] The blacksmith shop and other buildings began operating that same year.

Similar to François Vallé at Mine La Motte, Austin employed a mining manager. Josiah Bell supervised the work of the other miners who were assigned as miners, smelters, woodcutters, and carters.[140] A fuller complement of American and European miners at Mine à Breton would quickly usher in a faster transition away from the Native American mining practices than the gradual transitions taking place at Mine La Motte. Like Vallé, Austin also recorded early examples of the division of labor but rarely mentions who performed mining, dressing, or smelting work except for when he wrote specific instructions to John S. Brickey. For example, on February 22, 1815, Brickey was assigned to work the reverberatory furnace built at Mine à Breton. In his letter, Austin states, "I give you the following memorandum by which you will fully understand, what I wish you to take under your charge."[141] Austin provided Brickey with detailed regulations regarding what was to follow after the lead ore was extracted from the shaft. Brickey and the slave miners were to restock the wood, charge the furnace, and smelt the lead ore. Additionally, Austin instructed Brickey that "Negroes and Whites are to be prevented from visiting the furnace not suffered to remain about the furnace, they draw off attention of the hands."[142] These comments suggest that slaves worked in the mines or separated and cleaned the ores prior to smelting and that a division of labor existed that was more intense than the earlier models.

Austin and his miners prospected for lead ore by applying longstanding European prospecting approaches, which were similar to the amalgam used by early miners. They used their environmental knowledge to sense changes in odors, changes in the texture of the soil, and changes in the earth's color, all the while observing small specks of ore scattered about in their trenches or shaft. At times Austin noticed "grains of lead mixed in with the soil."[143] When English miners recognized these grains, they endeavored to sink their shafts deeper to access the ores and then hoist them to the surface using the European-style windlass. More importantly, like those who came long before, Austin's associate miners had the same familiar and intimate knowledge of the mining environment. In 1799, Mine à Breton was the location where Austin planned to sink his first shaft. Before doing so, Austin cleared the surrounding surface of all the alluvial soil. Austin describes "the first shaft reaching a depth of eighty feet," and the shaft was indeed vertical. He noted, "The clay at Mine à Breton was thirty or forty feet thick"; and it was into this that he sank his shaft, but he "stopped at the limestone rock." He did, however, dig tunnels into some of the "crevices that contained massive ores."[144]

Once the shaft was cleared of soil, miners put in place a windlass and bucket tied to a rope, which was a common English and American practice. Then a shed was built to protect the pit and workers from rain, wind, and cold weather. The shed was also a good place to keep wheelbarrows and to store the miners' tools. One miner would dig and place the earth into the bucket, and another miner controlling the windlass from the top would raise the filled bucket to the surface.[145] There were tunnels extending in several directions from the shaft.[146] It is possible that as many as ten miners could manage one shaft.[147] By the end of the eighteenth century, Moses Austin had clearly introduced modern large-scale mining techniques.

Accompanying these new technological applications to mining was the need for a large number of slaves. In addition to the American and English lead miners, Austin carried with him a number of his slaves, and although it is not clear how many slaves he transported to Spanish Louisiana, by the early nineteenth century he had purchased more slaves for his operations. However, after his arrival at Mine à Breton, Austin hired black slave labor from local slave owners. Then in 1814, Austin was in a position to purchase fifty-two slaves including men, women, and children. It is also worth remembering that the French colonists had been using black slaves in the lead mines for over fifty years, so when Austin arrived, he merely expanded a system that was already in place. Austin had his slaves mining and carting lead ore as it accumulated around the shaft. The slaves used carts to transport the ore to a nearby stream to wash it before smelting. Similar to the divisions of work in Europe and Africa, it is probable that Austin used female slaves and slave children to wash the ore in nearby streams.[148] Agricola speaks of male miners doing all the work from prospecting to smelting, without a mention of women having a role near the mines. In Africa, older males did the mining and smelting; however, young boys accompanied their elders to the mines to learn the craft. Neither women nor young girls participated in any mining activities. It was considered harmful to the mining and smelting venture if women were nearby as men conducted their work.[149] Slave miners proceeded to separate any remaining spar and rubble by breaking it into small pieces with handheld hammers called buckers. The fist-size pieces were then placed in a buddle (trough) to be sifted in water. They then underwent a second washing to separate them from the lighter foreign matter in another buddle for sorting the ores. To assist in the process, another miner or dresser agitated the buddle with a hoe so that the lighter material would float away.[150] Slaves repeatedly cleaned the lead ore in preparation for the smelting process.[151]

The construction of the furnace house was completed and put to blast in January 1799. Austin's reverberatory furnace is a matter of considerable interest, as it appears to have been used to smelt lead ore as well as lead ashes. Schoolcraft described Austin's furnace as the only one in the district. Austin mentioned only one furnace which he termed *reverberatory*. Reverberatory furnaces in Europe were built of bricks. It is not clear whether Austin was capable of making his own bricks or whether he purchased them from New Orleans. Most likely, Austin purchased the bricks from New Orleans and had other instruments shipped in from Philadelphia, which included a fireplace, a grate, and the ash pit. The smelting period required to roast the lead ashes was between six and twelve hours.[152] As mentioned above, the introduction of innovative furnaces made a significant contribution to the mining district. It allowed lead ashes to finally be reclaimed. What Vial and Luna first attempted with the large amount of lead ash heaped and scattered near the old furnaces would now allow mine owners and miners to begin to see additional profits.

With care, the ash furnace made possible the reclaiming of the ash heaps, and each month it could produce sixty thousand to ninety thousand pounds of lead.[153] When Schoolcraft visited the lead mines, he observed and documented Mine à Breton as "the chief of the entire lead" mines. Contemporary European mining manuals also contain Schoolcraft's extensive illustrations of the furnace. Austin's furnace was twice as efficient as the log and stone furnaces that it would eventually replace. The furnace worked on the principle of an oven. It used hot air, and the ores did not have direct contact with the fire. By this method, Austin was able to obtain seventy-five percent of the lead from the ore. Schoolcraft elaborates on how the reverberatory furnace was also capable of "refining lead ashes," which were a mixture of impurities and lead that miners melted into zane.[154] As the molten material flowed into the iron pots, the impurities were ladled off the top. This mixture of impurities and imperfect lead was called zane and was equal in quantity to the lead. The lead was poured into molds; the zane was taken back to the log furnace where it was easily converted into lead by another smelting.[155]

Austin and Schoolcraft both speak of the reverberatory furnace as the primary European furnace of the entire lead district, but they fail to give a precise description or to designate the type of furnace. Austin mentioned only one furnace, which he termed a reverberatory furnace, while Schoolcraft gave an elaborate description of an ash furnace. Both speak as though there was just one efficient furnace in the district. But the ash furnace, which was a reverberatory furnace, smelted only lead ashes.[156] Ordinarily, an ash furnace was built of limestone, a very poor material for the purpose. A limestone ash furnace usually lasted only fifteen to twenty days; and with care, it might be made to last a full month during which time it was possible to produce 60,000 to 90,000 pounds of lead.[157] The introduction of the ash furnace was also a significant contribution to the lead mining district since it made possible the reclaiming of the ash heaps lying around the old log furnace sites.

Smelting began after the miners placed a layer of ashes and a layer of wood at the mouth of the furnace hearth. When the furnace was ready, the timber was ignited and the furnaces had to be heated thoroughly before opening the taphole, which allowed for a longer heating process. After the lead was melted, the smelter opened the taphole using a long iron tapping bar. With the taphole open, the melted lead was allowed to flow into the iron pot, and any impurities were ladled off the top by the smelter. This usually occurred six hours after the furnace was in full charge. The taphole remained open because longer periods of heat were required. For lead ore, the smelting time extended for as long as three days and nights. During the time of Agricola, a smelter's workday was twelve hours, and when replacements arrived, that smelter had to be ready to keep the process going uninterrupted.

Although manufacturing zane appear to be a tedious process, it was well worth Austin's time and effort because he sold it to glass manufacturers who made bottles in Frederick, Maryland, or Pittsburgh, Pennsylvania, where he secured business connections. Both locations were suitable markets for zane for making glass bottles.[158] For example, in 1784, when Johann Friedrich Amelung settled in Maryland, he proceeded to construct a glassworks factory. Amelung produced some of the most beautiful glass ever made in America. Similar to Austin, Amelung brought sixty-eight glass craftsmen, of German origin, and furnace equipment to the young United States. During the following decade, Amelung built housing for approximately 400 to 500 American and German workers. Similarly, in the late eighteenth century, James O'Hara founded the Pittsburgh Glass Works, where Moses Austin may have hoped to sell zane for their bottle works.[159]

Austin manufactured a considerable portion of the lead mined in Upper Louisiana into shot. Prior to Austin constructing his shot making enterprise, either frontier homes contained shot-making equipment, or one villager manufactured the shot for the community. He designated one building to make "patent shot of good quality" to conform with William Watts's English shot making method.[160] After erecting his shot tower, Austin hired one worker to manage the furnace and the kettles used to smelt and cast the lead. The lead was mixed with arsenic, which made it more fluid during the casting process. Arsenic also quickened the hardening process. After the worker thoroughly mixed the lead and arsenic, he dropped the lead into water through an iron or copper frame perforated with round holes according to the required size. To make the smallest shot, the elevation had to be about ten feet above the water, and for the largest shot, about a hundred and fifty feet. Austin recommended to manufacturers that, "a material difference of the height is required in the local climate," according to the seasons of the year.[161] After the molten lead shot cooled, it was glazed and polished.[162] One artisan could cast between four and five thousand pounds a day and polish the entire batch over the course of nine days. Polishing was accomplished by placing the shot into a wooden tub with a crank that workers turned. The constant motion caused the shot pieces to hit each other, making each piece smoother and ready for market.[163] When Henry Rowe Schoolcraft visited the region in 1819, he sketched the town of Mine à Breton, which at the time contained over forty wooden-frame structures.[164] Austin built his family home, which he called Durham Hall, and on a nearby hill above his quarters, he built housing for forty miners. In the sketch, all the homes appear to be located on hills and in valleys arranged in a sprawling geographical cluster.

Attend to Embarkation

Sometime after miners settled at Mine La Motte, their mining and smelting work increased to two seasons per year, which "coincided with the seasonal flow of lead bars" to frontier villages and cities as packhorses carried four lead bars per trip.[165] The activity along the Mine La Motte path to Ste. Geneviève was intense. At times, De Guire, La Chance, Nicollet, Matis, Chevalier, or Variat, along with at least two of Vallé's African slaves, transported pig lead using ox carts to François Vallé's home or to another location before shipment to New Orleans. Before Austin established another village at Herculaneum, Missouri, in 1809, Ste. Geneviève was the leading lead storage location and depot for both Mine La Motte and Mine à Breton.[166]

Shipping lead from Mine La Motte and Mine à Breton to New Orleans required Vallé and Austin to observe "the different stages of the water in the different seasons."[167] Travelers suggested that beginning in mid-July until early October, "embarkation should be attended with considerable detention."[168] In other words, so as not to be held back, the safest time to float downriver would be at very low levels of water. Merchants and miners understood that the "spring season journey from the mines to Ste. Geneviève and onto New Orleans would be slow and safe." Moving pig lead during autumn was also "encouraged for navigating the inland waterways." However, merchants were cautioned to sail by "late December [before the] ice formed, [closing] rivers, [which] greatly hinder[ed] navigation."[169] Following the arrival of Moses Austin, lead shipments to New Orleans and beyond increased.

As an additional advantage of the association formed with Vallé and De Luzières, Austin gained unlimited access to navigate the Mississippi River. At the time, only residents of the Spanish Territory could trade or transport goods between Ste. Geneviève and New Orleans. Thus, De Luzières provided Austin with the necessary passports for unlimited travel to and from New Orleans, which allowed Austin to ship his sheet lead and patent shot directly from Mine à Breton. Both articles flowed along navigable rivers to frontier societies. On April 12, 1801, at about ten o'clock in the morning, Moses Austin departed Ste. Geneviève for New Orleans with two small timber flat boats.[170] Each was loaded with the sheet lead and shot he and his artisans manufactured. Austin was careful to embark on his trip at a time that coincided with easily navigable conditions. Austin arrived at his destination twenty-three days later and decided, with John Merieult, to have the lead articles loaded on the schooner *Nancy* bound for Philadelphia.[171]

In conclusion, numerous accounts from Mine La Motte and Mine à Breton highlight the significance of lead ore, in the form of shot, sheet lead, and zane, to society and to the growth of settlement on the frontier. This chapter has chronicled the ability of merchants, farmers, and miners to establish Mine à Breton where they worked year-round and Mine La Motte where miners continued to employ seasonal mining practices. Like the French miners who adapted their mining practices to the Kaskaskia hunting seasons, the French came to understand how to exploit the seasonal diversity of their environment by practicing mobility and following a particular cycle. However, at Mine à Breton, Austin went one step further to manufacture lead articles on the frontier. The task of the numerous prospectors, miners, and smelters was the same: to construct processes to locate, extract, and manufacture this useful metal, using both Native American and European mining traditions to increase lead production.

The evidence suggests the continuing significance of mining to a late eighteenth-century frontier society. Clearly, lead ore was an agent of change. It transformed and influenced mining, smelting, and new settlement styles after American and English settlers carried their knowledge and technology into this frontier. Like early eighteenth-century French engineers De Ursin, La Renaudière, and De Gruy, De Luzières also envisioned a more efficient mining operation and a well-settled village in closer proximity to the lead mines.

Concern over the inefficiency of the furnaces and the wasted lead in the slag annoyed De Luzières, and he attempted to find a solution by luring Americans and Europeans into the Louisiana Territory to help establish a "foundry and forge" similar to those used in England. Similar to merchants, miners, and scientists in the United States who desired to increase mining profits, like Jefferson, De Luzières too believed that innovations and early experimentation should be developed to efficiently smelt lead ore in stone furnaces. De Luzières would have been well aware of the latest European innovations and technological improvements of the age associated with mining. Although cross-cultural mixing continued at the mining frontier borderland, earlier syncretic techniques began to disappear slowly in the shadow of new hegemonic instruments that accompanied American and European miners' settlement practices.

Although De Luzières and Austin wanted to discourage the La Motte miners from using the Kaskaskia-French amalgam, François Vallé and other miners there continued to produce over two hundred thousand pounds of pig lead between 1799 and 1803. The miners at Mine à Breton, who installed the latest European equipment, produced over three hundred thousand pounds of pig lead during the same period, a significant change from what De Gruy had produced during the 1740s.[172] When farmers, miners, and smelters crossed the Mississippi River to work in Missouri's early lead district, they carried with them new ideas about how to link the arts with commerce. They united a set of hybrid practices to form ornaments for society. Most significantly, this chapter suggests those who migrated to the mines in Spanish Louisiana promoted useful knowledge and technology in the mining environment. Local farmers and Austin, encouraged by the lead prospects, settled near Mine La Motte and Mine à Breton.

At the close of the Spanish period, the United States government asked Moses Austin to compile a report on the Missouri lead mines. He reported that miners had discovered ten additional mines in the southern mineral region; and that from 1802 to 1804, only twenty-five men had worked there, and for only a few months. The region remained well watered and residents described it as "pure and wholesome."[173] The next chapter focuses on the first geological survey conducted by Austin to determine the quality and quantity of the lead ores lying beneath the soil.

CHAPTER 3

Early Mineralogical Assessments and Emerging Science

Introduction

To prepare for the transfer of the Louisiana Territory from France to the United States following the Louisiana Purchase, President Thomas Jefferson and Congress wanted to learn as much as possible about the vast new territory—including the lead mines where Native Americans, French settlers, Creoles, African slaves, and now American settlers worked. In 1804, the newly appointed commandant to Upper Louisiana, Captain Amos Stoddard, traveled to St. Louis to establish the American government.[1] Stoddard had fought in the American Revolution and went on to practice law in Maine before returning to the army in 1798. He was selected to represent both France and the United States during the transfer of the Louisiana Territory in March 1804. When Stoddard arrived in the territory, he carried a request from Jefferson stating a desire to learn more about the lead mines located across the Mississippi River opposite the Illinois village of Kaskaskia, which he mentioned in his *Notes on Virginia*.[2] Specifically, Jefferson needed figures on "the number of hands employed" at the mines, the amount of lead "annually produced," and how settlers and merchants used the metal. Stoddard reached out to the merchant and miner Moses Austin of Mine à Breton, requesting him to write a prospectus or a "memorandum of the number, extent and situation, [and] the quality of the mineral produced" at the local mines. Moses Austin applied his mineralogical skills to conduct experiments on the lead ore, writing, "[I] found that the mineral" was of superior quality and that "in the hands of skillful smelters" they could produce "sixty and in some veins seventy percent" metal at Mine à Breton and Mine La Motte.[3]

Austin worked hastily to complete the mineralogical report entitled *A Summary Description of the Lead Mines in Upper Louisiana*, and in the middle of June of the same year, he asked Captain Meriwether Lewis to take "a copy of [the] dissertation on the Lead Mines in Upper Louisiana" to Stoddard before he[Lewis] left on his mission to explore the territory of the Louisiana Purchase. In a letter to Jefferson, Stoddard explained that Moses Austin "owns an extensive mine, situated about thirty-eight miles back of St Geneviève, which he has worked for some years past; and from his education and experience, I conceive him to be better calculated to give correct information on the subject than any other man in this quarter."[4] After Austin's report arrived in Washington, DC, on November 8, 1804, in his 1804 message to Congress Jefferson reported, "The lead mines in Louisiana offer so rich a supply of that metal as to merit attention."[5]

During this age of emerging science, the timing of the United States' purchase of the Louisiana Territory could not have been better. The president and natural philosopher, Jefferson, had already decided to launch the Corps of Discovery across the continent in search of a northwest passage to the Pacific Ocean. To lead the mission, Jefferson selected the young Virginian, Meriwether Lewis, for his gift of observation and knowledge that birthed a keen fascination with the western country's plants and animals. After accepting the assignment, Lewis in turn asked his friend William Clark to be a joint commander of the enterprise. Initially shrouded in secrecy, the expedition only became known after news of the Louisiana Purchase became public. In the meantime, to help Lewis prepare for the mission, Jefferson offered Lewis access to his personal library of scientific works, tutored him in geography and paleontology, and sent Lewis for training with other natural scientists in Philadelphia. Before leaving Philadelphia in 1803, Lewis purchased books on botany and mineralogy, a supply of medicines for the journey, and instruments needed for surveying the landscape. On July 5, 1803, the day after news reached the capital that France had formally agreed to sell the Louisiana Territory to the United States, Lewis quietly departed Washington, DC, with supplies and a letter for the commandant Captain Amos Stoddard in St. Louis.[6]

During this period, as mineralogical studies developed in the United States, natural philosophers also embraced scientific mineralogy for its utilitarian value. Moses Austin, Meriwether Lewis, and Thomas Jefferson were representative of late eighteenth- and early nineteenth-century natural philosophers who wanted environmental knowledge from familiar and unfamiliar places. Like other natural philosophers, they valued the utilitarian as an aspect of science to conduct further investigations into nature and thereby revolutionize the national economy. Jefferson desired far more than geographic knowledge; he also desired to learn as much as possible about the occurrence of lead ore in the mines he wrote about in his *Notes on Virginia*. As a true son of the Enlightenment, like his American and European counterparts, Jefferson had practical and scientific goals to obtain mineralogical information about the newly acquired Louisiana lead mines. He believed that Austin's mineralogical assessment would benefit the nation. Therefore, it was all the more important for him to ask Stoddard to reach out to Austin, the experienced miner from Virginia who with the help of a number of English miners successfully established a mining settlement at Mine à Breton. Whatever information Austin gathered about the deep-rooted lead mining and smelting enterprises in the Ste. Geneviève mining district—Jefferson understood that the new scientific knowledge would become useful for dissemination to both Congress and the public. Jefferson hoped that the new knowledge about the natural history surrounding the Louisiana Territory's lead mines would promote programs for gathering, systematizing, and eventually publishing this knowledge to encourage more Americans to settle in the region for the benefit of national wealth.[7]

When Thomas Jefferson took office, it marked a change and anticipation of a new century. Merchants must have been excited about potential opportunities of new progress and development to connect the United States' eastern coastline with the Mississippi River. One indicator came from the United States census. In 1800, the population of the United States reached 5,308,483 million, a thirty-five percent increase since 1790. The republic's population would continue to increase, nearly doubling by 1820 and doubling again by 1840. This rapid growth was powered by an unprecedented commercial and geographic expansion—especially in the West.[8] As the United States' population increased, Americans looked for the opportunity to purchase and settle on affordable lands adjacent to the Mississippi river in the west.

Simultaneously, extraordinary events began to unfold that would alter the geography of the United States and ultimately shape the nation's economic, social, scientific, and technological practices in the region of the Louisiana Territory. Like many Americans who desired to relocate west of the Mississippi river, Thomas Jefferson hoped that the Louisiana Territory would eventually be joined to the United States. It was thought that if Louisiana came under the jurisdiction of the United States this natural extension of the country would promote critical access to the valuable sources of the lucrative fur trade. Most significantly, he secretly harbored the idea that Ste. Geneviève's lead mining district "on the Spanish side of the Mississippi river, and not far from Kaskaskia," was one ideal location for landless Americans and immigrants to settle. Like Moses Austin and the English lead miner Josiah Bell, Jefferson also believed that those who migrated west would not only supplant the American agrarian tradition but also transfer new American and European mining techniques, which he deemed essential to the wellbeing of the nation.[9]

Additionally, to comprehend the nature and distribution of lead ores in the Louisiana Territory, Jefferson needed Moses Austin to identify and catalog the minerals, signaling a cultural sea change in efforts to improve a nation's economy. Since the arrival of the French settlers and mining engineers, it had been extensively believed that the lead ore found in the *country full of the mines* were superficial deposits. Therefore, due to this opinion, capitalists felt restrained from giving their full confidence and financial support to undertaking mining in a more efficient manner. The French empire was afraid to invest their means beyond an amount necessary, and chose instead to conduct partial diggings and excavations near the surface and hesitated to sink shafts and establish mining, smelting and lead manufactories on a larger scale. However, following Moses Austin's arrival and after he applied his mineral cataloging skills more Americans embraced the utility of mining in relation to producing results favorable to national industry.

A decade after the publication of Moses Austin's *A Summary Description of the Lead Mines in Upper Louisiana*, the American physician and mineralogist Dr. Archibald Bruce published the *American Mineralogical Journal*, which presented early nineteenth-century accounts of pioneers of American geology in a friendly and enjoyable narrative style. The first publication in 1810 (January) included a comprehensive list of the contributors including "the miner, the quarrier, the surveyor, the engineer, the collier, the iron master and even the traveler" who assisted in making geological observations. Additionally, much of the content in the *Journal* was devoted to prospecting and studying minerals of economic importance such as coal, iron, copper, and lead, as evidenced by *Description and Analysis of an Ore of Lead from Louisiana* by Dr. William Meade. Meade states, "Having been favored with a good specimen of lead ore from the mines at Ste. Geneviève on the Mississippi, and being desirous of ascertaining its constituent parts and productive quality, I submitted it to a careful analysis in the most accurate way. It is to be regretted that the country which affords this abundant and productive ore, has not as yet been described by any Mineralogist."[10] Austin and Meade each reveal the mineralogy of the lead mines where Native Americans, French, African slaves, and American miners extracted and smelted lead ore. They each engaged in illuminating the natural history of these particular lead mines and called for improving metallurgical methods. Most significantly, they both were engaged in the dissemination of knowledge regarding useful applications of minerals to serve society with lead products.

By the opening decades of the nineteenth century, Europeans and especially Americans migrating from the East to settle in the Louisiana Territory considered their way of thinking and identifying the earth as profoundly different from the ideas and practices of the peoples they encountered after their arrival in this vast new territory. While Austin's and Meade's mineralogical reports supported the economic value of lead, Austin's was the first American mineralogical survey of the lead mines. His survey exposes his confidence in European scientific advances and practices, which he believed had surpassed those of other civilizations. Notably, the mineralogical survey reveals the beginning of a desire to abandon the Native American techniques and seeks to encourage miners to transition to European methods of experimentation, mining, and smelting. Like Europeans who believed their prospecting instruments to be superior to others, Austin also alleged that applying his apparatus correctly would prove his techniques superior to the local miners' longstanding syncretic methods.

By the late eighteenth century when European states became increasingly interested in mining and metallurgical industries, a number of locales in the United States focused their attention on the study of mineralogy as well. Mining officials, merchants, miners, and national leaders all needed information about the location and properties of metallic ores such as lead, zinc, copper, and tin. In Europe, many trained physicians contributed to mineralogical knowledge, and in the United States, some physicians and laymen did the same.[11] While these changes affected mineralogy studies on the European continent, the mineralogist Moses Austin developed his skills after reading translated versions of European mining texts and British mining treatises. Unlike many mining engineers and mineralogists in Great Britain, Moses Austin, the experienced miner from Virginia, and the British miners who helped Austin to establish Mine à Breton did not have access to mining schools.

Like skilled English miners developing their skills as apprentices, Austin developed his skills at Connecticut lead mines as an apprentice and then as an owner of a Virginia lead mine. All of these experiences required that he study to become successful at extracting and smelting lead ores.[12] Also referred to as "practical men," mineralogists in England understood the nature and distribution of minerals in their mining areas. However, lacking institutional support and the ability to publish results, these practical men learned to depend on wealthy professionals, or like Austin, they used their business associations to exchange their mineralogical knowledge with one another.

By the late eighteenth century, United States' metals were being employed in a multitude of products. Cast-iron kitchen utensils—pots, pans, and kettles—were cast directly from blast furnaces, as well as fireplace andirons and pokers. Additionally, blacksmiths fashioned horseshoes, farming equipment, axes, and chains from iron. Kettles, boilers, roofing, and stills to manufacture beer and whiskey were designed and sold by coppersmiths. Finally, lead was employed to make ribs for windowpanes, to make sheet lead to roof homes and to sheath ships, and to manufacture musket balls. Following the publication of Austin's survey, red and white lead manufacturers seeking quality ores in great quantities began to purchase Missouri's pig lead. Austin's survey reveals that the constant need for minerals was responsible for the active study of these substances by the early American scientific community. These mineralogical reports were common. Each projected the idea that nature was a resource and commodity for sale.

After Jefferson presented Austin's assessment to Congress, both houses of government envisioned the Louisiana Territory lead mines as another means to decrease America's dependency on foreign products made from lead and to establish a dynamic domestic market for lead products. For most Americans, shot continued to be the most useful product made from lead for protection of personal property; however, sheet lead, white lead, and red lead were also beginning to gain economic value. To manufacture white lead correctly, artisans needed lead ore in its purest state—galena, free from any contaminants. The pigment white lead was prepared by casting the lead into sheets, rolling it up in a spiral form, and setting it to corrode in clay pots partly filled with vinegar.[13] White lead was used for painting wood and to plaster interior walls. It was applied alone or in conjunction with other pigments to serve as a base and to give paint body. White lead continued to be an important chemical product for making paint.[14] After 1670, red lead became an important product in Europe for the production of flint and crystal glass. Red lead was produced by moderately heating metallic lead in a reverberatory furnace, like the one built by Moses Austin at Mine à Breton. Following hours of heating, workers witnessed metallic lead transition to a vivid red-orange-colored soft powder that crumbled between their fingers.

As the Western region's population grew and new markets emerged, settlers began looking eastward for their manufactured goods. With escalating tensions in Europe and possible effects on overseas trade, northeastern merchants began to cultivate trade with western outposts along the American banks of the Mississippi River. However, many merchants began to encounter problems exporting and importing goods. In 1800, the best trade route to the American West was by water along the Atlantic coast into the Gulf of Mexico and then up the Mississippi River from the port of New Orleans, which was still under Spanish control. The other option was to ship goods westward along a series of connecting roads, rivers, and canals, which were all still in the formative stages.[15]

Crude Methods to Discover Ore

During the early eighteenth century, as Native Americans and French explorers traveled to the lead mines, explorers observed and documented the region's mineralogical information both in narrative and cartographical form. With an eye always on expanding the empire, the French also recorded many natural features such as rivers, hills, plant life, and animals, as well as minerals. Eventually, the French and their Native American guides endeavored to create a middle ground, centered on prospecting, extracting, and smelting lead ores, which developed into an amalgam of mining practices. By the 1740s, De Gruy would begin to critique miners' methods; however, by far the most extensive critique of the Kaskaskia-French amalgam was contained in Austin's *A Summary Description of the Lead Mines in Upper Louisiana*. Before the close of the eighteenth century, Austin would embark on similar mining expeditions with a plan to transfer new European science and technologies to extract and smelt lead ore, which would begin in earnest the erasure of the old syncretic practices. Like his forerunners, Austin planned to offer a more systematic scientific study of the mining district by providing a more accurate appraisal of the amount of lead ore available for consumption. Although the broad title of this first American mineralogical report, *A Summary Description of the Lead Mines in Upper Louisiana*, suggests this was a sketch of the lead mines, the account was also notable for its emphasis on changing all aspects of frontier mining, smelting, and manufacturing of lead products. Stemming from an era of enlightened civic pursuits to embrace new environmental knowledge, his analysis critiques miners and their methods while simultaneously stressing the need to install European methods.[16] Austin's mineralogical interpretations would become only one of the tools to supplant the longstanding syncretic practices.

Before the 1830s, when improvement societies began to organize local surveys to replace Native American methods, miners measured the amalgam against what they believed to be more sophisticated European technologies. The Austin survey reveals that American miners believed the amalgamated prospecting methods needed to be replaced with up-to-date scientific assessments of natural resources. As mentioned earlier, Austin's report started a critique of those practices influenced by Native Americans. When Austin first arrived from Virginia to survey the mines, he was surprised to see local miners discovering lead ore as they did. For example, when miners went to "discover ores," they used the method established during the early eighteenth century to follow environmental markings. Austin noted how miners examined "the soil, the slope of the hills, and trees," and in some cases sought to determine if lead ore was "attached to these natural markers."[17] Along the way he may have observed miners checking the color of plants to see if they could detect the presence of a vein running beneath the soil's surface.[18] Embracing European practices to master nature as manifested in scientific discovery and technological advancement, Austin wanted to learn if a vein of ore was a branch or a main lode. Therefore, he disdained the way these particular miners haphazardly worked their trenches.

Austin believed that local miners were unable to effectively exploit the lead resources because they not only lacked the proper European environmental knowledge but also lacked the European tools to determine the quality or quantity of lead veins successfully. When visiting Mine La Motte, Austin and his staff observed how many miners worked. He observed how "skilled miners would most likely have been astonished to see miners" in this territory applying such "crude methods to discover lead ore."[19] Local miners prospected for lead ore by digging numerous "scattered pits" or trenches "four to five feet deep until a good vein was found." Along these veins, they worked until their mining progress "was impeded by rock" or by water flowing into the trench. Whenever these obstacles emerged, miners abandoned their trench and began to search in another location. Even if the miner was ten or twelve feet below the earth's surface, "he usually quit and began digging a new pit" according to Austin.[20] These observations and notes in *A Summary Description of the Lead Mines* showed the American disdain for these practices emerging. Again, after visiting one of the mines and witnessing miners' prospecting methods, he referred to their actions as "inefficient," as their only tools for discovering ore were a "pick and wooden shovel," and he stated that they clearly had a need for European mining instruments.[21] Furthermore, he reported that miners "seldom dig deeper than ten feet" making it "impossible to determine whether the mineral terminates in regular veins or not; for when the miner finds himself ten or twelve feet below the surface, his inexperience obliges him to quit his digging and begin anew" at an alternate location.[22]

Although Austin refers to the local miners as "inexperienced," there were other factors that guided French Creoles to continue using their syncretic methods in the Ste. Geneviève mining district. In many cases, if miners were unable to form associations with a wealthy mine owner, then they were unable to construct masonry furnaces or water pumps to remove streams of water when encountered trying to dig beyond ten feet, or to build houses nearby to the mines. Miners who lacked the capital to prospect, extract, and smelt minerals using modern techniques had to join with entrepreneurs interested in mining the earth for its minerals to make money. To determine the extent of a vein, miners needed machinery to dig deeper and rope and cranks for a windlass to remove debris and lead ore from deeper shafts and to lower tools for workers.

Based on Austin's observations and his knowledge of European techniques, he recommended improvements after witnessing miners abandoning their trenches after only "scratch[ing]" the surface. And crucially, no matter how much lead ore miners discovered using their current techniques, Austin felt that they wasted time, noting that "one half of the miners' time is taken up in sinking new holes or pits."[23] For three years, Austin observed miners lacking the proper instruments to determine where the richest lead veins were located. And without the correct prospecting tools they failed to gauge the quantity and quality of the lead ores waiting to be extracted. He believed that with correct supervision and training, miners would become competent in using instruments specifically designed to guide their searches for ores. The tools he wanted to see miners carrying included iron hoes, picks, shovels, and crowbars to prospect for lead ore. Austin held that their current methods hindered miners from truly observing the nature, length, and depth of the veins. Austin's survey and analysis of mining practices illustrates why it was deemed necessary to transfer European science and technologies to the Louisiana territory and for miners to stop using the amalgam they had adopted from the Mine La Motte miners—a further demarcation between old and new mining technology.[24]

The Mining Region Extends Over 2,000 Acres

Various writers of the time saw the role of science as essential to the evolution of humanity and the ability of people to control nature. The power of Austin and his assistants to explore the secrets of nature using European tools and experiments contributed to the perceived "mastery of nature" and the gradual erasure of the Native American and French prospecting practices from the frontier. Because Austin conducted his survey prior to the first Virginia survey, his work highlights the importance of mineralogical surveys of geographic spaces and how those reports benefited early American miners and merchants much earlier than historians have argued.[25] Conducting an organized survey and cataloging of minerals after the Louisiana Purchase was not an easy task on the mining frontier. To conduct chemical and physical tests on minerals precisely, Austin had to rely on Europeans' manuals, procedures, and tools to effectively classify, analyze, and record exact measurements on lead ores.[26] Austin's early nineteenth-century analysis reveals how miners used small-scale laboratories to conduct field experiments to communicate their scientific findings about nature to the president, Congress, the public, and American naturalists.

Prior to Moses Austin and a small number of English miners successfully completing the region's first shaft near Mine à Breton, they had to understand the location and direction of the lead ore veins below the surface.[27] Because Austin is noted as the sole author of the survey, it is difficult to know exactly who assisted him with his extensive mineralogical analysis of a mining region that extended over 2,000 acres and included ten operating mines. Most likely, due to the expanse of the mining territory, he required the assistance of Josiah Bell and Timothy and Matthew Mullins who had recently immigrated to the United States from England. As skilled miners, Austin, Bell, and the Mullins brothers viewed science as a measure of their own civilization's achievements. Together they conceived a plan to conduct experiments on lead ore specimens they had collected from the ten mines listed in Austin's *A Summary Description of the Lead Mines*. While Austin determined the quality and quantity of the ores in his home laboratory, Bell and the Mullins brothers conducted field observations on veins of galena. To analyze and describe lead ore, Austin, the Mullins brothers, and Bell applied European systems of mineralogy and chemistry in both types of frontier laboratories.

By the 1770s and 1780s, fieldwork and laboratory experiments became more important for utilitarian geology in Europe. Similar to Abraham G. Werner who was associated with mining in Germany—a natural philosopher who frequently descended into the mines in the course of his work—Austin noted the importance of traveling to the mines to study all aspects of prospecting, mining, and smelting. Moreover, if Austin was unable to travel to the mines himself, then he had Bell's and the Mullins brothers' surveys to study.[28] Before Austin and his staff departed on their multiple expeditions to the numerous mines, they most likely packed a cart with an iron probe, shovel, and poll (pick on one side, hammer on the other), which was similar to what English miners would have used when prospecting the lead mines of Derbyshire.[29] In England, miners used the four-to-five-foot-long probe to "plunge into the ground in alternating places until a vein was struck."[30] After locating veins of lead ore, miners penetrated the earth using a pointed shovel (spade).[31] Austin, Bell, and the Mullins brothers applied these same practices as they searched for lead ore, and Austin hoped that more miners would learn to incorporate similar standards when prospecting for lead ore buried in veins beneath the earth's surface.

While traveling through the mining region, whenever Austin spotted miners digging trenches to expose the lead ore, he had difficulty evaluating the character or termination of the numerous veins.[32] So, Austin became the first miner in the

Missouri Territory to apply traditional European nomenclature to describe and name numerous lead veins, which informed natural philosophers and manufacturers back east about the district's value.[33] To distinguish differences in vein color and possible depth, Austin depended on the encyclopedias of European mineralogists. Using the tools and books he carried from Philadelphia, European environmental knowledge was transferred to Louisiana to assist him in redefining and promoting the region's wealth.

Austin would have been acquainted with a number of European natural philosophers writing about ores, and he was most likely familiar with the methods of Abraham G. Werner, a professor of mineralogy at Freiberg.[34] Werner used a descriptive approach to study and examine the external characteristics of minerals to determine the quality of lead ore.[35] According to Werner, minerals' external characters such as color, smell, and taste, needed to be correctly classified. He described eight principal colors: white, gray, black, blue, green, yellow, red, and brown. In addition, Richard Kirwan of Ireland, whose works were well known in the United States, published Werner's system of mineralogy in 1794. Kirwan's manuals would have been carried in the knapsacks of many miners and natural philosophers venturing into the North American frontier just as Austin and Meade had done.[36] In 1803, the Presbyterian minister Samuel Miller published *A Brief Retrospect of the Eighteenth Century*, which included chapters on the history of mechanics, chemistry, natural history (which at the time included mineralogy and geology), scientific associations, education, and separate discussions of several countries. Miller reported that many American mineralogists applied Abraham Werner's practice, which they learned from the work of Richard Kirwan. Miller also made mention of Kirwan's collection of minerals as "the best collection on earth."[37] Although Austin may have considered his practices superior to the longstanding syncretic methods, at contact, the Kaskaskia and French prospectors also used soil color and taste to characterize the environment to locate lead ore.

Like Europeans, Americans carried to the early frontier an understanding of the superiority of scientific thought and technological innovation.[38] Moses Austin and his English miners possessed a sense of their preeminence in inventiveness and organization. They understood their practices and tools to be superior and their knowledge of nature to be justified as they proceeded to control the lead mines by improving how miners located, defined, and assayed lead ore. Carrying their portable European-style laboratories and the Kirwan manual in their knapsacks to conduct chemical analyses on lead ores, Josiah Bell and Matthew and Timothy Mullins conducted numerous field trips between Mine à Breton and other mining sites.[39] At the conclusion of the month-long field surveys, Austin compiled their information into one report. In line with contemporary European scientific analysis, his mineralogical survey outlines the general layout of the mining district, the names of lead veins, and the quantity and quality of lead ores for the president, Congress, naturalists, and merchants in the United States.

Austin and his assistants began their survey in January of 1804 and completed the mineralogical assessments by February of 1804. The timeframe coincided with the phase when most miners ceased their extracting and smelting activities and remained close to home preparing their tools for the spring planting and mining season.[40] The short amount of time required Austin and his staff to work long days traveling between locations. Their plan was to combine the breadth of observations and chemical analyses of the lead ores in the mining district as well as provide a short history of each mine. The organization of the report is also telling. Instead of Austin beginning with the oldest, Mine La Motte, he chose to begin with his recently established mining settlement—Mine à Breton—revealing pride of place. Since Austin desired to promote the significance of his settlement to lead manufacturers and miners back east, he needed to communicate to others that the European-like mine shaft, the masonry smelting furnace, and the lead manufactory where he produced sheet lead and shot only existed at Mine à Breton. In addition, although Austin's staff conducted "a few experiments" at nearby mines, "because of the strong appearances" of quality lead ore at Mine La Motte, he explained the benefits of that mine as well.[41]

One striking illustration of the central position of science taking shape on the frontier is Austin interacting with nature using his utensils to experiment and measure lead ore. Clearly, the procedures to prospect or assay minerals outlined in European manuals also improved Austin's, Bell's, and the Mullins' understanding of the quality of Missouri's lead. To remain well versed in mineral classification, Austin and his staff depended on the continuous flow of reprinted British publications in the United States. To conduct experiments, Austin relied on other early American mineralogical reports. For example, in 1804, Benjamin De Witt produced an essay entitled "Mineral Productions of the State of New York."[42] Together, the reports suggest that Americans were eager to control and redefine the natural environment. Similarly, Austin recognized that economic benefits would come from a systematic charting of mineral resources, as focused miners used their skills to create a clearer picture of nature and "the number, extent and situation of the lead mines in Upper Louisiana."[43] Austin explained how the mining region "may be said to extend over two thousand acres of land." In each entry on a mine in the survey, Austin first described how miners "[worked] using a method that renders it impossible to determine whether the mineral terminates in regular veins or not." Thus, it appears that one of his primary goals was to define the lead veins at each mine. Austin and his assistants had to estimate the length and depth of each vein to determine how and where miners would construct shafts or European-like galleries, and then erect "machinery" to aid the recovery of larger quantities of lead.

Although Austin had built a laboratory at Mine à Breton, to guide his assaying activities, it must have been helpful to acquire a number of pocketsize manuals and portable laboratories from England. All of these mineralogical tools helped emerging scientists to quickly conclude the amount and purity of each mine's lead ore. English miners used these miniature manuals and laboratories to prospect and assay minerals when away from the main laboratory. What we are unable to gather from the survey is how Bell and the Mullins brothers corresponded with Austin, who remained at Mine à Breton, from the field. Since the surveys were conducted during the winter, timing of transportation would have been improved through densely forested regions by the elimination of leaves, growth, and insects. In addition, it was cold enough for people to easily travel along ice roads and trails. At times, it must have been difficult to pass over certain roads because of snow or mud, and these obstacles may have even damaged their equipment.[44] Because Captain Amos Stoddard gave Austin a short amount of time to complete the work, the miners must have endured some level of exhaustion. However difficult the process of collecting and analyzing samples may have been, they each understood that it was necessary to acquire information from the local mines so that Austin could incorporate their analyses into the report.

The European method of classifying the Louisiana lead mines served the president, Congress, natural philosophers, and merchants well. Clearly, the survey's outline of soil, lead veins, and rock descriptions helped solidify geological ideas about the quantity of lead ore. The mineralogical report provided Americans with an analysis detailing the "strong appearance of mineral—to render the mining business generally advantageous" to both the miner and farmer.[45] Another central element to the report's success was the use of instruments and equipment in the lab or in the field to represent the quality of lead ore meticulously.

With Austin's staff in place and the tools and organization completed, they were prepared to look at the mining environment and organize it according to European innovation. Amos Stoddard recognized Austin as a miner and, more importantly, as a chemist who understood geology. Most likely, Bell and the Mullins brothers carried with them to the United States similar skills and also understood how an accurate survey and estimate of the average quality and quantity of mineral produced annually would benefit lead production and promote the economy of the lead mining district and beyond.[46] As Austin visited other mines, he may have praised the final amounts of lead produced; however, similar to Europeans questioning African and Asian mining methods during the early days of exploration, Austin argued that the Kaskaskia-French amalgam produced small amounts of lead.[47]

Replacing Old Prospecting Practices in Low and Flat Prairies

Mine à Breton was originally designated as the mine on the forks of the Meramec River, and François Azau Breton was the first French settler to prospect and extract lead ores there in 1780. Mine à Breton is located a few miles south of the mines where Father Jacques Gravier and a number of Native Americans walked the network of trails to the rich lead mine near the Meramec River. Just a few miles south of both locations is where Moses Austin established the town of Potosi, Missouri, in the late 1790s. The mine was approximately thirty-eight miles west of Ste. Geneviève and adjacent to the Grand River fork west of the Mississippi River. The Grand River was navigable during the spring season, which made it the ideal location for shipping lead down to the port of New Orleans. Austin characterized the mine as being in "the greatest part—in an open prairie which rises nearly a hundred feet above the creek. The mine extends over two thousand acres of land; but the principal workings are within the limits of one hundred sixty acres."[48] Similar to those explorers who came before, Austin proceeded to organize his survey with mineralogical information to bolster the need for expanding the mines. To promote Mine à Breton as the chief mine of the Mississippi valley to Jefferson, merchants, and naturalists back east, *A Summary Description of the Lead Mines* not only highlighted new mining operations at Mine à Breton but also described the area's natural features such as the waterways, mountains, vegetables, and animals, as well as the mining operations at Mine La Motte.[49] However, Austin argued that it was only after replacing old prospecting practices in these low and flat prairies with European prospecting practices that he and his English miners and his slaves created an orderly way to work at Mine à Breton. Therefore, Austin also wanted to duplicate these changes at the other important mine, Mine La Motte.

Considered the oldest of all the mines in the district, Mine La Motte differed in every way from Mine à Breton. Austin's appraisal of Mine La Motte showed it to be more extensive than Mine à Breton. Similar to Mine à Breton, Mine La Motte was positioned on a river, the St. Francis River. Closer to Ste. Geneviève than Mine à Breton, Mine La Motte was just thirty miles to the southwest of the village. After characterizing the landscape as "low and flat," Austin also noted that miners still applied the old practices at Mine La Motte. He described how after "miners penetrated the ground to a depth of twenty-five feet ... water seepage often prevented additional mining except during the dry season beyond that depth."[50] He also noted that "when the miner finds himself ten or twelve feet below the surface, his inexperience obliges him to quit his digging and begin anew" without understanding or viewing a vein's character or shape.[51] Austin's descriptions of miners' techniques at Mine La Motte are by far the most extensive critique in his survey. According to Austin, because the miners there adopted the old syncretic mining methods, he was motivated to apply European prospecting practices to more clearly describe and study the lead ore veins. For example, most of the mineral at both mines was found in what Austin termed "regular veins" of varying sizes.[52]

After the publication of Georgius Agricola's prospecting, mining, and smelting manual in 1546, it remained the leading textbook for miners and metallurgists for nearly two centuries. At a time when families, guilds, or towns held most industrial processes secret, Agricola thought it important to publish every practice and improvement that he observed and considered of value. Since the time of Agricola, Europeans had borrowed the term *vein* from its use to describe animals. Just as animals' veins are distributed through all parts of the body, allowing blood to flow from the liver throughout the whole body, so too the mineral veins traverse the whole globe. Agricola suggested that from a geological standpoint, mineral veins indicated canals in the earth or channels of ore. More particularly, it was believed that these veins were present in the mountainous and hilly regions, where water runs and flows through the veins.[53]

Closely adhering to Europe's natural philosophers' instructions in their manuals, Austin decided to communicate to his readers the underground patterns of *regular* veins at Mine à Breton and Mine La Motte. In *A Summary Description of the Lead Mines*, he uses the terms *rake-vein* and *flat-vein* to distinguish between two types of veins of lead ore deposits. Europeans considered rake-veins to be true veins because they were "narrow, and often mixed in with limestone." As mentioned earlier, De Gruy reported how he usually observed lead ore enclosed in limestone, and Austin also observed a similar pattern to the rake-veins near limestone as having "a thickness of at least a foot." Furthermore, he noticed how the veins widened "three to four feet in width" in most pits or shafts and how at greater depth the veins at Mine à Breton and Mine La Motte contained an abundance of lead ores.[54] Comparable to miners in Great Britain, Austin observed the lumps of ore "to have been rounded into common gravel," and he assumed that flat-veins held small beds of ores connected to rake-veins and were only productive to a certain distance from the body of a vein. Often after miners working these flat-veins removed three or four feet of earth, they would reach gravel lead ore intermixed with "sand rock," which could be easily broken using their picks.[55] Austin's survey highlighted the "regular vein" more than any other vein structure, which not only confirms his understanding of European terminology but also his ability to apply their methods to prospect deeper into the earth. He could clearly view where mineral veins terminated. Austin's use of the term *regular vein* most likely means rake-vein, since both appeared near limestone. At both mining sites where Austin prospected, he noted that rake-veins were the most common form in which lead ore occurred. Additionally, his reference to "quartz and pyrites" notes that galena was nearby.[56]

The classification of mineral veins also included a description of the external characteristics, and in *A Summary Description of the Lead Mines*, Austin dedicates an extensive section to the qualities of the lead ores at Mine La Motte where he observed three distinct qualities of ore: gold-colored fossil mineral, gravel, and "fine steel grain" mineral. The gold-colored fossil mineral, yet another type of mineral, was a term that had been applied to rocks and minerals since the medieval period. The meaning stemmed from the Latin word *fossilis*, which referred to anything dug out of the earth. The metallic luster of gold fossil was pale with a "brass yellow hue," which often caused miners to confuse it with gold.[57] One hundred years earlier, the French explorers Des Ursin, La Renaudière, and De Gruy also described the galena as "rounded pebbles or gravel ore" that they discovered in small, detached lumps "buried beneath three to ten feet of the earth's soil." According to Austin, gravel ore could be seen immediately under the soil and was intermixed with the "lumps" of lead "the size of a fist weighing from one to fifty pounds" that he observed in the flat veins.[58] Believing in their scientific instruments to determine the location of rich veins, Austin and his assistants became excited when he first recognized the dark blue "fine steel grain" mineral at Mine La Motte.

Known as galena, or potter's lead ore, it was the most common of all the lead ores and appeared as grains shaped like moderate-sized cubes. This was the same silvery lead ore that the Mississippians prized for local use and distant trade. To create ancient glitter, they would crush it; and this type of galena was cherished because it could be easily broken with the blow of a hammer. Native Americans traveled over land and river networks to deliver their crushed galena to tribes located along the Mississippi and Ohio Rivers. Because of galena's richness, Native Americans used it with very little preparation as a glaze for coarse pottery and exchanged it with other Native American tribes for products not available in the Mississippi valley.[59] Europeans classified this galena "as soft unless it was mixed with iron making it harder."⁶⁰ Since the lead appeared like colorless glass and had a refracting quality, in *A Summary Description of the Lead Mines*, Austin, Bell, and the Mullins brothers referred to these specimens as carbonate of lead.⁶¹ With the discovery of deep veins where carbonate of lead lay hidden, Austin outlined in his report the need to improve prospecting and mining techniques to exploit it.

At Mine La Motte, Austin detected that the greatest portion of carbonate of lead could be found embedded in a thick stratum of marl clay adjacent to limestone.⁶² To access carbonate of lead more efficiently, Austin called for local miners to abandon their trenches and start digging deep mine shafts using European machines. Since the carbonate of lead was free of contaminants, glass manufacturers craved it as a key ingredient to manufacture quality red lead to fabricate glass. In certain lighting, it appeared colorless, transparent, and glass-like—all natural qualities for producing flint glass. Flint glass is different from all other glass due to its superior quality, luster, and purity. Flint glass was considered to be a better glassware, perfected in England, because powdered flint was added to the glass formula to improve clarity. The glass was primarily produced in the United States and Britain from the 1820s through the 1860s. At a later date, a lead compound was added to the formula, and it was found to give the glass much more clarity, resonance, and weight. Soon after, the powdered flint was dropped from the glass mixture, but the name was entrenched in vernacular. After experimenting on the particular galena type from Mine La Motte, Austin discovered the amount of lead registered between forty-five and eighty-three percent.⁶³ Austin's experiments prompted him to write in his report that Mine La Motte was a "gold mine descriptive of its wealth" and capable of furnishing vast quantities of lead.⁶⁴

After applying their new European skills, Austin and his staff brought the Louisiana Territory lead mines to the attention of the president and Congress. In addition, after natural philosophers and merchants read about the dark blue "fine steel grain" galena found at Mine La Motte, all understood this type of lead would be useful primarily for manufacturing red and white lead.⁶⁵ A decade later, when the New York glassmaker and geologist Henry Rowe Schoolcraft toured the lead district, he would state, "when this ore is piled near mines, the reflection of the light makes it appear white, and the unfamiliar workers unacquainted with this ore might readily mistake it for silver."⁶⁶ Schoolcraft's description of the structures and the various character traits of galena was similar to Austin's, and he noted "the external luster of this type of galena was found to be resplendent and specula to glimmering" and refracted like glass.⁶⁷ Austin described it as a pinhead-to-chestnut-size gravel mineral intermixed with the soil. Like Austin, Schoolcraft, after visiting Mine à Breton and Mine La Motte, argued in his 1819 publication of

A View of the Lead Mines of Missouri that if local miners possessed the tools and machines to prospect and mine correctly, they could extract carbonate of lead from the rich veins more efficiently than when using the traditional syncretic methods. To classify mineral veins and describe the external characteristics of lead ore accurately in *A Summary Description of the Lead Mines*, Austin most likely set up a laboratory with instruments to conduct experiments on the lead ores he and his assistants observed. Austin's mineralogical report added to the landscape not only European terminology to classify galena and carbonate of lead but also a laboratory "fitted up with furnaces, instruments and apparatus" to characterize lead ore.[68] He used European chemists' tools such as furnaces, cisterns, pots, and utensils for examining features of the natural landscape to identify veins and lead formations and to determine their ores' quality.[69]

Found by Experiments That The Mineral . . .

The observations documented in *A Summary Description of the Lead Mines* were all an effort to discourage the syncretic methods and simultaneously promote new environmental knowledge about the mining landscape. While the early Native American and French practices to test the quality of lead ores were closely associated with human taste and touch, the European method appears to have been more distant and required tools. In all probability, to test the quality of the galena gathered from the field, Austin and his assistants constructed a laboratory. The chemical analyses performed by Austin and his staff and other local miners represent a critical moment in early American lead mining history, signaling the further disappearance of the Kaskaskia-French amalgam from the mining district.

A broader significance of the report and the experiments conducted is an acknowledgment that beyond applying a miner's sensory skills, Austin used other European instruments and measurements to achieve his goal of redefining and controlling the region's most precious natural resource. While the tools and practices appeared useful and beneficial, in the hands of Austin and his associates, the laboratory instruments contributed to the erasure of the traditional Kaskaskia-French amalgam. For example, Native Americans had a method to wash and analyze lead ore prior to smelting. If clay covered the ores, it was necessary to wash before smelting. If gangue minerals adhered to the galena, Native Americans hand-cobbled the ores with a small pointed hammer known as a pickwee. Gangue minerals are the substances that are extracted along with the desired minerals or ores because of their close association and nearness to veins, and these minerals are considered commercially worthless. Therefore, Native American, European, and American miners understood that any gangue had to be separated out after the extraction process. Also, when miners extracted galena that was heavily coated with white carbonate, they understood that the smelting process would be slowed. Although Native American smelters may have done some preliminary washing and hand-cobbling, it is unclear how long they were aware of these practices. But, because French Creole and Native American smelting processes were labor intensive, there is no doubt that along with prospecting, extracting, and assaying, smelting required considerable judgment.

As earlier mining engineers had understood, the most important step after extracting lead ores was for the miner to assay the ores correctly.[70] Similar to La Renaudière and Des Ursin, who conducted an assay on the lead ore at Mine La Motte, Austin and his field assistants understood the importance of determining the amount of metal contained in the lead ores from the mines they visited. If overheated, the metal could be lost in fumes; if underheated, the metal would be lost with the slag and furnace accretions. To analyze these samples, Austin combined fieldwork with lab work. Testing metals required a set of tools, such as the cupola, an absorbent, shallow vessel. After acquiring a small sample from each mine, they proceeded to carefully assay the ores.[71] Austin's acceptance of the superiority of his tools, instruments, and precise observations highlights a way of thinking about and perceiving the natural world. Most significantly, the early nineteenth century marked a period when many Europeans and Americans were confident that because of the emerging sciences they had surpassed all other civilizations; and Austin, natural philosophers, and miners desired to comprehend and control the natural world. To do so, they believed that the best means of redefining the mining region was not only to characterize nature but also to use the latest European apparatus to conduct experiments on lead to measure its quality.[72]

Armed with new scientific tools, Austin and English miners began the informal process of transferring European assaying practices to the frontier. The laboratory at Mine à Breton may have been in the same building as the reverberatory furnace near the sheet lead factory.[73] In addition, Austin and his staff carried portable or pocket laboratories when traveling between the various mines throughout the district.[74] Both the stationary and pocket laboratory included a variety of European instruments to test minerals.[75] The trained chemist who worked at the medical school in Philadelphia, James Woodhouse, was also devoted to mineralogical science. A number of years after Woodhouse became a physician in 1792, he created his *Young Chemist's Pocket Companion: Connected with a Portable Laboratory* in 1797.[76] Woodhouse also spent the opening years of the nineteenth century conducting analysis on lead ores at the Perkiomen Creek mines near Philadelphia. With these instruments and manuals in their hands, Woodhouse, Austin, Bell, and the Mullins brothers found by experiments on lead ores that they were able to produce new knowledge about their mining environments with technical specificity.[77]

The design of the portable laboratory reveals the care by which miners conducted their experiments, and also offers insight into the topography of the mining area where they traveled. Miners depended on these instruments to help them assay lead ore at the mining site, and packed in a "small oak box, the size of a book, the field laboratory was designed for the miner who travels from one mine to another." Manufacturers designed the boxes to protect the instruments by "lining the compartments with green velvet," and they "covered it with leather" so that the boxes would not become damaged as the carts or wagons passed over the mining region's rough and often hilly topography. Upon opening boxes, miners discovered the iron or ivory blowpipe, the glass triple magnifier, and the lamp furnace all housed in separate compartments to further protect each instrument.[78] When used, the design of the blowpipe eliminated the sensation of miners holding a piece of metal "between the teeth and lips" for a long time. This simple blowpipe with a tapering iron tube at the end and a concise opening allowed Austin and his assistant miners to blow air "under relativity high pressure." Miners used the blowpipe, which acted as a portable bellow, and mimicked the large furnace at Mine à Breton, to perform chemical tests.[79] The *Compendious System of Mineralogy & Metallurgy* states that "when the blowpipe was applied to compact or common galena," it melted and gave off a "sulphureous odor" and produced a button-size piece of metal at the bottom of the crucible.[80]

Natural philosophers and miners used other tools to assay their minerals. Most would have followed the chemical analysis techniques outlined by Woodhouse and others.[81] Closely adhering to the Woodhouse model, Austin conducted experiments with great care and skill. Any error made during the assaying process would have been multiplied when the bulk ore was worked at a later time, and would have resulted in losses to the owners. Since most encyclopedias were cumbersome and difficult to carry between mining sites, Austin provided his field assistants with the Woodhouse pocketbook-sized manual and instruments, which acted as a field laboratory to conduct their investigations. The pocket laboratory was not intended for "a person who always resides at one and the same place."[82] Because of time constraints and obstacles on the frontier, the miners used both the Woodhouse manual and instruments to assay ores.

By the late eighteenth century, advances in optics also aided miners with their experiments. A triple magnifier able to produce "seven magnifying powers" and one lamp furnace were both stored in separate compartments of the pocket laboratory. The magnifier enabled Austin to enhance his observations. With the triple magnifier, miners could view small and middle-sized "crystals with specula, splendent or sometimes rough surfaces that appeared uneven and splintery."[83] Miners used optics to detect whether foreign matter was attached to the minerals. In these cases, the miner would have to take extra care when cleaning the mineral in preparation for assaying. The mineralogist could clearly distinguish the structure and metallic parts of lead ores. The blowpipe and lamp furnace would sufficiently perform in a few minutes and "with very little expense the assay, which otherwise would require large vessels like a reverberatory furnace," for the miner when the mobile iron or clay furnace was not available. Traveling along the rough trail, field assistants seldom had an opportunity to carry portable iron or clay furnaces. They may very well have been content with the portable laboratory and apparatus, "which are sufficient for the most part [for] such experiments as can be made on a journey," to the local mines.[84] With the blowpipe, the triple magnifier, and the lamp furnace, Bell and the Mullins brothers could supply Austin with the analytical results on the lead ores from all ten mines to which they traveled and which are covered in *A Summary Description of the Lead Mines*. The eighteen-page report included approximately seven pages devoted just to the mineral analyses that Austin collected from his assistants.

Two more brief examples of the types of instruments miners armed themselves with to survey the lead mines reveal the transfer of European scientific practice to the frontier. Field assistants carried a separate portable laboratory that contained a washing trough to rinse mineral ores with water to separate the ore from "adherent rock" before conducting chemical assays on the minerals. Troughs came in multiple sizes and were very common in laboratories. Woodhouse recommended that miners use "one of a moderate size for field use." He provided instructions for building troughs according to the following dimensions. Troughs were to be twelve and a half inches long by three inches wide at one end and one and a half inches wide at the other end, and should slope down from the sides and the broad end to the bottom, where it is three quarters of an inch deep. Troughs were commonly made of smooth, hard, or compact wood containing no pores where minuscule grains of ores "may conceal themselves." Finally, there was also a separate laboratory box to house different acids to test the mineral bodies. The most common were niter of vitriol, vinegar, borax, and common salt.[85] The tools Austin and his assistants carried across the mining landscape represent how they came to view science and especially technology as a measure of their own civilization's achievements.

The analyses conducted on lead ore at Mine La Motte proved that the lead ores were of a superior quality to those at Mine à Breton. Austin tested the ores to determine the true percentage of metal. In 1818, glass manufacturer Henry Rowe Schoolcraft also conducted an analysis on Mine La Motte's galena. Schoolcraft "found by experiments that the mineral in the hands of skillful smelters" would produce sixty or seventy percent metal. Austin also believed that if "fifty men under a proper manager" worked "with a good smelting furnace," the mine "might produce five or six hundred thousand pounds of weight of lead per [year]."[86] As mentioned above, Schoolcraft confirmed what Austin had already discovered—that Mine La Motte appeared to be like a gold mine.[87] According to Schoolcraft, the amount of sulphur associated with lead ore was important to miners. Although Austin fails to mention veins of sulphur or the amount of sulphur contained in the lead, Schoolcraft noted, "The ores of La Motte contain an unusual portion of sulphur.... I draw this inference from its refractory nature." Schoolcraft's experiments taught him that when the ore was heated, "the more sulphur there is driven off, the brighter it grows."[88] In effect, the analyses of Austin and Schoolcraft concurred that with proper management and the transfer of skilled smelters to Mine La Motte, the gross production of the mine would increase.

The highly anticipated Austin report produced new and more scientific descriptions of the lead region's features. By the beginning of the nineteenth century, American and English miners began to specify their understanding of European environmental knowledge, which most likely further influenced changes to the mining district. They took up all manner of questions, including topics we would now assign to medicine, engineering, travel writing, or even ethnography. Science was welcomed in both urban and frontier spaces. Everyone who could do so read about the sciences. Notably, near the lead mines of Missouri, experiments and demonstrations were public events. Newspapers regularly published accounts of experiments, new technologies, and medical news.[89]

Like Austin, eighteenth- and early nineteenth-century men of science did not specialize in or make their living by science. They had other occupations or careers. Like Thomas Jefferson, they were presidents, clergymen, members of Parliament, farmers, tax administrators, ship captains, noblewomen, printers, and even miners or smelters.[90] On the early North American mining frontier, Austin and his field-workers connected European scientific instruments with a landscape that had originally been defined according to Native American and colonial French practices. Austin's report centers on the development of science in these spaces. By producing new kinds of environmental knowledge, the survey would become a tool for improvement predicated on the importance of applying enlightened views to resolve extracting and smelting problems.

On November 8, 1804, Thomas Jefferson provided Congress with copies of Austin's report on Louisiana's rich lead mines. During the next five years, the Missouri Territory's lead industry was one of slow but substantial growth.[91] The European techniques of prospecting, naming of veins, and conducting assays enabled Austin to determine with certainty the percentage of lead contained in the ores. As word spread in the form of his mineralogical survey, others were encouraged, which prompted additional explorations and experiments on newly discovered deposits; new towns were settled, and the manufacture of lead on an important scale began. After Austin completed and sent his survey of the mines to the president, the analysis in *A Summary Description of the Lead Mines* alerted merchants in frontier and urban spaces about the types of galena discovered after the United States acquired the Louisiana Territory.

Additionally, the first mineralogical survey performed by Austin and his associates made a significant contribution to the young United States, as the volume of lead that flowed from the mining district to frontier and urban factories increased. The vast frontier of minerals would continue to be examined, exploited, and manufactured for the next several decades. Austin's observations and experiments would eventually be refined for Philadelphia's merchants, who desired additional mineralogical and chemical data about Missouri's lead. The environmental knowledge produced by Austin and his associates was transferred to glass manufacturers who coveted quality lead ore for the manufacture of red lead.

By 1809, Joseph Brown and John A. Storch both migrated from Philadelphia to St. Geneviève, Missouri. Brown was a land surveyor who worked for the government defining treaty lines. He surveyed the state line between Missouri and Arkansas. Although there is no information related to their assaying activities, the *Missouri Gazette* announced that Brown and Storch planned to erect a permanent laboratory to assay minerals, establishing a business for "analyzing ores of every description." Brown and Storch subscribed to the idea that anyone possessing any kind of ores could have them "ascertained for five dollars per experiment," with the precondition that they send no more than one pound per visit.[92] As the Missouri Territory became more populated with miners, they would discover new deposits of mineral that needed to be assayed. Following the publication of Austin's analysis, news of the region's mineral wealth spread from the frontier to eastern manufacturers already engaged in the production of shot and red and white lead.[93]

Learning That This Country Would Furnish Lead Sufficient for the United States

Austin conducted his survey during America's era of early industrialization, a period when American leaders considered that independence would depend on cultivating the nation's natural resources. Decades before the Revolution, colonists mined iron, copper, lead, zinc, and salt. At the turn of the nineteenth century, frontier miners and urban merchants well understood the natural relationship between the newly acquired lead mines and manufacturing. Austin and other miners examined the properties and changes in the composition of minerals, which enabled them to contribute to the young United States' growth.[94]

There is a close link between the timing of Austin's report and industrial development along the Eastern Seaboard of the United States. In 1794, Tench Coxe listed over fifty classes of assembled goods. He cataloged "many articles indeed" as being "manufactured in the city of Philadelphia, in the boroughs, and in the counties of Pennsylvania."[95] Local officials as well as politicians encouraged development in and beyond Philadelphia. For example, shot, sheet lead, red lead, and white lead were useful products to both frontier and urban places. There was also a growing desire to discontinue importing both lead products and glass from London into the United States. The increasing friction between the old and new countries during the early 1800s provided added impetus to commercial independence for the younger one. As tensions continued to rise, Congress passed the Non-Importation Act of 1806 restricting importation of glass and other commodities. The restrictions were not enforced until the Embargo Act of 1807, so merchants continued to accept large orders of white and red lead, litharge, paints, drugs, and chemicals from London, as Great Britain gladly continued to flood American markets with these products.

On March 8, 1809, the *Missouri Gazette* announced that Moses Austin had established at Herculaneum a "shot manufactory," saying that "the situation is especially adapted for the purpose of having a natural tower of stupendous rock, forming a precipice of about 130 feet." The village of Herculaneum was located on the west bank of the Mississippi, about thirty miles below St. Louis. Frederic Billon noted, "After the transfer of the country [west of the Mississippi River] to the United States, and the extensive development of lead mineral throughout all this region back from the river, two enterprising Americans, Colonel Samuel of St. Louis, and Moses Austin of Ste. Geneviève, perceiving the advantages of this point for an extensive lead business from its nearer proximity to the mines than

Ste. Geneviève, then the only point of shipment on the river, purchased from Jonathan Kendall, on January 9, 1809, a tract of land at the mouth of Joachim Creek and immediately laid off their plat of the town" (figure 4).[96]

The *Gazette* noted that because of the nearby lead mines and the accessible harbor, the Missouri Territory would be able to "supply the Atlantic states on such terms as will defeat the competition."[97] In addition to the closeness of the mines, Herculaneum provided a good landing for boats in the nearby Joachim Creek, which afforded an excellent harbor.[98] Neither Austin nor Kendall feared new inventions or establishing new frontier manufactories. Like their Northeastern merchant counterparts, they were also aware of the escalating tensions in Europe and their potential threat to overseas trade. Moreover, they began to explore expansion possibilities with western outposts from Lake Erie to St. Louis and beyond to sell their shot and sheet lead.

Rising almost two hundred feet from the riverbanks were perpendicular cliffs. The natural elevation allowed shot makers to drop the molten lead, and this meant they did not have to invest substantial capital in the construction of shot towers. Still, John Maclot erected a tower to manufacture patent shot at Herculaneum, and the *Gazette* reported that he had been successful at "casting shot comparable to the best English quality in Herculaneum." Maclot, who recently appeared in the Missouri Territory came from France with plans to erect the first shot tower west of the Allegheny Mountains in 1809 on the south bluff of the town. Workers at the shot tower melted lead on a crest of the cliff and while still molten, it was dropped into receptacles at the bottom. By the time the molten lead had reached the receptacles, the lead had hardened into small round shots. On November 16, 1809, the *Missouri Gazette* wrote "The proprietor of this establishment has commenced casting shot equal to the best English patent." Maclot planned to further excavate the rock to render the fall to be about 130 feet, which he believe would enable him to manufacture ten thousand pounds of the largest size shot per day.[99] A visitor to the region noted how the energy with which capital turned to this form of production at this early date on the frontier was remarkable.[100]

On June 20, 1811, the *Gazette* reported that the lead mines in the "district of Ste. Geneviève promoted a spirit of industry [that was] everyday manifesting itself among the people of this territory." The fur trader William Ashley began to experience a decrease in the value of peltry and increasingly turned his attention to lead mining.[101] It was also reported that captains of several large boats no longer left Ste. Geneviève but instead departed from Herculaneum, "the larger mineral establishment." These boats carried more laborers to supply a quantity of lead "sufficient for at least one half of the consumption of the United States." Local

FIGURE 4. View of Herculaneum and the Shot Tower, sketched by Charles-Alexandre LeSueur, April 10, 1826.

citizens hoped that if the trade in lead became successful, it would lessen "the imports of this article into the United States from foreign countries, and decrease our dependence on them, as well as give activity and life to the trade of this territory." Now they needed artisans with scientific knowledge to turn their attention to quality galena in the Missouri Territory in order to begin the production of red and white lead locally.[102]

The Austin report encouraged manufacturers of red and white lead in Philadelphia to search for domestic natural resources instead of imported ones. By 1804, the Wetherill Brothers of Philadelphia, "learning that this country [would] furnish lead sufficient for the United States' consumption," would slowly decrease their flow of white lead and red lead from Great Britain and begin to manufacture similar products using galena from the Missouri Territory.[103] The Wetherill Brothers had a long established association with the London shipping, warehousing, and leading color manufacturer and supplier to the trade firm Brandrams, Templeman & Co. The association with this firm endured over many years. Now the Wetherills were considering a way to make glass, and dry white and red lead to sell from their

126 | Early Mineralogical Assessments and Emerging Science

Philadelphia store. To manufacture similar products, they made plans to secure from Mine La Motte the best quality of pig lead for their Philadelphia red and white lead factory.[104]

In 1809, five years after the publication of Austin's report, frontier towns also began to witness change as there was a widening use of lead. Merchants and artisans erected factories to produce white and red lead in Cincinnati, and glass was being manufactured in the town of Pittsburgh. Painters and glassmakers depended on manufacturers who understood Austin's assays. Manufacturers had to acquire the highest quality of galena for the production of white lead for paint and red lead for glass.[105] Red lead was a primary ingredient used to prepare enamels for potters' glazing. However, a greater quantity of red lead was used to manufacture glass, "which increased its luster, weight, and strength." With the correct amount of "lead combined with other raw materials, glass became more pliable to work and easy to mold into a variety of ornamental forms."[106] Similar to Austin's building shot and sheet lead factories and applying new scientific knowledge on the frontier, factory owners and their workers who manufactured glass, red lead, and white lead were required to have a level of scientific acumen to correctly measure and control the process.[107]

Additionally, the report encouraged Christian Wilt and Joseph Hertzog, two Philadelphia merchants, to consider establishing a white and red lead factory in St. Louis. On October 26, 1811, Christian Wilt, the Philadelphia merchant who had recently migrated to St. Louis, published an article in the *Missouri Gazette* stating, "This is the proper country for the establishment of manufactures of lead such as shot, red lead, and white lead, which in the vicinity of the mines can be carried on better than any other location. Persons have been sent on this summer to erect buildings and commence the manufacture of red and white lead."[108] In the same edition, the paper outlined an estimate of lead shipped from the region at two billion pounds, for total revenue of $200,000. Soon, Wilt would establish the first red and white lead factory on the edge of the frontier in St. Louis.

During the early nineteenth century, as the eastern markets became flooded with British shot, the Philadelphia shot factories suspended operations. As a result, Joseph Hertzog, Wilt's uncle in Philadelphia, decided it was a good time to begin manufacturing red lead in St. Louis to supply nearby frontier towns. In addition, he visualized possibly selling red lead to potters located along the Ohio River. Wilt and Hertzog also hoped to extend their sales to Pittsburgh where five glass factories had recently been erected, all using red lead.[109] Austin's chemical analysis of the Mine La Motte lead ores encouraged cultivators of science in Philadelphia to produce red lead, a key ingredient for manufacturing glass.

Wilt and Hertzog made plans to manufacture red lead and sell it to factories in Pittsburgh and Cincinnati.[110] Both merchants were aware that in addition to glassmakers, there was also a growing number of pottery artisans located along the Ohio River whom they could supply with quality red and white lead.[111] Both men apparently lacked the experience required to manufacture either product; therefore, to help construct and operate the factory, Hertzog hired two Philadelphian artisans.[112] Joseph Henderson was employed to color the lead since he had some chemical expertise. The factory also needed a skilled smelter, and John Sparke, a former employee of Samuel Wetherill, the owner of the largest lead factory on the East Coast, was hired as the grand smelter charged with the job of preparing the bar lead to manufacture red and white lead.[113]

In his job coloring lead, Joseph Henderson was an example of an early nineteenth-century American worker who applied the European sciences to manufacture red and white lead. Henderson carried his chemical experiences from Philadelphia to St. Louis to transform the lead that flowed from Mine à Breton and Mine La Motte to the first frontier urban white lead and red lead factory. During Henderson's leisure time, he studied Aikin's *Dictionary of Chemistry and Mineralogy* and Nicholson's *A Dictionary of Chemistry*. Using both works, Henderson conducted experiments to work out a formula to manufacture over four tons of red lead a week. He hoped to "discover some valuable processes for making sundry profitable articles," and he worked out a new system to manufacture red lead through the use of heat alone. Henderson believed that if his chemical tests were successful, his invention would make it unnecessary to pound, grind, and wash the lead, thereby saving time, and that he could manufacture red lead according to the demands of the market. He also assumed that with his new discovery, the weekly output would not be limited to four tons of red lead but that the quantity of manufactured lead could be regulated according to the demand of both the urban and frontier markets. Although making several attempts, Henderson's experiment was unsuccessful; his attempts represent the importance of scientific experimentation on the frontier.[114] The ability of Henderson to conduct research by applying European systems according to Aikin's and Nicholson's chemical methods during the early nineteenth century on the frontier tells of the growing importance of red and white lead to society. Merchants in St. Louis, Ste. Geneviève, and Herculaneum sent their products east along the Ohio River or down the Mississippi River to New Orleans, where they were loaded on ocean vessels bound for Philadelphia, New York, and Boston.[115]

Lead Colic Has Existed from Time Immemorial

The same year Jefferson announced Moses Austin's mineralogical report to Congress, the French physician François Victor Merat published *Dissertation sur la colique metallique*. Merat conducted his observations and research while at the bedsides of colic patients in Charity Hospital (Hôpital de la Charité) in Paris. In 1812, he published a follow up to the original report—*Traite de la colique metallique*. Both works followed a long tradition of medical writing on lead poisoning including physicians such as Hippocrates, Samuel Stockhausen, and Bernardino Ramazzini who endeavored to shed some light on the history of colic's relationship to the preparations of lead. In order to understand the actual state of knowledge of this disease in the late eighteenth and early nineteenth century, Merat conducted a number of critical experiments on animals and humans. While working at Charity Hospital in Paris, he examined what effect drinking water impregnated with lead had not only on animals but also on humans, and he studied the symptoms associated with lead colic. He stated that "persons have sometimes been strongly incommoded [inconvenienced] by having drank water."[116]

At the turn of the century, physicians continued to articulate their concerns about lead poisoning and the risk associated with artisanal and other industrial activities related to working with lead ores, lead compounds, and associated products. Although there was most likely limited access to literary reading circles among villagers living near Mine La Motte and Mine à Breton, what Merat documented at Charity Hospital in Paris was experienced firsthand by families living near Old Mines. The villagers living in close proximity to these lead mines reveal their understanding about the chemical effects of lead compounds on their water supply. Far removed from the city, in this rural setting where seasonal and year-round mining, smelting, and metallurgy increased, villagers voiced their concerns about developing lead poisoning from polluted water near the mines.

In April 1804, the cleaning and smelting of lead ores occasioned a complaint by the people of Old Mines where fifteen families were dependent on a nearby spring located a mile above the settlement for their water supply. In 1802, the families had reminded those who would listen that they were already forced from their land. At that time, Amable Partenay Mason, who had recently moved from Illinois to Ste. Geneviève and eventually to Potosi, Missouri, to mine, had polluted the stream at Mine à Breton. Mason placed a pigpen and a buddle for lead-ash washing above the villagers' lots. As a result, they explained, "the well water near the mines was impregnated with sulphur and other minerals and was both unpalatable and injurious."[117] According to the families issuing the complaint, in the opening years of the nineteenth century, mining in Upper Louisiana still affected the environment. Those families became concerned that substances were mixing into the water that approximately forty people relied on for drink and possibly for fishing. The families were not only concerned about raising and maintaining cattle, they were also troubled by the sulfur and arsenic polluting their drinking water. In the mining district, prior to smelting, lead ores had to be washed. Miners and slaves washed the ores in a buddle, or a pond, which the complaining families believed held waste from Mine à Breton. Over time, the buddle that Amable Partenay Mason built would eventually hold gallons of potentially toxic liquid and definitely toxic sludge polluted with arsenic, mercury, and sulphur.

Their understanding of the chemical effects of lead compounds on their water supply ultimately led the fifteen families living near Mine à Breton to sign an official complaint. Later they learned that the mine manager had granted permission to Mason to pollute the Old Mines water supply by building a buddle above the settlement. When the villagers requested Mason use a site below the village instead, he refused. When they applied to the government official at Mine à Breton, he told them to "tear down anything that Mason might construct that threatened their security."[118] Instead, the villagers decided to petition Amos Stoddard to send an American to survey the situation. Stoddard advised them that he would have to learn more of the district before he could take action. Eventually officials ignored the complaint, which forced the villagers to start a new establishment at Old Mines. The environmental effects of smelting also increased the quantity of waste in the form of lead ash and tailings about the mining settlement, and vapors floated across the landscape affecting the vegetation and water supply.

Back at Charity Hospital in Paris, Merat collected medical evidence that he believed would "convince any persons who make lead" of the dangers inherent in handling the compounds associated with making lead products. Merat went on to mention many cases of lead colic in workers including painters, plumbers, potters, glaziers, workers in glass, gilders, chemists, and miners who were "often attacked with the most severe colics, sometimes succeeded by death, from having only handled saturnine preparations, or even from being placed within the sphere of their emanations."[119] However, by 1837, physicians also began to note that manufacturers of white and red lead, grinders of colors, sheet lead makers, and manufacturers of shot were affected through contact with the compounds associated with these products. What distinguishes Merat—and later, Louis Tanquerel des Planches—is his attention to lead poisoning in miners, chemists, and those working with pigments.[120]

Within months after Christian Wilt's arrival, Andrew Wilt and Joseph Henderson, who were lead coloring experts from Philadelphia, joined Christian Wilt and began to color lead for sale in markets between St. Louis and New Orleans and Pittsburgh. Close to the completion of the mill, Christian Wilt realized the need for additional help; therefore, on September 6, 1812, he decided to write to his Uncle Hertzog in Philadelphia. He stated, "Our mill is nearly done and, in a week, I allow [the mill] will be going for certain" and needing extra hands "I have engaged a Negro," Claude Henry, a free African American, for "twelve dollars a month for ten months." In the same letter, Wilt also requested that his Uncle Hertzog in Philadelphia arrange to purchase a number of slaves at "a relatively low cost" from Maryland to aid in the production of red lead. In the letter, Wilt told his Uncle to purchase the slaves from Maryland because slaves were selling at four hundred dollars compared with five hundred in St. Louis.[121] Andrew Wilt, Henderson, Henry, and potentially a few Maryland slaves would perform the hazardous and difficult job of coming in contact with lead and its vapors while coloring lead. Further complicating their work and possibly absorbing lead particles into their bodies, they used furnaces, cisterns, pots, and utensils to transform pig lead to red lead mechanically and chemically.[122]

In addition to Wilt desiring his Uncle to purchase slaves from Maryland because of the savings, it is possible that Wilt understood that the Upper South slave state began manufacturing chemicals in 1810. Stimulated by changing trade patterns brought about by the Napoleonic Wars, the Maryland Chemical Works factory began to produce alum, pigments, and dyes to supply Baltimore's early cotton mills. Combatants' seizures of enemy and neutral merchant vessels reduced supplies of imported chemicals and finished goods made with or from chemicals. Concurrently, the expanding domestic manufacture drove up demand for chemicals such as alum, used as a mordant to set dyes in cloth. In a number of those factories, owners hired free blacks and whites, and they rented slaves from their owners. At the Chemical Works, it was also the practice either to own slaves or to hire other masters' slaves from throughout the city. Between 1780 and 1820, enslaved African Americans were engaged in many kinds of work dealing with new masters who never held slaves before acquiring them in Baltimore. When the Baltimore chemical business began, owners erected a new factory and purchased slaves as operatives in these new chemical plants. For example, the records of the McKim and Sons Chemical Works provide a rare, in-depth portrait of the operations of industrial slavery in Baltimore and of the importance of training and keeping slave workers productive.[123]

During the 1804 transition from Spanish Louisiana to American Louisiana Territory, there were early debates about continuing slavery and opening the territory to the interstate slave trade. This would create a market for the growing surplus of slaves in the old Upper South—Maryland, Delaware, and Virginia— and set the stage for the Missouri controversy that erupted fifteen years later. As slaves migrated with their owners to the Missouri Territory, St. Louis city officials were compelled to adopt ordinances between 1808 and 1811. These ordinances were designed to restrain slaves from drinking, holding public gatherings, and mixing socially with free blacks and whites. Like African American slaves contributing significantly to Baltimore's early development, in the Missouri Territory, free and enslaved blacks contributed to further establishing this new United States territory. Free and enslaved blacks had long mined lead, cleared land, planted and cultivated crops, hunted and trapped, and engaged in domestic service. Now with the installation of the first white and red lead mill west of the Mississippi River, free and enslaved blacks learned the art of chemically turning pig lead into red lead. Therefore, Christian Wilt's request of his uncle to purchase slaves is significant because his factory not only employed free white and black workers but also would eventually have slaves working together with free people. Together they would perform the hazardous and unhealthy jobs of grinding, washing, and coloring lead at the mill.[124]

The lead coloring experts, Andrew Wilt and Joseph Henderson, were responsible for manufacturing the pigment red lead, which is a peculiar oxide of lead, that required coloring experts to understand calcining the lead by roasting or exposing the ores to strong heat. The process also called for Wilt and Henderson to rake, grind, strain, and dry it for its final conversion to minium or red lead. A number of times throughout his letter book, Christian Wilt writes, "Henderson has begun to mash [grind] lead yesterday and will commence coloring next week." According to the Scottish physician Andrew Ure's *Dictionary of Arts, Manufactures and Mines*, published in 1837 and subsequently translated into almost every European language, including Russian and Spanish, to manufacture red lead experts required calcining the lead, by roasting or exposing to strong heat upon a reverberatory hearth with a slow fire. Prior to starting the coloring process, workers had to allow the lead to go through a number of calcinations and watched for the lead to change into dusty grey ashes. During the process, Wilt and Henderson continually raked it, which also prevented the formation of lumps in the batch as it oxidized.[125] The calcined mass was then ground to a fine power in a mill before it would have been separate into lighter and heavier particles by suspension in a water. To further purify the particles workers would have strained it with a stream of water, to carry off the finely reduced powder or smooth paste, which were then deposited in tanks. After sprinkling the particles with water, workers carried the powdery material to the mill for grinding, which exposed their bodies to poisonous particles and vapors.

Thereafter, the powder was dried, and ready to be converted into red lead or minium by putting the dried substance into iron trays, 1 foot square, and 4 or 5 inches deep. These were then piled up in the reverberatory hearth overnight, for fuel economy, allow the substance to absorb additional oxygen, which then becomes partially red lead. When the lead was ready to be colored by the absorption and combustion of oxygen with the metal, it took over a twenty-four-hour heating process.[126] Wilt writes, "Andrew has washed about three tons of lead," which involved continuing the stirring and raking between four and six hours exposing the fresh surfaces of lead to convert the entire batch into a dusty oxide. Finally, this partially red lead was stirred and raked under the low calcining heat once and again, and form a marketable red lead. Under the skillful eyes of Andrew Wilt and Henderson, the color changed to red as they raked the substance out onto the main room brick floor, sprinkled it with water, and continued raking until the entire batch turned color.[127]

The grinding of lead was the most dangerous part of the process. The environment in which the lead color men and grinders worked usually produced an atmosphere "charged with lead particles" and "minute almost impalpable dust produced by" grinding the lead, "which cover[s] the workmen who breathe and swallow the lead polluted air." In April 1803, while conducting his observations, Merat documented the fate of Jean B. He described him as a "forty year old housepainter with a sanguine temperament, and strong constitution" who was carried into Charity Hospital in Paris, France. Merat said that Jean B. arrived "senseless" and experiencing "excruciating pains in the abdomen . . . ; the limbs were agitated with violent convulsions," and he "died a few hours afterwards." Jean B. was just one of the color grinders affected with lead colic carried into Charity Hospital. A little over twenty years later, Louis Tanquerel des Planches, who also cared for patients exposed to lead at Charity, noted, "Color grinders employed in this part only, handling neither brush nor color, are often attacked with colic. Indeed, no part of the lead trades is so dangerous, so liable to cause disease, as that of color grinding." The symptoms that Jean B. exhibited were exactly what Christian Wilt penned to his Uncle on November 1, 1812. A careful reading of Wilt's letter book shows that this was the first instance of his brother Andrew becoming ill; Wilt states, "Andrew was taken with a kind of bilious fever a few days ago. . . . The fever is continual upon him and he complains of pains in his bowels. I hope he may recover." Additionally, as Andrew continued to "wash about three tons of lead," on November 8, 1812, Christian Wilt penned to his Uncle an update about Andrew's health: "Andrew has been very low all the last week, with a bilious fever conjunctly a bilious cholic [colic], but he is rather better this morning."

The symptoms exhibited by Jean B. in Paris and Andrew Wilt in the frontier St. Louis lead mill were similar in nature to the Devon colic, a condition that affected people in the British county of Devon in the seventeenth and eighteenth centuries. In Devonshire, the primary drink of the people was cider, and a connection was found between cider drinking and colic. The symptoms included severe abdominal pains that could lead to death. In the 1760s, Dr. George Baker hypothesized that the lead in cider was to blame, and he believed that the symptoms of colic were similar to those of lead poisoning. He pointed out that lead was used in the cider-making process both as a component of the cider presses and in the form of lead shot, which was used to clean them. He also conducted chemical experiments to demonstrate the presence of lead in Devon apple juice.[128]

At the turn of the nineteenth century, physicians in Europe and settlers in Upper Louisiana articulated their concerns about lead poisoning and the risk to nature associated with lead ores, lead compounds, and lead mining products. The experiences of villagers living near Old Mines, of tradesmen at Charity Hospital in Paris, and of lead grinders at the St. Louis red lead mill reveal their shared understanding about the chemical effects of lead compounds on their water supply and their bodies. Far removed from the city, the frontier became a space where seasonal mining, smelting, and metallurgy increased concerns about developing lead poisoning.

As word of the Missouri Territory's mineral-rich region spread east, the *Missouri Gazette* announced the arrival of Henry Marie Brackenridge, a gifted mineralogist, and requested that he "be protected and given guidance during his two-year residence," in the territory.[129] When Brackenridge finally made his way to the lead mining district, he observed the methods miners employed to extract and smelt lead ore. He asserted that although "the principal employment of the inhabitants is agriculture, the greater part is, also, more or less engaged in the lead business. This is a career of industry," he continued, "which is open to all, and the young in setting out to do something for themselves usually make their first assay in this business."[130] Brackenridge estimated "from the best information he could gather," the annual production of lead by the number of workers at each site, reporting that fifteen workers at Mine à Breton produced 50,000 pounds, and forty workers at Mine La Motte produced 100,000 pounds of lead.

The estimated output of lead from the Missouri Territory from 1800 to 1829 was 73,000 tons of metal valued at $4,188,000. The amount of shot manufactured during the same period approximated an annual average of $50,000. The lead product sent from Herculaneum alone during the eighteen months ending in June 1818 was 668,350 pounds. The value was $46,784, and the exports of bar lead during the same period were valued at $126,294. In 1819, Schoolcraft estimated that 1,130 persons including miners, blacksmiths, smelters, woodcutters, and carters were engaged in the lead business for a portion of the year.[131]

When Captain Amos Stoddard presented President Jefferson's request for additional information about the lead mining district to Austin, the natural philosophers and businessmen alike desired to understand the mineralogical and chemical properties of lead ore for national development. Austin and his British miners' knowledge and skills were regarded as valuable and indeed necessary to emerging science and national growth to aid manufacturers in providing the comforts and luxuries of nineteenth-century life to American citizens. The miners' experiences with chemistry and mineralogy would encourage the flow of lead from Mine à Breton and Mine La Motte to frontier and urban spaces.

Austin's analysis highlights an example of the convergence of early nineteenth-century environmental knowledge with technology. By using European experiments to describe and assay lead ore, he promoted the region's lead ore quality, which aided in the development of lead manufactories between 1801 and 1813. The primary mining sites were Mine à Breton and Mine La Motte. Although failing to recognize the Native American influence on mining, Austin observed one hundred and fifty miners extracting mineral using "primitive methods" by digging shallow pits and trenches. When Schoolcraft traveled through the same district in 1818, he recorded that over one thousand workers were employed in mining by "digging trenches." He also listed the primary methods miners used to smelt lead ores as being "in the common log hearth furnace, and the British reverberatory furnace at Mine à Breton," a significant change from local miners who used the Native American log furnace method as described earlier.[132] Therefore, a mix of methods continued into the early republic.

When Samuel Miller published *A Brief Retrospect of the Eighteenth Century* in 1803, he sketched out the "revolutions and improvements in science." In his discussion on mineralogy, he believed that Americans were handicapped by the newness of the country, and that it was necessary to conduct more field examinations to discover the country's vast mineral wealth. Austin too was fully convinced that mineralogical science influenced the development of the United States economies and industries. Regarding Austin's galena report, Miller stated, "a larger proportion of the growing wealth of our country will hereafter be devoted to the improvements of knowledge and especially to the furtherance of all the means by which scientific discoveries are brought within popular reach." Miller believed Moses Austin and Joseph Henderson represented future prospects of frontier American science. Certainly, Austin, Bell, the Mullins brothers, Henderson, and Andrew Wilt all implemented European instruments and analytical techniques that can be seen as the impetus for the replacement of the Native American amalgam that American miners practiced.[133]

CHAPTER 4

Unhealthy Spaces Fitted Up with Furnaces

Introduction

The year 1819, when Henry Rowe Schoolcraft published his scientific observations and illustrations based on his survey of Mine La Motte, marked one hundred years after the Kaskaskia and European miners Marc Antonine de la Loire des Ursin and Philippe de La Renaudière created their hybrid mining methods. During his visit to the same lead mines, Schoolcraft observed miners using the same hybrid methods while newly settled Americans were applying European approaches. After closely describing each technique, Schoolcraft stated that the European mining technology was "superior to those in use under the French and Spanish governments." He also noted that there was still "ample room for improvement." He portrayed the mines as places where "the population had increased, and the progress of settlement [has] made advances in civil refinement, mechanical arts, and useful inventions." Schoolcraft watched how mines "worked in a more improved manner by [employing] a greater number of miners." He believed that to satisfy red and white lead manufacturers back east, more "improvement was needed in raising ore and smelting it." According to Schoolcraft, the rapid settlement of Americans in the region would also change the character of the environment from a space of "barbarism to refinement."[1] In addition, as Alexis de Tocqueville crossed the countryside, he too observed how the invasion of American settlement practices caused the gradual diminishing presence of Native American tribes, stating, "their huts were replaced with the civilized man's house."[2] Tocqueville's observations also provide a lens for understanding why American settlers desired to replace the "primitive" Kaskaskia-French mining amalgams with their new prospecting, extracting, and smelting methods. The invasion of Americans carrying new mining technologies, about which Schoolcraft wrote, caused the gradual conversion of the Native American and French mining hybrid to the innovative American system.

The opening decades of the nineteenth century mark a time when Americans continued their search for scientific information. Schoolcraft, boosters, and naturalists believed that the "mine country" was in need of miners possessing "scientific knowledge, practical skill, and industry, which characterized the best [worked] European mines."[3] Guided by Enlightenment confidence, travelers, geologists, and miners who made their way west in the early nineteenth century to survey the Missouri lead mines wanted to promote the region's resource potential. Acting like American boosters and "improvers," visitors to the mines believed the mining district to be incomplete and only expected it to become complete with the introduction of more advanced mechanical interventions.

Before Americans installed their machines, they envisioned the possibility for progress and began to rename the lead mining amalgam as "primitive, uncivilized, simple," or "inefficient."[4] Natural philosophers and medical doctors like Henry Marie Brackenridge, Henry Rowe Schoolcraft, and Dr. Lewis Fields Linn similarly endorsed the American desire to "civilize" the mining district.[5] Just as the early eighteenth-century natural philosophers Des Ursin, La Renaudière, and De Gruy carefully surveyed and documented the quality of the lead ore at the Native American lead mines and their practices on the western side of the Mississippi River in *le pays des Illinois*, later, emerging scientists and miners sought to improve on current extracting and smelting skills, thereby increasing the production of lead products for future economic growth. This chapter looks at the role and the motivations of these improvers.

Americans' confidence in improvement has received a good deal of scholarly attention, but mostly as part of the story of agricultural development or western expansion, and mostly for the later nineteenth century. In the case of mining, Brackenridge and Schoolcraft both supported progress by explaining how altering settlement procedures and adopting European mining devices could increase the scale of what lead miners produced. In addition, by the 1830s, two geologists, George W. Featherstonhaugh and Thomas Clemson, confirmed earlier prophetic voices by noting how new mining shafts and machines and a steam furnace had increased the amount of lead that flowed from the ground into society.

Accompanying the conversion from Native American and French hybrid mining to American systems were material changes. The growth in the number of villages and towns with settlers living near mines created the need for doctors to ply their services among mining communities. By 1816, doctors began examining miners, diagnosing miners with lead poisoning, and documenting early accounts of the region's unhealthiness in connection with new production techniques. While doctors' medical observations framed the mining settlement as unhealthy and dangerous, visitors and geologists continued to promote the region as healthy, which signaled a reinterpretation of the meaning of improvement.[6]

During the opening decades of the nineteenth century, doctors' medical observations and treatment practices as well as miners' health reports testify to the high incidence of unhealthy settlements associated with new extracting methods. Together these reports raised new questions about land unhealthiness as lead production increased in scale. Americans' efforts to improve or "civilize" the mining district meant digging deeper shafts, working with lead particles in underground tunnels, and standing next to above ground furnaces that spilled fumes into the air—which collectively made settlers and doctors recognize and label the environment as unhealthy for settlement.[7]

If Mining Were Carried On in a Proper Manner

The 1811 travel account of Henry Marie Brackenridge and the 1819 mineralogical observations of Henry Rowe Schoolcraft both commented on the mining practices of early nineteenth-century Americans. The language of Schoolcraft and Brackenridge reflected an emerging American disdain for what appeared to be unsophisticated manners. For example, when Schoolcraft observed the manner in which miners at Mine La Motte discovered ore using a pick and wooden shovel, he labeled it as primitive and inefficient. In fact, they were both critiquing the Kaskaskia-French amalgam that dated back to the early eighteenth century. When Brackenridge, the botanist and mineralogist, traveled through the Missouri Territory's mining district, he wrote down his concerns after seeing miners discovering lead ore as they did. Brackenridge stated that "if mining were carried on in a proper manner," other than by "scratching the surface of the earth," mining profits "might increase."[8] When Schoolcraft, the glassmaker and mineralogist, traveled to the same mines, he also surmised that more "skilled miners would most likely have been astonished to see miners" applying such "crude methods to discover lead ore."[9] When it came to resource extraction and production, references to miners using "crude tools" also meant they lacked any superior knowledge and skill.[10] Like Pierre-Charles de Hault de Lassus de Luzières and Moses Austin of the late eighteenth century, Brackenridge and Schoolcraft desired to see miners apply superior devices that would eventually lead to forgetting the Native American traditions.

American visitors to the mines assumed that miners lacked the ability to fully exploit the region's lead ore. They also lacked the proper equipment and knowledge.

Similar to European colonizers, Americans believed that the agility required for digging trenches was no substitute for excavating deep shafts with more advanced equipment to access richer veins of lead ore. Improvers no longer wanted to see miners using Native American woven baskets to haul ore. Instead, Brackenridge, Schoolcraft, and others envisioned miners transporting lead ore in iron buckets and wheelbarrows and installing windlasses to raise ore hidden beneath the soil. Brackenridge, Schoolcraft, and other improvers thought American techniques should represent the most progressive equipment and "advances made to the useful arts," as was the practice in England.[11] The ultimate proof of their assumptions about achievement was their dependence on European mining technologies.[12]

The transition from old traditions to new European practices was a gradual one. Although Brackenridge witnessed miners to some degree using European equipment to dig their trenches, he remarked that miners built "primitive" log furnaces to smelt galena. Brackenridge noted how "Creole miners never smelted any other way than by throwing the lead on log heaps" and how directly next to each miner's diggings was a "smelting furnace, and the ore is smelted on the spot." Similar to previous observers, Brackenridge explained to his readers how he envisioned seeing multiple masonry furnaces, similar to the reverberatory furnace at Mine à Breton, which he stated "was introduced by Austin since the Americans took possession of the country."[13] By 1819, there was progress. Eventually, as more European immigrants and Americans made their way to the mining district, they transferred their skills to Mine La Motte. Schoolcraft reported that more miners employed European mechanical apparatus such as windlasses to hoist ore. Schoolcraft also commented that miners depended on English smelters to help build their stone furnaces.

Improvers suggested that local miners did not have enough technological acumen or intelligence to operate machines effectively; therefore, the evidence suggested that miners continued to block change by adhering to the amalgam. Schoolcraft believed that importing additional miners from England who could dig deeper shafts, conduct experiments to determine the quality of the ore, and build furnaces would expand production. Promoters of the "civilizing project" perceived the current approaches to mining as slowing progress and the ultimate creation of a mining district. They alleged that it was incumbent on new miners to seize the opportunity to extract the region's mineral assets more efficiently. Like improvers who promoted the acculturation of Native Americans, Brackenridge and Schoolcraft extended the same ideology beyond lead production to considering how to change miners' behaviors.

In addition to criticizing miners' methods, Brackenridge and Schoolcraft disparaged their manners. Brackenridge likened miners' behavior to that of "savages."[14] He noted that the "manners of the miners and the persons engaged in the mining business are barbarous."[15] He continued, "a few years ago there was a collection of worthless and abandoned characters, the different mines were scenes of brawls and savage ferocity."[16] The language suggested that miners, who improvers viewed as too lazy to exploit the resources, needed transformation as well. Civilization also applied to the "wild behaviors and work habits" of miners.[17]

To overcome what improvers considered miners' indolence required supervision and training. Since local miners were "unacquainted with the utility of machinery," according to Schoolcraft, miners needed more information to operate European machines correctly.[18] This was a change from earlier managerial procedures. During the late eighteenth century, mine owners appointed one miner to supervise the mining activities at Mine La Motte. In addition to the mine manager supervising mining activities, Schoolcraft also implied that the overseer should be held responsible for educating all miners according to the latest procedures. The supervisor monitored extracting lead ore, ensured the proper construction of log furnaces and that they were well supplied with fuel, and assigned miners to transport ore to the furnaces from the mines. Schoolcraft understood the European methods, which he connected to production, stating, "It was evident that miners under a proper manager in this country will furnish lead sufficient for the consumption of the United States."[19] In the end, Schoolcraft concluded that if English miners with managerial skills immigrated to the Missouri Territory mining district, they could direct miners, install machines, and thereby increase the flow of lead from the ground into society.

During the eighteenth century, European and American miners had envisioned bringing trained miners from Europe to replace the Native American and French methods. Similarly, in the 1830s, Lewis Fields Linn, the physician in Ste. Geneviève, Missouri, tried to incorporate new techniques rather than just calling for change.[20] Linn had always been interested in medicine and had completed some of his studies in Louisville, Kentucky. Following the War of 1812, he joined his brother as a surgeon of the Missouri Militia, and after the war he went to Philadelphia for further medical training before returning to Ste. Geneviève in 1816.[21] Dr. Linn recognized that the mining and smelting methods used were "crude, out of date, and highly wasteful of the lead content." According to Linn, the methods in use were inferior to the latest developments in Europe. Linn devised a plan with "great hopes of bringing back some highly skilled workmen" from England and France, after making an investigation of their latest mining methods, in order to remake Mine La Motte like Mine à Breton.[22]

In London, to search for skilled miners Linn published two pamphlets stating "at least one hundred lead miners" were needed in America and "extolling the virtues of" the district. He stressed how the region needed workers possessing the latest knowledge. He planned to have samples of the ores sent to chemists for careful analysis; and he anticipated installing "crushing and stamping mills so miners could pursue their work for more than half time they had been unable to work because their ores could not be processed" into manageable pieces. Dr. Linn also expressed hopes of building "lead furnaces of the most approved kind."[23] There is no record of how many immigrants Dr. Linn's promotional tour attracted to Missouri; however, there was a notable increase in the number of American, English, and French settlers who made their way to Mine La Motte.

Just months after Linn returned to the Missouri mines from Europe, the cultural and material sea change in mining methods that the improvers wanted to see began in earnest. Two miners, Henry Henriod and Ferdinand Rozier, arrived from France. Their short but informative journal describes a number of gradual alterations to mining work, as well as to settlement style. Both miners crossed the Atlantic and arrived at Philadelphia before boarding a steamer to New Orleans. Henriod and Rozier departed New Orleans on a steam-powered riverboat that carried them up the Mississippi River to Ste. Geneviève, Missouri. Eventually, they settled twenty-five miles west of Ste. Geneviève in the newly established village of St. Michael adjacent to Mine La Motte.[24] Henriod and Rozier were among many settlers who immigrated to the mining region over new roads and by canal systems from Philadelphia to New Orleans on steam-powered boats, most likely the same boats that transported increasing amounts of lead to the growing national markets.[25]

Daily, from Monday to Saturday, Henriod and Rozier departed the village of St. Michael for Mine La Motte to extract lead ore. Their ritual marks a transition from early practice. Before settlements were established near the mines, miners planted their crops, and then traveled two days to the lead mines where they extracted lead for two to four months before returning to Ste. Geneviève to harvest their crops.[26] After the harvest, miners returned to the mines either to extract more lead or to smelt the ores they previously recovered. Now that a new settlement surrounded Mine La Motte, Henriod and Rozier could quickly travel to their shaft to continue "burrowing further into the east-west hole in search of lead ore" by clearing away what Henriod called "pretty yellow and red earth." They applied their iron tools to "bore through the lengthy cold and wet shaft and tunnel," encountering rock and the continuous flow of water.[27] Like most early miners, Henriod and Rozier prepared for these obstructions "by loading into their cart necessary utensils" such as iron shovels, pickaxes, sledgehammers, and steel gads that were driven into crevices of rocks, or into small openings made with the point of the pick to loosen and detach protuberances.[28]

By the 1830s, miners added other utensils to their array of tools. Henriod and Rozier included a "compass, stretch cords, a ladder, and the windlass."[29] When miners in Missouri began using the compass to guide their search and measure lead ore veins is unclear; however, in Europe, miners used the compass to measure shaft depth and map the length of newly discovered veins of lead.[30] Since the time of Agricola, miners had used a system of measuring the depth of a shaft using a compass and a number of stretched cords. The cords helped the miners determine the depth of their newly dug shaft with greater accuracy. Cords also aided miners in determining the length of tunnels that ran perpendicular to the original shaft. However, in Henroid and Rozier's journal, there is no mention of them digging additional galleries or tunnels connected to their original shaft. In Henriod and Rozier's case, the cords allowed them to measure the length of veins beyond the surface and helped them to build the correct size ladder, which they could easily descend and ascend in proper fashion. Daily Henriod wrote in his journal how "the first line of attack when using" new extracting devices "was to climb down the thirty-foot homemade ladder into the shaft."[31] Additionally, since miners no longer extracted from trenches, Henriod or Rozier installed a windlass above their shaft's opening.

The windlass enabled Henriod and Rozier to remove ore, dirt, or water more easily from the depths of the earth. During this period, windlasses hauled a small amount of material; therefore, the use of windlasses was limited to prospecting, sinking, beginning shafts, and raising ores. To assemble the windlass, first miners cut timber, and then attached a cord purchased in Ste. Geneviève. Once in the shaft or a connecting tunnel, Henriod and Rozier tried to "straggle between stones to the linking reservoir, to cross a small fissure to continue their search for lead ores."[32] When Henriod discovered lead ore, he would return to the opening of the shaft to obtain the tools Rozier had lowered in the iron basin connected to the windlass. Clutching the sledgehammer, pick, and gad, Henriod began to separate rock, earth, and lead ore from the tunnel's sidewalls. At times miners screamed to their partners, as when Rozier yelled, "I see an opening which seems to go under the rock." To which Rozier replied, "Clear away as much as you can and I will descend later to clean the rubble away." After Henroid ascended to the surface, Rozier descended into the shaft, and continued working the tunnel. Rozier noted how he "filled five to seven iron basins of rubble, opened an additional cave, and began to load another four basins of rock, earth, and lead ore" to be raised to the surface.[33] Daily Henriod and Rozier worked in this fashion, unless they were sick or injured.

The daily experiences of Henriod and Rozier reinforced Schoolcraft's call for improvement and Dr. Linn's efforts to transfer new techniques from Europe to the Missouri mines. Henriod and Rozier's daily work represents the change from trench mining to excavating "shafts in imitation of some practical miners" from Europe, who were "rewarded with the most perfect success." Henriod and Rozier's acquired skills and equipment provided them with the ability to dig multiple shafts to a "depth of about sixty feet and locate sulphuret of lead vein" before discovering additional "horizontal veins upwards of one foot thick."[34] Sulphuret of lead is the ore most commonly found as lead sulphide (PbS), or galena. Galena was the heavy and shiny grey metallic ore that Native Americans and early French explorers and miners prized, extracted, and smelted. They worked with galena at both Old Mines and Mine La Motte, and it was usually cherished for the small amounts of silver it contained.[35] The modern mining style and the associated steps practiced by Henriod and Rozier show their abilities with European tools, but their actions also indicate the continued waning of the Kaskaskia-French amalgam. In addition, largely because of the high demand for quality pig lead by Americans back east, the United States planned to conduct more environmental and scientific investigations into Missouri's galena.

After Moses Austin and Henry Schoolcraft wrote their mineralogical reports in 1804 and 1819, respectively, no other survey was conducted in Missouri until the 1830s. As miners started to employ their new equipment to open numerous shafts, geologists and merchants back east wanted a clearer assessment of the district's geological structure. In 1834 and 1838, respectively, George W. Featherstonhaugh, the leading English geologist, and Thomas Clemson, the American "mineralogist, chemist, and geologist of great merit," completed their evaluations of the district in 1835 and 1839 respectively.[36] On July 12, 1834, Lieutenant Colonel J. J. Albert of the US Topographical Engineers instructed geologist Featherstonhaugh to "personally inspect the mineral and geological character of the highlands . . . of that elevated country lying between the Missouri and Red Rivers." However, during his travels, Featherstonhaugh also took time to also observe the lead deposits in Missouri, which he recorded as "the most ancient diggings which had been carried on since the French had possession of the country." While touring the region he visited Mine La Motte where he first received satisfactory evidence in the lead shafts. Featherstonhaugh's geological survey helped to build interest in science as a valid means to describe the environment.³⁷ During the same period, the Geological Survey of Virginia was conducted. The Virginia survey offered an analysis on minerals and soils for further agricultural development. In similar fashion, improvers suggested that natural philosophers apply chemistry and geology to reveal "the great wealth which lies buried in the earth" in Missouri.³⁸ Featherstonhaugh and Clemson used their environmental knowledge and conducted experiments to confirm Austin's and Schoolcraft's tests, assuring the quality and quantity of Mine La Motte's lead for commercial production.³⁹ Their reports, Clemson, Featherstonhaugh, and Linn were included in the account of the mining district along with Charles Gregoire, agent to the mines, which provided readers with detailed descriptions of the geological features of the lead mines and the surrounding environment. Clemson, Gregoire, and Linn's descriptions were all part of *Observations of the La Motte Mines*, a pamphlet written by Linn and Clemson, published in 1839 to promote the recent improvements completed at the mines.⁴⁰ Widely read, this account acted as a scientific text that resembled a travel narrative about a landscape of abundant quality galena, while at the same time advertising settlement opportunities.

In 1838, when Clemson visited Mine La Motte, Henriod and Rozier offered him the opportunity to work together with them in their shaft. His experience with Henriod and Rozier confirmed his critique of old versus new mining and his advocation of new smelting furnaces. Observing Henriod and Rozier, Clemson stated, "a reform in the whole system of mining is taking place and I believe the changes would henceforward be conducted upon acknowledged principles which will enable them to contribute powerfully to the national resources."⁴¹ Henriod and Rozier, Clemson noted, "sunk [their] shaft to a depth of about one hundred and ten feet, when I was there, and very obligingly let me down into it, and gave me every aid and facility in examining their works, which enabled me to observe the very curious [underground] structures they built."⁴² In addition, Clemson was not only impressed with the depth of the shaft but also with the European-style galleries Henriod and Rozier constructed.⁴³

Clemson carefully watched how Henriod and Rozier prospected for the origins of "subterranean structures."⁴⁴ They created underground tunnels, which exemplified how such miners applied their skills and were "observant and experienced in both the arts and sciences."⁴⁵ Miners employed their compasses to guide them to valuable mineral veins and to manage tunnels successfully. Like European miners, Rozier and Henriod merged their knowledge of surveying and arithmetic to sink shafts and to construct underground tunnels in Missouri. Clemson was satisfied with the lead veins he studied alongside Rozier and Henriod, and he decided to turn his attention to watching how miners cleaned and crushed lead ores before smelting. Rozier and Henriod, and others, harnessed the force of nearby rivers and streams by installing water-powered mills to accomplish these processes.[46]

Recall how the Kaskaskia and French also managed water to clean lead ore, before hammering large chunks of galena into smaller pieces with their stone and iron hammers prior to smelting. Now, Henriod and Rozier gathered stones and controlled the rivers and streams located throughout Mine La Motte to power machinery. The abundance of limestone and sandstone on the property made it feasible for miners to manufacture millstones large enough to grind and crush lead ore. Mills at Mine La Motte were small and included a waterwheel on the lower floor to access the water's edge. These major developments in frontier mining helped Mine La Motte to become what improvers had predicted, a landscape where settlers ploughed their farms and mined the earth. By the late 1830s, Mine La Motte expanded into a twenty-four-thousand-acre property consisting of farms and mines, and the old log and ash furnaces continued to be used throughout the first third of the nineteenth century.[47]

During Clemson's visit, Charles Gregoire informed him of a new type of smelting furnace. By 1836, the Scotch hearth furnace was put into operation at Mine à Breton, Old Mines, and Mine La Motte. This furnace type, having its early progenitors in England, was introduced into the region by Major Manning of Webster.[48] The Scotch hearth furnace centerpiece was an open box, several feet square, originally with a back and sides of stone, but later generally of less corrosive cast iron. After slaves sorted and cleaned the lead ore of debris, it was then crushed into small chunks and placed into the furnace where it mingled with the fuel, generally wood or charcoal. The molten lead flowed from the box down the hearth groove to the melting pot and was ladled into pig molds. The Scotch hearth eliminated the need to have the log furnace in conjunction with the ash furnace. Following the initial blast period, the ash residue that gathered in the furnace box bottom was drawn out and placed on the new charge of fuel material and ore, and the process was initiated again in the same furnace.[49] Although the Scotch hearth furnace did not eliminate the other furnace types, its use spread in the lead regions of the Mississippi valley. Until the advent of large-scale lead production after the Civil War, the Scotch hearth played a major role in Missouri lead production.[50]

Following the initial use of the Scotch hearth furnace, Gregoire, Mine La Motte's manager, contributed to a pamphlet published Clemson and Linn to attract more migrants, stating, "For smelting of this lead a new furnace has been erected."[51] This was another major evolution at Mine La Motte: miners who once smelted lead ores using two types of furnaces—log and ash reverberatory—were now powering a Scotch hearth that eliminated the need for an ash furnace. With the Scotch hearth in place, miners could obtain a higher percentage of lead by reintroducing the remaining lead to the same furnace for a second firing. Rozier stated, "I fired the furnace today but I cannot say now with what success."[52] However, he most likely did not do so badly, for Charles Gregoire stated, "I shipped on board the steamboat *Jubelle* 1428 pigs of lead weighing 85,016 pounds, being lead taken from Rozier's" Scotch hearth furnace.[53] Before Clemson completed his report of 1838, he wrote, "to render these mines more productive" and improve the mining district, "it would be necessary to bring in at least one hundred lead miners and smelters from Europe for five or ten years."[54] The American geologist witnessed and endorsed the developments new miners were achieving—after conducting scientific observations using new utensils, they installed new, more efficient furnaces adjacent to their village, further transforming the mining district in line with improvers' specifications.[55]

When Moses Austin settled at Mine à Breton, it was the only village located in the district. Prior to 1820, the combined populations of the counties surrounding the lead mines was 11,613; in 1830 the population was recorded as 18,648. The majority of settlers who came to Missouri migrated from Kentucky, Tennessee, and Virginia. Steamboats now plied upriver more frequently carrying settlers interested in lead mining.[56] As the scale of lead production increased, and the population grew, new villages and towns were established.[57]

During the 1830s, Ste. Geneviève and Herculaneum were the primary towns where miners shipped their pig lead to New Orleans and eastern cities. Both towns were between twenty-five and thirty-five miles away from Mine La Motte. Desiring to establish a settlement in closer proximity to the mines and to begin year-round operations, settlers developed the village of St. Michael. The village contained "fifty houses, several stores, and a post office all located at the center of the richest farming district" where miners worked individual farms.[58] Since Mine La Motte was only two miles north of the village, its miners devoted their time to farming as well as mining and smelting. In conjunction with the modifications to daily routines, the village enabled miners to work year-round instead of following the former seasonal schedule.

Besides making periodic discoveries of thousands of pounds of lead ore, Henriod and Rozier typified those miners who maintained farms. Prior to departing for the mines, Rozier and Henriod spent their early morning hours "cleaning orchards and calf-pens, organizing the tobacco and vegetable gardens, making hay in the nearby forest, or repairing tools." By the late morning, they made their way "across the small patches of farmland, through the adjacent forests, up the hillside path, and across the Mine Creek" to their mine shaft.[59] In conjunction with the changes to mining systems and equipment, Henriod, Rozier, and the community of miners now completed their daily "household, animal care and farming chores" before and after mining tasks.[60]

In 1839, Thomas Clemson summarized the changes that followed the incursion of European techniques at Mine La Motte. Clemson observed how the inhabitants spent time as "both farmers and miners who were either directly or indirectly engaged in the mines year-round." He further noted how miners took advantage of numerous agricultural benefits, "but that [the] mines [were] regarded as the principal object of prosperity to the surrounding country." Clemson echoed what early travelers and miners had previously predicted—"the extraction and reduction of lead ore to supply the nations' growing demands would depend on year round operations."[61] As dangerous year-round extraction of lead ores from the earth, washing and crushing of ores in nearby water-powered mills, and long hours of smelting by miners such as Henriod and Rozier generated the material flow of pig lead into society, the memory of Native American and French fusion of techniques was washed away.

Visual Depictions Confirming the Disappearing Amalgam

The departure from the traditional amalgam, called for by Brackenridge and Schoolcraft, and the systems employed by Henriod, Rozier, and Gregoire contributed to the increase in lead production at Mine La Motte. In addition to Featherstonhaugh's and Clemson's accounts of the numerous cultural and material changes that occurred during the 1830s, the French naturalist and painter Charles-Alexandre Lesueur, a French visitor, illustrated how the mining landscape had been transformed. After leaving Philadelphia, Lesueur traveled west along the inland rivers to Ste. Geneviève, Missouri. When he arrived at the mines, Lesueur started to draw vivid images to capture where miners lived and where miners worked. His catalogue is depicted in his *Drawings and Sketches*.[62] Lesueur's leisurely journey along the Ohio and Mississippi Rivers to the early nineteenth-century La Motte mines allowed him the opportunity to record the transformation of a once picturesque landscape to a landscape of ingenuity.

On March 22, 1826, Lesueur sketched a watercolor view of the homes adjacent to Mine La Mott, which at the time were centrally located.[63] Lesueur carefully penciled the contours of the landscape reshaped by miners according to their needs. Directly behind the village, Lesueur united the hills and the sky at the horizon, which draws the viewer's attention to the abundant trees lining the hilltops in the background; however, the settlement's foreground appears to be practically devoid of trees, an indication of the amount of timber settlers used to construct their homes, build their windlasses, and fuel their furnaces. The barren forest, scattered trees along the hillside, and miners' dwellings are clustered together just beyond the fence. Signs of domestication are evident; smoke-filled chimneys and a clothesline are symbols of care (figure 5).

Perhaps Lesueur's are a few of the known drawings of early American lead mining in Missouri at Mine La Motte with African American lead miners completed by a French naturalist. During Lesueur's trip to the Missouri mines, he also sketched another watercolor view depicting enslaved miners along with their equipment in the early nineteenth century.[64] He commented, "When I approached the mines there was opened to my view a large space of cleared ground" where "three miners [were] surveying and working the mines."[65] Included in this sketch are more than seventeen windlasses blanketing Mine La Motte and one of the three miners clutching an iron pick (figure 6).

Lesueur captures the silencing of Mine La Motte's Native American and French mining amalgamation and the eventual fate of traditional mining practices. He reveals over seventeen Euro-American-style windlasses in operation, which emphasizes the dominance of new technologies and practices. His images reiterate the material implications of the changing frontier industry—as Native American miners determined to control their own future and decided to leave the mining region voluntarily. As more American settlers arrived to mine lead, the Shawnee and the Delaware Indians chose to relocate one hundred miles west to the Ozark region. After the Shawnee had lost their lands in Pennsylvania and Kentucky, they eventually settled in the Ohio Territory. Likewise, the Delaware migrated into the Ohio territory from Pennsylvania. Both tribes ultimately crossed the Mississippi River and lived not far from the lead mines in Spanish-occupied Ste. Genevieve.

The immigrant Shawnees and Delawares used the Mississippi River as a political and economic route of communication engaging with French traders and American miners in the area. Moses Austin's son Stephen remembered Shawnee and Delaware Indians frequenting his father's Mine à Breton store to do business as "he traded with the Shawnees and Delawares as their friend." Stephen noted that as a child he played with the Shawnee children when his "father had hundreds of them at his home at the Lead mines."[66] Although most settlers considered their ways to be superior to those of their Shawnee and Delaware neighbors, they traded and borrowed their customs from them. Louis Lorimier, the commandant at Cape Girardeau embraced an amalgamated lifestyle that was a mixture of Native American and European customs. As well, in St. Louis, during the

FIGURE 5. View of Mine La Motte Village with twelve houses and a clothes line, sketched by Charles-Alexandre LeSueur, March 22, 1826.

FIGURE 6. View of Mine La Motte with three miners among the lead pits and windlasses, sketched by Charles-Alexandre LeSueur, April 10, 1826.

eighteenth century, Native American culture remained very noticeable as French and eventually American immigrants learned to cultivate Indian corn, squash, and beans. For example, the daughter of Francois Vallé sent a friend a sample of Native American handiwork in 1794. Some even used expressive words and idioms in their daily speech, while other became interested in their decorative arts.[67] However, with the arrival of the nineteenth century and more American settlers crossing the Mississippi, the close-knit French, Shawnees and Delawares community faced separation as the later once again would face forced migration. The 1825 Treaty of Prairie du Chien resulted in the Shawnees ceding their Missouri lands and accepting a tract of land along the Kansas River, in present day Johnson County, Kansas.[68]

In effect, what improvers advocated had come to pass. They believed that "civilizing" the area required the separation of American and Native American mining behaviors.[69] Discarding any obligation to the assimilation policy of earlier times, like European colonizers, Americans disparaged native environmental knowledge and technical skills in their pursuit of increased lead production. Their language and actions reflected a desire to control the mines by installing sophisticated and complex machines. Similar to the travelers who carried the improvement ideology to the mines, American settlers viewed the Kaskaskia-French amalgam as having no place in the modern, Americanized mines. The expulsion of the Native Americans west of the Mississippi began in earnest with the *Indian Removal Act of 1830*, although many Native Americans living near the lead mines had been offered land in southwestern Missouri over a decade earlier.[70] By the 1820s, the Native American and French methods had disappeared from Mine La Motte. The discourse was about superiority and inferiority and the need to import new mining labor, equipment, and skills to satisfy the American lead industry being established back east.

Fitted Up with Furnaces, Instruments, and Apparatuses

Before Missouri miners began shipping their pig lead to the East Coast, they made numerous lead products for both rural and urban consumers. As discussed in chapter 3, Moses Austin and his English miners not only extracted lead ore but also produced shot, sheet lead, and a key component for making glass bottles, zane.[71] After establishing settlements at Mine à Breton and Herculaneum, Austin manufactured a considerable amount of shot, which he sold throughout the Mississippi and Ohio valleys. Austin went on to build a special room to house large limestone tables to mold sheet lead, which he shipped to New Orleans where it was used for sealing shipping vessels, covering roofs, lining cisterns and bathtubs, and molding pipes for the conveyance of water. Settlers and city dwellers also used sheet lead to make small boxes to preserve food items. Finally, Austin utilized lead ashes to make zane, which glassmakers in Pittsburgh coveted to make bottles.[72] The changes to mining and settling habits made it possible to extract, produce, and ship shot, sheet lead, and zane to merchants year-round. The nineteenth-century transfer of technologies to Missouri's lead mines not only made it a new mining district but also influenced where lead products would eventually be made.

Notably, a correlation existed between the arrival of miners carrying new apparatus to the Missouri mines and the establishment of the Wetherill lead factory in Philadelphia. By the opening decade of the nineteenth century, Samuel Wetherill had learned about the abundance of quality lead at the Missouri mines through Austin's 1804 geological survey.[73] With this new knowledge, Wetherill hoped to stop importing lead products from England and to begin making the same goods in the United States. Most importantly, there was an emerging market for white lead to produce paints and for red lead as a key ingredient in flint glass.

Samuel Wetherill, the founder of Samuel Wetherill & Sons, was born in Burlington, New Jersey, in 1736. Being the eldest in a large family, at the age of fifteen, he was encouraged to relocate to Philadelphia, and became apprenticed to the homebuilder Mordecai Yarnall. In addition to learning the technology of building, Wetherill was also initiated into purchasing, bookkeeping, and other business methods until he completed his apprenticeship at the age of twenty-one in 1757. Thereafter and until 1783, Wetherill, acquired a wide range of interests in a number of businesses, community activities, and service to others as a Free Quakers preacher. In the same year, 1783, Wetherill also began to establish his paint and hardware business, particularly such goods as nails, hinges, saws, glass, and red and white lead. By 1789, Wetherill had shipped a large quantity of white lead and other paints to New Orleans. 1789 was also the year that confirmed the interest of the company in a letter to the firm of Brandrams, Templeman & Co. in London writing "We are Druggists as well as Oil and Colour Men," over the signature of Samuel Wetherill and Sons.[74] By this time, Samuel had married Mordecai Yarnall's daughter, Sarah, in 1762, they had three sons, Samuel Jr., born in 1764, Mordecai, in 1766, and John, in 1772, and they were all the "sons" of the company.

Before American manufacturers endeavored to build a laboratory, most would make trips abroad or to other factories, while others, like Samuel Wetherill and his sons remained in Philadelphia and chose to send weekly correspondence across the Atlantic to London to purchase lead and glass products from the firm of Brandrams, Templeman & Co. The firm was a leading color manufacturer and supplier to the trade. From 1782 to 1803 they were known as Brandram, Templeman, & Jacques, and then Brandrams, Templeman, & Co. from 1803–1819. The Wetherills maintained detailed letter books starting in 1789 that contain numerous requests for various products and raw materials; they also often requested information regarding the latest innovation in the manufacture of lead products.[75] In one letter, Wetherill placed an order for instruction manuals prior to beginning production of his own red and white lead.[76] To outfit a laboratory to make products in the United States, first Wetherill needed to understand how to design a laboratory with the appropriate apparatus; and second, he needed to access quality pig lead from the mines to maintain consistent production.[77] Finally, by the early nineteenth century, Samuel Wetherill and his sons began to build his factory in Philadelphia to produce red and white lead.[78] Eventually, he made contact with a mine owner and a lead agent to secure a steady flow of pig lead. Thereafter, Wetherill began to decrease the number of lead products he purchased and turned his commercial gaze on establishing deep connections with Mine La Motte's agent.

Similar to the way miners adopted English techniques, the Wetherills adopted their processes to make red and white lead.[79] Like Samuel Slater, who transferred fresh ideas about factory operations to America, the Wetherills' company followed the English plan to design, build, and operate their lead laboratory.[80] The Wetherills intended the rooms to be "fitted up with furnaces, instruments, and apparatus."[81] These tools as well as knowledge of science and art were necessary to reintroduce the pig lead to millstones, tools, furnaces, and cisterns once it arrived in Philadelphia.[82] By grinding, raking, melting, and washing the warmed lead, they carefully observed the yellow, brown, and white dusty substance transform to white or red lead constituted for commerce.[83]

At the Wetherill factory, workers prepared white lead by casting the lead into sheets, rolling it up in a spiral form, and setting it to corrode in clay pots partly filled with vinegar. To produce red lead, the Wetherills moderately heated lead in a reverberatory furnace until the lead turned into a vivid red-orange soft powdery substance.[84] Similar to the Missouri miners' application of new European tools to discover, extract, and smelt lead ore in their workspaces, the Wetherills fitted their workspaces with European furnaces, cisterns, pots, and utensils to convert pig lead to red and white lead.[85] Innovations at the lead mines also changed the way that East Coast lead products were manufactured.

To produce quality products, the Wetherills required a superior grade of pig lead year-round. Therefore, on May 26, 1812, Samuel Wetherill wrote a letter to

Benjamin Morgan, a New Orleans lead agent, stating, "Having learned of the quality lead being extracted and produced I would like to request two separate shipments of 100 and 200 tons of Mississippi valley pig lead."[86] Satisfied with the quality of pig lead he received from Morgan, in January 1813, Wetherill then turned to Henry Thompson of Baltimore, another customer of Morgan, "to provide him with fifty tons of Missouri pig lead" from Mine La Motte.[87] Within a few days, Samuel Wetherill also told his Baltimore agent, John Kipp, to send fifty tons of pig lead "as soon as navigation opens as we are out of lead."[88] Again, in June of 1815, Wetherill secured from Benjamin Morgan another fifty tons of pig lead.[89]

In addition to creating business alliances with Morgan and others, the Wetherill sons corresponded directly with mine owners to arrange multiple shipments of pig lead. And after several attempts from afar, on March 30, 1835, John P. Wetherill, Samuel's brother, was at the lead mines in Galena, Illinois, to arrange for a steady flow of pig lead in person. In 1839, John made a another trip to Galena, Illinois, and while Samuel Jr. remained in Philadelphia he reminded him by letter "not to fail to visit Mine à Breton and [Mine La Motte] and acquire all the information you can connected with its advantages."[90] The Wetherills' ongoing communications with, and two ventures to, the Missouri mines not only emphasize the quality and abundance of Missouri's pig lead for making red and white lead, but also show how miners were now capable of conducting year-round operations. Back east Wetherill learned to depend on the flow of pig lead from the mines to New Orleans and Philadelphia and into his urban factory.

During the early nineteenth century, miners used wagons to transport pig lead from their furnaces to two primary shipping points on the Mississippi River—Ste. Geneviève and Herculaneum, Missouri. At both depots, they hoisted pig lead onto keelboats or steamers bound north to St. Louis or east to Philadelphia.[91] Moving pig lead around the country was connected to changes in shipping patterns and newly constructed transportation routes between Philadelphia and Missouri. Scarcely 100 miles of canals existed in the United States in 1816, and roads in most parts of the country were uneven wagon paths that became impassable during wet weather; if the wagons made it, customers often received their glass products broken. The construction of canals made shipping more secure, easier, faster, and less expensive. Canals connected the Missouri mines with Philadelphia by way of the Mississippi, Ohio, and Monongahela Rivers—primary inland waterways.[92] These extensive channels of water provided an alternative means over which quantities of pig lead traveled from the region "abounding with mines" to Philadelphia's harbor. Skilled navigators guided their barges, keelboats, and steamers loaded with pig lead, wheat, hemp, and tobacco to eastern depots. The Mississippi River provided an umbilical connection with the rest of the world. It joined the mining district and its villages to St. Louis, Canada, New Orleans, and the American East. The river continued to be a primary connector, for keelboats were the chief method of transporting heavy cargoes of lead from the Mississippi valley to New Orleans.[93]

Pig lead also journeyed from the mines directly down the Mississippi River to New Orleans. With the aid of high water and a fast current, boats departing Ste. Geneviève or Herculaneum in February or March traveled five to six miles per hour. They often reached New Orleans usually within two to three weeks. Major voyages of these large boats took place twice a year; such frequency of connection was similar to travel between the Atlantic colonies and England in the seventeenth century. After steam boats arrived on the Mississippi in 1817, the time to New Orleans was shortened to around seven days.[94] On arrival in New Orleans, the pig lead had to be moved from the keelboat or steamboat to a much larger ship heading to Atlantic coastal cities. So, pig lead often remained in port for ten days more before being placed "aboard a good vessel bound for [the] Gulf of Mexico" and beyond.[95] Upon leaving the Gulf, vessels carrying pig lead rounded the Florida Peninsula and arrived in Philadelphia eighteen days later. When the pig lead arrived in Philadelphia, porters hoisted it onto horse-drawn wagons and delivered it to the Wetherill factory.[96]

While the Wetherills made their own red and white lead according to English specifications, miners mastered new production machinery, and navigators embraced new forms of transportation. These multiple changes eventually diminished the likelihood of the Midwest becoming the principal production site for most lead products. The mines and laboratory workspaces became part of a larger system of production where pig lead traveled east, and new commodities made from lead traveled west. Inevitably, as larger amounts of lead moved from the earth into society, the new techniques miners employed also increased the amount of particles and fumes that flowed over the district.[97]

The Business of Smelting is Both Unhealthy and Deadly

Changes to mining and smelting production also altered the way visitors, mine agents, and mineralogists defined the health of the region. They each wrote about how new machineries increased the amount of pig lead shipped back east, not simply as specialists but also as advocates proud of the region's resource potential. Brackenridge's, Schoolcraft's, and Clemson's texts were calculated. They did much to highlight the benefits of the new equipment, but also, they diminished how unhealthy the region had become for settlers. At the same time, though, doctors and miners began to draw attention to the unhealthiness of the settlements near the mines.

During the early decades of the nineteenth century, Brackenridge, Schoolcraft, and Clemson combined improvement ideology with civic boosterism to promote Mine La Motte. Their inquiries were of deep interest to the country in terms of business advantages. In a speech to the townspeople at Fredericktown, Clemson expressed his hope to see miners and settlers experience the immediate development of the mineral capital of the region. The three early observers did not hesitate to promote improvement, which they believed would produce results favorable to national industry. They described the region as healthy and planned to convince skilled miners and smelters to migrate to Missouri.

Brackenridge and Schoolcraft described Mine La Motte and its twenty-four-thousand-acre property as healthy but minimized how lead production affected the environment and its inhabitants. For example, although Brackenridge noted how "the business of smelting is considered unhealthy," he only mentions how "animals raised near the furnaces are frequently poisoned by licking the ore or even the stones."[98] Schoolcraft also failed to call attention to the effects of vapors, fumes, and lead particles emanating from the furnaces on the local crops or miners' health. Instead, like Brackenridge, he reported how "the tract is well watered by several springs of the purest water ... and is decidedly admitted to be the healthiest of Missouri."[99] Since production increased in scale, and because the Missouri mines were capable of supplying the nation with abundant amounts of lead, they chose to promote only the positive aspects of improvement. By the nineteenth century it was a well-established fact that lead colic, or miners' sickness, was produced by the deleterious action of lead on the animal system, and men and animals often died of the same disease. Brackenridge and Schoolcraft, both men of science, and miners all understood the effects of lead poisoning from the first monograph on occupational diseases.[100]

Admitting to the fact that there was a correlation between human and animal health problems and lead production, the first Missouri state legislature enacted a law in 1821 necessitating the fencing of furnaces. Fencing was to be at least ten yards from the furnaces and maintained for as long as six months after the furnaces were last used. Actually, French miners had, as a custom, fenced mines and furnaces long before this Missouri law. They were indeed concerned about their own crops and animals suffering. The 1821 law was necessary. By then, as new miners populated the region, they tended to disregard environmental concerns and the health of both miners and local residents, and they had to be coerced to take these commonsense precautions.[101]

Although early promoters at Mine à Breton did not inform miners and farmers about the possible dangers of living near furnaces, Brackenridge did hint that a poisonous haze coated the landscape. Although there was only one brick furnace at Mine à Breton, he observed a considerable number of oxides spilling into the atmosphere. Brackenridge commented how "unhealthiness arises from the fumes of the furnace in which are quantities of arsenic and sulphur."[102] He also "perceived a peculiar taste in the air as vapors emanated over the landscape."[103] Austin told Brackenridge that "none of the miners experienced attacks" of lead poisoning. This is extraordinary since miners working the reverberatory furnace would have been exposed to vapors that would have caused injury to their health. Additionally, Brackenridge and Austin would have been aware of the effect of lead oxides on workers.[104]

During Schoolcraft's visit, he also failed to write about how handling lead ores affected miners. To boost the attraction of skilled workers, Schoolcraft only informed his readers about how the fumes affected animals. Schoolcraft attributed "mine sickness wholly to the quadrupeds." He observed cats and dogs experiencing "violent fits [which] never fail in a short time to kill them." Moreover, he watched cattle "licking about the old furnaces and falling down." Schoolcraft's medical observations attributed the cause of animal deaths to the inhaling of "sulphur, which is so abundantly driven off in smelting lead."[105]

Promoting the region required boosters to highlight the landscape's resource potential. However, they also reported the number of tributaries of clean water, which meant healthy land. Schoolcraft noticed these springs, but also reported how a small number of rivers became unwholesome. Still he did not associate this unhealthiness with the increase in lead production or its effects on settlers. He outlined how the vapors originating from the furnaces flowed over the landscape and impregnated the drinking water with lead particles, which only animals absorbed into their systems. Schoolcraft's account suggests that as technology evolved there was an ecological impact affecting both animals and vegetation. Schoolcraft's medical observations called attention to the unhealthiness of the water supply, but he did not report cases of miners getting sick. His goal was to promote the populating of the "healthy" country with skilled miners and farmers, but in doing so he played down the potential unhealthiness of the district.

During the early nineteenth century, settlers frequently used the healthiness of land as a guide to settlement practices; therefore, visitors to the mines chose to minimize the use of the term *unhealthy* in their reports.[106] By 1839, in *Observations on the La Motte Mines*, Clemson would also combine straightforward civic boosterism with notions of improvement and land cultivation. Like Brackenridge and Schoolcraft, he not only promoted the region's supply of galena but also published that "about one-third of the [land] is first rate farming land."[107] He linked the significance of mining and farming to the development of Mine La Motte, calling "to the greatest extent . . . and to invited miners, and those engaged in kindred operations, to make it the theatre of their future exertions."[108] Clemson underscored the numerous tributaries that he believed would not only "afford ample power to create and power" mills to stamp and crush lead ores into manageable pieces for smelting but also provide water for crops.[109] Clemson also understood that early nineteenth-century migrants, miners and farmers, searched for new lands in the midst of a "healthy country."[110] Although Brackenridge, Schoolcraft, and Clemson failed to connect the impact of the new devices on the health of miners and smelters, their observations did begin to change the perception of the mining settlements' healthiness among settlers and doctors.

By the early nineteenth century, the multiple furnaces at multiple mining sites melting large quantities of lead ore contributed to the growing unhealthiness, and deadliness, of the mining district. More settlers migrated to settlements near the mines; and physicians migrated to the region to service miners and smelters affected by the lead particles or vapors. In 1843, Dr. Hardage Lane of St. Louis, Missouri, published the first medical report of lead poisoning while he was president of the Medical Society of Missouri. He published his account in *St. Louis Medical and Surgical Journal*. Lane also had experience with the 1849 cholera epidemic, and told patients to wear tourniquets on both the arms and legs.[111] In 1816, Lane treated a thirty-year-old at the lead mines who seemed to have the usual symptoms of *colica pictonum*—lead poisoning.[112] Lane documented, "He was called to see a [man] who seemed to be laboring [from] the symptoms of *colica pictonum*."[113] Lane wrote that the man was "vomiting" and experiencing "pain and discharge of blood from the bowels, which continued for two to three days." Following the man's death, he made "an early post-mortem examination, when the mucous membrane of the whole alimentary was found in a highly inflamed condition, and necessarily engorged and thickened." Lane noted, "The most remarkable feature of the case is that there was found in the small intestine ashes, cinders, slag, and fourteen shot."[114] Dr. Lane continued, "This discovery at once accounted for the symptoms of lead colic, and the fatal termination of the miner."[115] Lane also recalled how he learned that "some slight difficulty had existed between the boy and his master and it was presumed that he had attempted [committed] suicide by taking the articles found in the bowels." According to Lane, "This case clearly establishes the fact that poisoning may be produced by lead in substance."[116] Lane uncovered what boosters tried to hide. He revealed how the scale of lead production and the application of new methods affected human health, which may have alerted settlers who considered migrating to these settlements. Additionally, Lane's autopsy revealed the first known case of suicide among enslaved African Americans in the Missouri Territory.

By the time of Lane's autopsy on the thirty-year-old African American slave, there was already an understanding that lead in all forms was "poisonous when taken in any quantity."[117] In England, after smelters became increasingly exposed to lead oxides from inhaling or handling galena, they had "bowel complaints, paralytic symptoms, and other maladies."[118] At Mine à Breton, families clearly understood the chemical effects of lead compounds on their water supply, and they eventually signed a complaint attempting to stop the owner from polluting their drinking water. Finally, while preparing lead for coloring in St. Louis, Andrew Wilt became ill and complained of pains in his bowels due to a bilious cholic [colic]. In all the aforementioned cases, individuals were exposed to the poisonous effects of lead production. Additionally, both Brackenridge and Schoolcraft observed animal sickness, linking it to lead at the Missouri mines. Ultimately, these occurrences suggest that the perceptions of settlers, miners, and physicians about the unhealthiness and deadliness of the mining and smelting district were examples of early environmental consciousness brought about by the increase in manufacturing of lead products.

The incidence of an African American male slave gorging himself with a large amount of lead ashes and swallowing shot reveals slaves' understanding of the power of lead to poison. Possessing this chemical knowledge allowed the thirty-year-old slave to take control of his own life and commit the ultimate act of self-destruction. In addition to this report of self-destruction observed by a frontier physician, the voices of specific slaves contemplating self-destruction or witnessing other slaves committing self-murder come to us from slave narratives.[119] One way to understand slave suicide is as a form of defiance. Another way to contemplate the history of suicide within the context of slavery in the Missouri Territory is to go beyond the resistance model and examine slave self-destruction through slave suicide ecology—what Terri Snyder terms as the emotional, psychological, and material conditions that fostered suicide among slaves. Despite the abundance of perceptions of slave self-destruction in the American context, trying to understand one suicide in response to enslavement remains difficult. To be sure, resistance, kidnapping, forced migration, rape, brutality, starvation, natal alienation, and family separation all gave slaves comprehendible motives for suicidal responses to their owners and their condition. However, without their own words to explain their actions, we must look at the conditions surrounding their decisions.

In the years leading up to Missouri statehood, the slave Scipio of the last territorial governor, William Clark, shot and killed himself three years after having attempted suicide by ingesting lead ashes, slag, and shot. Again, in April 1835, a male slave who was sold down the river also attempted suicide. First, he tried to cut off both his legs; however, after failing at that, he cut his throat. Failing at his second attempt, he walked a short distance to drown himself in the river.[120] Slaves might have understood that even death by suicide was preferable to life under slavery and considered death as a means of transmigrating and returning home to Africa. They prepared for death in a ritualistic manner or with materials they believed would aid their journey.

When Dr. Lane performed his post-mortem examination on the slave at the lead mines, it was revealed that the alimentary canal was inflamed, engorged, and thickened. After exposing the small intestine, he found ashes, cinders, slag, and fourteen shot. Possibly, after the slave and his master had an argument, or after the slave learned that he was going to be sent down the Mississippi River to New Orleans with the next shipment of sheet lead, he planned his self-destruction in a ritualistic manner. Using materials closely associated with his existence that he believed would aid his journey to his ancestors, the slave began taking the articles that Dr. Lane recovered from the his bowels during the autopsy. In other cases, and in other places, before killing themselves, newly imported slaves would put on or remove all their clothes, place food and water nearby, or wrap chains around their waists; they believed that these actions and objects would sustain them through the transmigration. Since ecology is the branch of biology that deals with the relationships of organisms to one another and to their physical surroundings, examining this thirty-year-old miner's act of self-destruction in the context of his emotional and physical environment allows for a more nuanced exploration of how lead became a tool in the powerful act of suicide, revealing on another level the unhealthiness of lead and slavery.[121]

As miners and agents expanded their activities, more miners became stricken with lead poisoning, and doctors continued to service the community. Dr. Stephen Skeel migrated and settled in the middle of the lead region while Clemson was compiling his report. In Skeel's medical observations, he recorded smelters suffering from severe attacks of lead poisoning and noted a high incidence of the sickness among miners between 1838 and 1840.[122] Skeel recorded over 100 cases of the disease. He treated "miners and cleaners of the mineral who handled [lead ore] or inhaled the flying particles that mingled with the saliva."[123] One evening Skeel "was called to visit the 45-year-old E. B. Harris, who had been three days sick, pulse slow and full, tongue white, pain in the stomach, belly, back and collar bones; no vomiting and but little sick stomach, obstinate constipation, great thirst, but drinks but little." After examining the miner, Skeel wrote, "It is mine sickness, has had the disease before, two years ago." Skeel not only diagnosed the disease but also prescribed a plan of treatment for his patients, stating, "We have taken great pains to ascertain its nature, and the best plan of treatment, adapted to particular cases."[124]

Skeel also treated John Perkins who "had been troubled with bowel complaints while working in the neighborhood of the Perry Mines," a short distance from Mine La Motte.[125] After staying with Perkins for some time, Skeel concluded, "that the disease must be caused by mineral; that the action on the bowels was, probably, kept up by the particles of mineral acting mechanically on the alimentary canal." Skeel developed a plan of action beginning with opium, "which relieved the pain and procured rest." The following morning Skeel "gave him two drops of croton, and one ounce of castor oil" and left Perkins's side to allow the medicine to take its course.[126] Following the treatments, Skeel was happy to report "my patient was relieved when Mr. Perkins discharges were completely filled with particles of mineral." Skeel also reported that the bowels of Mr. Perkins contained two handfuls of lead particles. The fact that the excrement contained such visible quantities of lead apparently shows that miners were exposed to a high level of galena.

The increase in the number of miners and smelters who suffered from lead poisoning during the early nineteenth century coincided with the increase in lead production as new procedures and machines continued to arrive. Dr. Skeel, who lived in the lead country, documented a high incidence of lead poisoning among the Missouri miners at the same time that Clemson was reporting the healthiness of the country. Skeel's observations, diagnoses, and treatments also represent how settlers began to redefine the locality in terms of unhealthiness associated with lead production. While boosters wrote pamphlets featuring the district's galena resources and praised the discovery of additional veins of lead, Skeel outlined in his medical report how the landscape was a dangerous place for settlement.

Dr. Skeel also treated miners who faced the danger of working in thirty-five-foot-deep shafts. In February 1838, Rozier wrote in his journal about how he became incapacitated for a number of months after a mining accident that occurred in his twenty-six-foot-deep shaft. In the process of filling his basin, Rozier stated that "with the velocity of a bomb, the rock, earth, and lead ore laden basin weighing from 150 to 200 pounds fell on top of my body, knocking me to the floor of the shaft."[127] Rozier wrote that he sustained broken bones, "in my right arm with such violence [that the bone] penetrate[d] the flesh causing a three-inch opening and my fingers shut from the pain coming from the nerves, muscles, and tendons."[128] After Rozier returned home, Henroid went to search for Dr. Skeel, who diagnosed Rozier as having a "great number of bruises and contusions as far as the marrow."[129] Rozier's accident represents the growing need for medical doctors in the community to help guard workmen against occupational hazards associated with new methods. In addition to blasting accidents, miners often slipped from damp ladders in the mines and emerged with broken arms, legs, or even necks. If mines were not properly drained, some fell into the water and were drowned, and others were crushed by cave-ins.[130]

Skilled miners recorded their medical observations as they redefined their own workspaces not only in terms of prospective resource capital but also in terms of danger. Like physicians' diagnoses, miners assessed their environment in terms of healthiness and unhealthiness. Although Rozier's journal shows that lead production became integral to the settlement and development of the region, his account also reveals how the incursion of new European technologies increased the dangers of mining in deep shafts. Brackenridge's and Schoolcraft's accounts of excessive fumes and animal sickness and Dr. Skeel's accounts of treating lead poisoning in miners reveal that people recognized at the time that the unhealthiness was linked to lead smelting. These changes in awareness were examples of early recognition of the environmental consequences of the increased production of lead. These observations alerted doctors and prompted them to study how changes in technologies affected the health of workers in other trades.

As mining work expanded to supply the increasing markets for lead products back east, urban doctors began to research and publish their medical observations of various trades. In 1837, Dr. Benjamin W. McCready of New York conducted a study on occupational diseases among New York City's manufacturers. McCready examined the health problems of agricultural workers, laborers, seamen, factory operatives, professionals, and literary men. He also discussed housing and the "general conditions of life" stemming from poverty and unhealthy cities. McCready fails to report on the health of miners or smelters in America, but he does connect Missouri's lead production with the health of painters who handled white lead. He states, "Painters are in the habit of constantly employing the mineral which has long been known for its poisonous qualities, and it is to this that they are unhealthy in their appearance." McCready instructed painters to do their work more quickly so that they would not breathe in the lead oxides spilling into an enclosed atmosphere. McCready explained how painters would be "affected with dizziness and head-ache, and often with nausea and vomiting."[131]

McCready's observations were similar to accounts of how lead distressed miners and smelters in the Missouri mining district. It is unclear why McCready chose not to visit the lead mines to report on the dangers of lead mining and smelting. McCready may have ignored these ailments because Americans were becoming regular users of lead products, and commercial ambition erased concerns about miners' health.

In conclusion, by the early nineteenth-century, guided by Enlightenment confidence, boosters and improvers desired to civilize the mining district by adopting more advanced mechanical interventions. As travelers, geologists, and miners made their way west to inspect the lead mines, they combined regional description and regional promotion as a way to recognize regional resource potential. Most significantly, in the mining district, after American miners installed their modern structures, settlers began to assess the healthiness and unhealthiness of their farms and settlements.

Prior to publication of Cincinnati physician Daniel Drake's *A Systematic Treatise, Historical, Etiological, and Practical, on the Principal Diseases of the Interior Valley of North America* in 1850 an atlas of health and place that shaped agricultural settlement practice, Missouri doctors and miners were giving greater attention to a region's "healthiness" or "unhealthiness" in agricultural mining landscapes.[132] Early observers and doctors understood and acknowledged that animals suffered from lead poisoning, but Drs. Lane and Skeel diagnosed settlers who were also suffering from lead exposure. Promoters, doctors, and miners were recognizing that the change in settlement practices, with miners living closer to the mines, was resulting in a high incidence of lead poisoning among the Missouri miners. As advocates praised the region by spreading glowing reports about vegetation, soil type, and quality galena, doctors dispatched observations about the region as dangerous and unhealthy. Although boosters' civic responsibility was to promote improvement, hidden within their positive assertions, they wrote about the effects of lead on the surroundings and the population.

As medical examinations revealed the harmful effects of lead at the Missouri mines, they spurred more thought and reflection about the healthiness of the district and about urban trades. The attentiveness to mining workplaces seems to be consistent with a tendency to connect disease with rural workplaces. But even as travelers and physicians often described these open spaces as "healthy country," the area surrounding Mine La Motte became "unhealthier" with the transition to European technologies. Miners and animals were sometimes sick and were sometimes in good health. Henriod and Rozier persisted in documenting their surroundings as being "healthy" or "unhealthy."[133]

When the miner, the digger, and the cleaner each handled lead, they inhaled airborne particles that mingled with their saliva. Swallowing a considerable portion of lead consequently affected many villagers and workers.[134] Some miners escaped related sicknesses, but those engaged in the smelting of lead ores were often attacked. The replacement of the relatively small-scale Kaskaskia-French amalgam mining system with European furnaces, triggered human suffering from increased exposure to lead in the neighborhood of the mines. Although, to date, there is a lack of evidence revealing the effects of lead production on early Native Americans and French miners, it appears that Drs. Lane and Skeel assumed that lead poisoning from mining and smelting only occurred after the Europeanizing of the industry. Their accounts of experiences with nineteenth century sick miners provided another way to show how early Native American and French mining history needs further study to learn if there was infact less lead poisoning using the old traditional methods.

CONCLUSION

Nearly two hundred years after representatives of the Illinois Confederacy guided Father Jacques Gravier, Antoine de la Mothe Cadillac, and French settlers to Old Mines and Mine La Motte in Missouri, the historian Reuben Gold Thwaites met the Ho-Chunk chief Spoon Decorah at his home in Wisconsin. The Ho-Chunk, also known as the Winnebago, are a Siouan-speaking Native American people whose historic territory includes parts of Wisconsin, Minnesota, Iowa, and Illinois. Born the year following the Louisiana Purchase and shortly after Moses Austin migrated from Virginia to Upper Louisiana's mining district, Decorah reminisced about the mining practices of his people. He explained that before the arrival of the Americans, Winnebago miners "dug lead for their own use, but most of them got it out to trade off to other Indians for supplies of all sorts" such as "lead for bullets, sometimes giving goods for it and sometimes furs," long before the Black Hawk War of 1832. Chief Decorah also noted that there were in fact many Native Americans "at work in this way, nearly all the time during the summer months, and during the fall and spring months when hunters would go down to the mines" to make "lead mining their regular work."[1]

It appears as if Decorah wanted to remind Thwaites that North America was not so much a new world but an old one, the product of millennia of Native American experiences and interactions with the environment. Decorah's account of Native American mining relays a story of Native Americans struggling to find ways to incorporate European people, their tools, and their ideas into Wisconsin and Missouri. Decorah wants Thwaites to understand that the invasion of Americans and their European technologies hindered Native American mining practices from continuing as they had in the past. Decorah may be suggesting that European technologies and the Americans that arrived from the Atlantic World may have physically displaced Native American lead mining practices from the mines but not from the memory of the Winnebago. According to Thwaites, for most of the interview, "Decorah recalled his story from memory," and at times he required the assistance of his nephew Doctor Decorah. Clearly, it appears as if Thwaites responded well to what he heard as he listened to Decorah's mining recollections, stating, "the narrative presents the Indian view of these historical events, thus providing insight into what [Native Americans] were thinking and talking about, in their campfire reminiscences of early experiences with the white man."[2]

Scholars have written important Native American histories. However, many of us who write about indigenous peoples usually overlook early instances of frontier mining, as much of the scholarship focuses on the well-known nineteenth-century gold rushes, without integrating Native American, European, and Euro-American mining practices into early American history. Analogous to mining and smelting in early European, Asian, and African settings, from prehistoric times, Native American peoples mined, melted, and traded crushed and melted lead throughout North America. Since early settlers lacked the tools and capital to dig deep shafts or lacked the bricks to construct furnaces, they adopted from the miners of the Illinois Confederacy the practices of trench mining to extract and log furnaces to melt lead ore. The syncretic methods they employed together, highlighted in this study, show how the Kaskaskia-French amalgam forged new alliances to make and trade lead ore. Without Native American influence, European-style mining development would have unrolled far more slowly. Correspondence from French mining engineers highlight Native American miners acting not only as guides for Europeans to their mining sites but also as skilled miners who developed their own environmental knowledge and techniques to prospect, extract, and melt lead ore on this early mining frontier.

Similar to the early eighteenth-century narratives that describe the significance of lead to all people living on the frontier, and specifically to the French and Native Americans, Chief Spoon Decorah also recounts for Thwaites the importance of lead to the Southwestern Wisconsin Winnebago peoples. Additionally, Decorah reminds us that the French settlers were not the first to prospect for lead in this region and that Native Americans had been extracting, smelting and trading lead they procured from their mines for centuries. It is unclear whether during the interview Decorah described Winnebago mining and smelting methods or examples of intercultural mining and smelting exchanges with Europeans, or whether Thwaites failed to include important aspects of indigenous mining environmental knowledge in the published account. However, it is clear that if we examine Decorah's late nineteenth-century mining recollections in combination with early eighteenth-century French natural philosophers' recollections and mineralogical reports, the non-precious metal lead shines a bright light on understanding an aspect of Native American history and the history of American frontier expansion and settlement.

What we have gained by tracing the development of lead mining and smelting—from precontact Native American processing, to the Ste. Geneviève district as a central French and Spanish lead depot, and finally to the Louisiana Purchase and American settlement of the frontier—is an understanding that these activities were neither wholly French, American, British, nor Native American, but rather a complicated amalgamation, or hybridization, of all. The Kaskaskia-French mining techniques that continued during the early lifetime of Chief Spoon Decorah, albeit in Missouri, underscore how indigenous peoples and European immigrants worked together, sharing their environmental knowledge and technological experience on the eighteenth century mining frontier. As European scientific knowledge, technology, commercial initiatives, and cultural mandates arrived, together these aspects worked to marginalize the earlier practices and put them on a new Euro-American footing, and the early amalgamation broke down.

Examining the multiple convergences of colonial management, landscape change, health impacts of lead, forms of government, and the desires of a young United States in this story about Native American and settler mining practices broadens our understanding of early colonial and Native American history in the *country full of mines*; it also directs our attention to another form of "middle ground" near the Mississippi River. Similar to Decorah's descriptions of miners working the spring, summer, and fall months and making lead mining their regular seasonal work, settlers in *le pays des Illinois* wrote about the seasonal nature of lead mining that continued among the Kaskaskia and French settlers. Combined with cyclical mining, the technological evolution—from Native American practices to an amalgamation of techniques and ultimately to the transfer of European technologies to the Missouri lead mines—created an efficient and year-round American standard of management and practice at a number of mining settlements. These transitions—from the cooperation between the French and Kaskaskia to create a relatively small-scale mining system respecting natural cycles, to the hegemonic relationships of Europeans and Americans over African slaves and the landscape—would become a means to dispel the earlier cooperative mining practices while creating industrial systems that ultimately contributed to the removal of Native Americans from the area.

Before new Euro-American mining operations would be installed at both Mine à Breton and Mine La Motte, French accounting records, Spanish census reports, and scientific works together revealed the role of scientific and technological development in Missouri, and later during the well-known nineteenth-century gold rushes. As American immigrants and settlers arrived in Upper Louisiana and began to establish their government, the first United States mineralogical assessment of the new territory's lead mines would focus American attention on the role of science in westward expansion. Such evidence and events forge a consistent analytical thread that reveals the messy world of eighteenth- and nineteenth-century scientific advances and the uneven ways environmental knowledge and practices circulated in the Atlantic World. Making lead mining the central theme is also to give attention to the history of the environmental history of mining on the American frontier to examine the reciprocal and ecological effects of lead on human and animal activities.

During a recent phone conversation with a museum supervisor, I asked about Native American contributions to lead mining and smelting. The supervisor appeared to be unaware of this early Native American mining history and their contributions that began over three hundred and twenty years ago.[3] Following our conversation, I also learned about the Joseph Lead Company documentary *Lead to Metal*, produced in 1950, in which St. Joseph Lead proposed that the history of lead mining began with the arrival of Europeans in 1720.[4] In the film, the company shows the French transferring their mining knowledge and skills to the *country full of mines*, extracting, smelting, and refining lead ore, without mentioning the Native American presence. The film confirms that Native American contributions to mining were not so much silent as silenced.

Endeavoring to uncover a narrative where Native Americans, Europeans, and Americans were not in opposition to each other, I hope this study elucidates another perhaps little-known set of alliances. What I have learned is that long before the technological exchange reached across the Atlantic to the United States, Native American and European miners engaged in lengthy and complicated interactions regarding their environmental knowledge and their respective methods. Europeans learned to mine and smelt lead ore according to Native American methods, and in turn, indigenous miners adopted European tools to extract and refine lead ore. Successively, Creoles and Americans also adopted the Kaskasia-French amalgam. For over one hundred years, miners interacted to create a cross-cultural dialogue that involved a hybrid of mining techniques that shaped their attitudes about each other during multiple encounters on the mining frontier. The proceeding chapters, rather than positing a precolonial past of ecological harmony, show that Native Americans, despite technological limitations in comparison to Europeans, engaged in a form of environmental manipulation. I hope that my study opens a window into how historians might bridge both cultural and material environmental history to reveal what impact intellectual change had on the mining environment through the lens of technology.

Nearly every region of the United States in every era of its history offers examples of Native American communities embracing mining opportunities. Whether the Pueblo in New Mexico, the Mississippian in Missouri, the Huron in Canada, the Winnebago near Lake Superior, or the Cherokee in Georgia, indigenous communities forged or adopted unique mining systems. We have much to learn about their motives, but many of the cases suggest they were seeking metals for exchange with one another or with Euro-Americans. By moving outside of well-established frameworks, historians can go beyond the longstanding separation of the American historical landscape to recognize where Native Americans and Europeans or Euro-Americans often engaged in lengthy and complicated convergences that connect them with the emergence of modern America. I hope that this study will challenge American historians to look for and recognize additional stories about other cultural and material exchanges between Native Americans, Europeans, and Americans. To appreciate fully the significance of the intercultural dialogue that shapes America's past, I believe we must demand for scholars to listen to Native American points of view. Only then will we be able to more fully integrate Native American experiences into the broader story of early American history.

NOTES

INTRODUCTION

1. Thomas Jefferson, *Notes on the State of Virginia*, (London: J. Stockdale, 1787); Douglas L. Wilson, "Thomas Jefferson's Library and the French Connection," *Eighteenth-Century Studies* 26, no. 4 (1993): 669; Donald W. Meinig, *The Shaping of America: A Geographical Perspective on 500 Years of History, vol. 2, Continental America 1800–1867* (New Haven, CT: Yale University Press, 1993).

2. Richard W. Judd, *The Untilled Garden: Natural History and the Spirit of Conservation in America, 1740–1840* (Cambridge: Cambridge University Press, 2009), 28–29.

3. Thomas Jefferson, *Notes on the State of Virginia*; Edmund H. Fulling, "Thomas Jefferson. His Interest in Plant Life as Revealed in His Writings—II," *Bulletin of the Torrey Botanical Club* 72, no. 3 (May 1945): 248–70; Joel Martin Halpern, "Thomas Jefferson and the Geological Sciences," *Rocks & Minerals* 26, no. 11–12 (November 1951): 601–2.

4. Reuben Gold Thwaites, ed., *The Jesuit Relations and Allied Documents: Travels and Explorations of the Jesuit Missionaries in New France, 1610–1791: The Original French, Latin, and Italian Texts with English Translations and Notes* [. . .], vol. 65, *Lower Canada, Mississippi Valley: 1696–1702*, (Cleveland, OH: Burrows Brothers, 1900), 103, 171, http://moses.creighton.edu/kripke/jesuitrelations/relations_65.html; Richard White, *The Middle Ground: Indians, Empires, and Republics in the Great Lakes Region, 1650–1815* (Cambridge: Cambridge University Press, 1991), 67–74.

5. Silvia Figueirôa and Clarete da Silva, "Enlightened Mineralogists: Mining Knowledge in Colonial Brazil, 1750–1825," in "Nature and Empire: Science and the Colonial Enterprise," ed. Roy M. MacLeod, *Osiris*, 2nd Series, vol. 15 (2000): 174–89; Roy M. MacLeod, ed., "Nature and Empire: Science and the Colonial Enterprise," *Osiris*, 2nd Series, vol. 15 (2000).

6. Robert H. Lamborn, *A Rudimentary Treatise on the Metallurgy of Silver and Lead: Containing a Description of the Argentiferous and Plumbiferous Minerals* [. . .] (London: John Weale, 1861), 132.

7. Frederick W. Hodge, ed., *Handbook of American Indians North of Mexico*, vol. 2 (New York: Pageant Books, 1960); Richard A. Bice, Phyllis S. Davis, and William M. Sundt, *Indian Mining of Lead for Use in Rio Grande Glaze Paint: Report of the AS-5 Bethsheba Project Near Cerrillos, New Mexico* (Albuquerque, NM: Archaeological Society, 2003); Lucy Eldersveld Murphy, *A Gathering of Rivers: Indians, Métis, and Mining in the Western Great Lakes, 1737–1832* (Lincoln: University of Nebraska Press, 2000), 77–100.

8. Robert Boyle, *The Works of the Honorable Robert Boyle in Five Volumes* (London: Thomas Birch, 1772), ii, 32; Henry Rowe Schoolcraft Papers, Container 82, Manuscript Division, Library of Congress, Washington, DC; Miriam Hussey and Wharton School Industrial Research Unit, *The Wetherill Papers, 1762–1899; Being the Collection of Business Records of the Store and White Lead Works Founded by Samuel Wetherill in the Late Eighteenth Century* [. . .] (Philadelphia: Industrial Research Dept., Wharton School of Finance and Commerce, University of Pennsylvania, 1942), 1–11; Ruth Schwartz Cowan, *A Social History of American Technology* (New York: Oxford University Press, 1997), 45–46.

9. D. J. Rowe, *Lead Manufacturing in Britain: A History* (London: Routledge, 2017).

10. Georgius Agricola, *De Re Metallica*, trans. Herbert Clark Hoover and Lou Henry Hoover (Reprint, New York: Dover, 1950), Book I of XII, https://www.gutenberg.org/files/38015/38015-h/38015-h.htm; O. C. Harn, *Lead: The Precious Metal*, (New York: Century, 1924); Robert B. Gordon and Patrick M. Malone, *The Texture of Industry: An Archaeological View of the Industrialization of North America* (New York: Oxford University Press, 1994); John Opie, *Nature's Nation: An Environmental History of the United States* (Fort Worth: Harcourt Brace, 1998), 223–25.

11. Hodge, *Handbook of American Indians North of Mexico*; Bice, Davis, and Sundt, *Indian Mining of Lead for Use in Rio Grande Glaze Paint*; Murphy, *A Gathering of Rivers*, 77–100.

12. E. N. Hartley, *Iron Works on The Saugus: The Lynn and Braintree Ventures of the Company of Undertakers of the Ironworks in New England* (Tulsa: University of Oklahoma Press, 1957); Arthur C. Bining, *Pennsylvania Iron Manufacture in the Eighteenth Century* (Harrisburg: Pennsylvania Historical and Museum Commission, 1973); Ronald L. Lewis, *Coal, Iron, and Slaves: Industrial Slavery in Maryland and Virginia, 1715–1865* (Westport, CT: Greenwood Press, 1979); John N. Ingham, *Making Iron and Steel: Independent Mills in Pittsburgh, 1820–1920* (Columbus: Ohio State University Press, 1991); Duane A. Smith, *Mining America: The Industry and the Environment, 1800–1980* (Niwot: University Press of Colorado, 1993); Charles B. Dew, *Bond of Iron: Master and Slave at Buffalo Forge* (New York: W. W. Norton, 1994); Timothy J. LeCain, "Moving Mountains: Technology and the Environment in Western Copper Mining" (PhD diss., University of Delaware, 1998); Kathryn Morse, *The Nature of Gold: An Environmental History of the Klondike Gold Rush* (Seattle: University of Washington Press, 2003).

13. Turner included a "miner frontier" but fails to see the significance of mineral exploitation in frontier Missouri. Frederick Jackson Turner, *The Frontier in American History* (New York: Henry Holt, 1920), 9–10, 183–84. For a discussion of prospectors and

miners pioneering the Far West during the California gold rush, see Rodman Wilson Paul, *Mining Frontiers of the Far West, 1848–1880* (Albuquerque: University of New Mexico Press, 1963), 1–11.

14. Paul, *Mining Frontiers of the Far West*; Rodman W. Paul and Elliott West, *Mining Frontiers of the Far West, 1848–1880* (Albuquerque: University of New Mexico Press, 2001). Other historians like Otis E. Young Jr., the noted historian of western mining, direct attention to the lead mines of the Mississippi valley. Otis E. Young Jr., *Western Mining, an Informal Account of Precious-Metals Prospecting, Placering, Lode Mining, and Milling on the American Frontier from Spanish Times to 1893* (Norman: University of Oklahoma Press, 1970); Duane A. Smith, *Rocky Mountain Mining Camps: The Urban Frontier* (Bloomington: Indiana University Press, 1967); Mark Wyman, *Hard Rock Epic, Western Miners in the Industrial Revolution, 1860–1910* (Berkeley: University of California Press, 1979); For mining technologies transfer, see Andrew C. Isenberg, *Mining California: An Ecological History* (New York: Hill and Wang, 2005), 24.

15. Elliott West, *The Contested Plains: Indians, Goldseekers, & the Rush to Colorado* (Lawrence: University Press of Kansas, 1998), 26–48; Stephen Aron, *American Confluence: The Missouri Frontier from Borderland to Border State* (Bloomington: Indiana University Press, 2009), 3–26; Andrew R. L. Cayton and Fredrika J. Teute, eds., *Contact Points: American Frontiers from the Mohawk Valley to the Mississippi, 1750–1830* (Chapel Hill: University of North Carolina Press, 1998); Peter C. Mancall and James Merrell, *American Encounters: Natives and Newcomers from European Contact to Indian Removal, 1500–1850* (New York: Routledge, 2007).

16. Susan R. Martin, *Wonderful Power: The Story of Ancient Copper Working in the Lake Superior Basin* (Detroit: Wayne State University Press, 1999).

17. Murphy, *A Gathering of Rivers*, 77–100.

18. Judith Ann Carney, *Black Rice: The African Origins of Rice Cultivation in the Americas* (Cambridge, MA: Harvard University Press, 2002); Max Edelson et al., "AHR Exchange: The Question of 'Black Rice'" *American Historical Review* 115, no. 1 (February 2010): 123–71.

19. Jack D. Forbes, "Frontiers in American History and the Role of the Frontier Historian," *Ethnohistory* 15, no. 2 (Spring 1968): 203–35; William Cronon, George Miles, and Jay Gitlin, "Becoming West: Toward a New Meaning for Western History," in *Under an Open Sky: Rethinking America's Western Past*, eds. William Cronon, George Miles, and Jay Gitlin (New York: W. W. Norton, 1992), 3–27.

20. Frederick Jackson Turner, "The Significance of Frontier in American History," Annual Report of the American Historical Association 1893, 199–227; For Turner's idea of a frontier community development, see Meinig, *The Shaping of America*; David J. Weber, "Turner, the Boltonians, and the Borderlands," *American Historical Review* 91, no.1 (February 1986): 66–81; See also Jeremy Adelman and Stephen Aron, "From Borderlands to Border: Empires, Nation-States, and the Peoples in Between in North American History," *American Historical Review* 104, no. 3 (June 1999): 814–41; Aron, *American Confluence*.

21. Daniel H. Usner Jr., *Indians, Settlers, & Slaves in a Frontier Exchange Economy: The Lower Mississippi Valley Before 1783* (Chapel Hill: University of North Carolina Press, 1992), 6–9.

22. Murphy, "To Live Among Us: Accommodation, Gender, and Conflict in the Western Great Lakes Region, 1760–1832," *Contact Points*, 291–98; See also Colin G. Calloway, *New Worlds for All: Indians, Europeans, and the Remaking of Early America* (Baltimore: Johns Hopkins University Press, 1997).

23. For discussion on Native American and European contact, accommodation, and exchange as it existed during the early fur trade, see White, *The Middle Ground*; Usner, *Indians, Settlers, & Slaves in a Frontier Exchange Economy*.

24. Meinig, *The Shaping of America*, 4–22; Peter J. Kastor, *Nation's Crucible: The Louisiana Purchase and the Creation of America* (New Haven, CT: Yale University Press, 2012).

25. Carl Ekberg, *Colonial Ste. Geneviève: An Adventure on the Mississippi Frontier* (Tucson, AZ: Patrice Press, 1985); Walter A. Schroeder, *Opening the Ozarks: A Historical Geography of Missouri's Ste. Geneviève District 1760–1830* (Columbia: University of Missouri Press, 2002).

26. Amos Stoddard to Thomas Jefferson, June 16, 1804, in United States and Thomas Jefferson, *Message from the President of the United States to Both Houses of Congress, 8th November, 1804* [...] (Washington, DC: William Duane & Son, 1804). See also David B. Gracy, *Moses Austin: His Life* (San Antonio, TX: Trinity University Press, 1987), 95–118.

27. Description of the Lead Mines in Upper Louisiana, Thomas Jefferson Communicated to Congress November 8, 1804, in American State Papers, 8th Congress, 2nd Session Public Lands: Volume 1 Page 188.United States Congressional Documents and Debates, 1774–1875, House of Representatives, 8th Congress, 2nd Session, Thomas Jefferson to the Senate and House of Representatives, Communicated to Congress November 8, 1804, *American State Papers: Public Lands*, 1:13.

28. Rachel Laudan, *From Mineralogy to Geology: The Foundations of a Science, 1650–1830* (Chicago: University of Chicago Press, 1993), 102.

29. I. Bernard Cohen, *Benjamin Franklin's Science* (Cambridge, MA: Harvard University Press, 1990); I. Bernard Cohen, *Science and The Founding Fathers: Science in the Political Thought of Jefferson, Franklin, Adams, and Madison* (New York: W.W. Norton & Company Publisher, 1997); Joyce Chaplin, *The First Scientific American: Benjamin Franklin and the Pursuit of Genius* (New York: Basic Books, 2006); Tom Shachtman, *Gentlemen Scientists and Revolutionaries: The Founding Fathers in the Age of Enlightenment* (New York: St. Martin's Press, 2014).

30. Paul focuses on American technological and scientific innovations and the progress of economic and political institutions that made mining possible. Paul, *Mining Frontiers of the Far West*; Otis E. Young Jr., historian of western mining, directs attention to the lead mines of the Mississippi valley. Young, *Western Mining, an Informal Account*; Smith, *Rocky Mountain Mining Camps*; Wyman, *Hard Rock Epic*; For a different type of ecological destruction, see Isenberg, *Mining California*, 24.

31. Daviken Studnicki-Gizbert, "Exhausting the Sierra Madre: Mining Ecologies in Mexico over the Lougue Duree," in *Mining North America: An Environmental History Since 1522*, eds. John Robert McNeill and George Vrtis (Oakland: University of California Press, 2017), 19–46. Timothy J. LeCain, *Mass Destruction: The Men and Giant Mines That Wired America and Scarred the Planet* (New Brunswick, NJ: Rutgers University Press, 2009). Richard V. Francaviglia, *Hard Places: Reading the Landscape of America's Historic Mining Districts*. (Iowa City: University of Iowa Press, 1997).

32. Laudan, *From Mineralogy to Geology*.

33. Laudan, *From Mineralogy to Geology*.

34. Mary C. Rabbitt, *Minerals, Lands, and Geology for the Common Defense and General Welfare*, vol. 1, *Before 1879: A history of public lands, federal science and mapping policy, and development of mineral resources in the United States* (Washington, DC: U.S. Government Printing Office, 1982); Steven Stoll, *Larding the Lean Earth: Soil and Society in Nineteenth-Century America* (New York: Hill and Wang, 2002); David I. Spanagel, "Great Convulsions and Parallel Scratches: The Era of Romantic Geology in Upstate New York," *Northeastern Geology and Environmental Sciences* 17, no. 2 (1995), 179–82; David I. Spanagel, *DeWitt Clinton and Amos Eaton: Geology and Power in Early New York* (Baltimore: Johns Hopkins University Press, 2014); George W. White, "Early Geological Observations in the American Midwest," in *Toward a History of Geology: Proceedings*, ed. Cecil J. Schneer (Cambridge, MA: Massachusetts Institute of Technology, 1970), 415–425.

35. Gracy, *Moses Austin: His Life*.

36. James Edward and Harold Dorn, *Science and Technology in World History: An Introduction* (Baltimore: Johns Hopkins University Press, 2015).

37. Rachel Laudan discusses the importance of the Wernerian Radiation, who were the pupils of Abraham Werner. Lauden, *From Mineralogy to Geology*, 102.

CHAPTER 1

1. August Reyling, *Historical Kaskaskia*, 1963, Illinois History and Lincoln Collections, Special Collections Division of the University of Illinois Library, Urbana-Champaign, accessed September 19, 2020, https://libsysdigi.library.uiuc.edu/OCA/Books2009-06/historicalkaskasooreyl/historicalkaskasooreyl.pdf.

2. Daniel Hechenberger, "The Jesuits: History and Impact: From Their Origins Prior to the Baroque Crisis to Their Role in the Illinois Country," in *Journal of the Illinois State Historical Society (1998–)* 100, no. 2 (Summer 2007), 85–109, https://www.jstor.org/stable/40204675; Thwaites, *The Jesuit Relations*, vol. 65, 103–105.

3. Reuben Gold Thwaites, ed., *The Jesuit Relations and Allied Documents: Travels and Explorations of the Jesuit Missionaries in New France, 1610–1791: The Original French, Latin, and Italian Texts with English Translations and Notes . . .* , vol. 63, *Lower Canada, Iroquois: 1667–1687* (New York: Pageant Book, 1959): 65, 101; Pierre Margry, *Discoveries and*

Settlements of the French in the West and in the South of North America (1614–1754): Memories and Original Documents Collected and Published (Paris: Imprimerie D. Jouaust, 1876–86), 5:408; Ruby Swartzlow, "The Early History of Lead Mining in Missouri, Part I," *Missouri Historical Review* 28, no. 3 (April 1934): 184–94.

4. White, *The Middle Ground*, 50–93.

5. Records of settlers carrying 200 pounds of lead and 71 pounds of musket balls and trade relations with Native American. Colonial supplies; Indian goods; lead for bullets; New Orleans butchers. See, Kaskaskia Manuscripts and Parish Registers, Administration of the French Colonies, Series B: Letters to Envoys, April-October, 1746, 83:17. Orders and dispatches from King and Minister to La Jonquiere, Lenormant, and Vaudreuil. Illinois Historical Collections. See accounts in N. M. Miller Surrey, ed., *Calendar of Manuscripts in Paris Archives and Libraries Relating to the History of the Mississippi Valley to 1803* (Washington, DC: Carnegie Institution of Washington, Department of Historical Research, 1926–28), 297; Dunbar Rowland and Albert G. Sanders, eds. and trans., *Mississippi Provincial Archives, 1702–1729*, vol. 2 (Jackson: Press of Mississippi Department of Archives and History, 1927), 170; Natalia Maree, *Kaskaskia under the French Regime* (Carbondale: Southern Illinois University Press, 2003), 29, 43; Francis Jennings and George Irving Quimby, "Indian Culture and European Trade Goods: The Archaeology of the Historic Period in the Western Great Lakes Region," *Ethnohistory* 18, no. 1 (1971): 71; J. Joseph Bauxar, "The Historic Period," *Illinois Archaeology*, Bulletin no. 1 (1973): 40–58.

6. Jefferson, *Notes on the State of Virginia*; William E. Foley, *The Genesis of Missouri: From Wilderness Outpost to Statehood* (Columbia: University of Missouri Press, 1989), 14.

7. Hodge, *Handbook of American Indians North of Mexico*, 481, 847.

8. Aron, *American Confluence*; James E. Davis, *Frontier Illinois* (Bloomington: Indiana University Press, 1998), 24–29; J. Joseph Bauxar, "History of the Illinois Area," in *Handbook of North American Indians*, vol. 15, *Northeast*, ed. Bruce G. Trigger (Washington, DC: Smithsonian Institution, 1978), 594–601.

9. Raymond E. Hauser, "The Illinois Indian Tribe: From Autonomy and Self-Sufficiency to Dependency and Depopulation," *Journal of the Illinois State Historical Society* 69 (May 1976): 127–38; Robert T. Bray, "The Missouri Indian Tribe in Archaeology and History," *Missouri Historical Review* 55 (April 1961): 213–25; Carl H. Chapman and Eleanor F. Chapman, *Indians and Archaeology of Missouri* (Columbia: University of Missouri Press, 1983), 99–117.

10. William M. Denevan, "The Pristine Myth: The Landscape of the Americas in 1492," in *Annals of the Association of American Geographers* 82, no. 3 (September 1992): 369–85, https://www.jstor.org/stable/2563351.

11. Andrew Hurley, ed., *Common Fields: An Environmental History of St. Louis* (St. Louis: Missouri Historical Society Press, 1997).

12. Denevan, "The Pristine Myth: The Landscape of the Americas in 1492;" William Cronon, *Changes in the Land: Indians, Colonists, and the Ecology of New England* (New York: Hill And Wang, 1983), 49–51.

13. Thomas E. Emerson and R. Barry Lewis, eds., *Cahokia and the Hinterlands: Middle Mississippian Cultures of the Midwest* (Urbana: University of Illinois Press, 1991); Hurley, *Common Fields*; Schroeder, *Opening the Ozarks*.

14. William R. Iseminger, "Culture and Environment in the American Bottom: The Rise and Fall of Cahokia Mounds," in *Common Fields: An Environmental History of St. Louis*, ed. Andrew Hurley (St. Louis: Missouri Historical Society Press, 1997), 38–57; William R. Iseminger, "Relationships Between Climate Change and Culture Change in Prehistory," *Illinois Antiquity* 24 (Spring 1990): 2–4; Walter A. Schroeder, "Environmental Setting of the St. Louis Region," in *Common Fields: An Environmental History of St. Louis*, ed. Andrew Hurley (St. Louis: Missouri Historical Society Press, 1997), 13–37.

15. James J. Parsons, "Raised Field Farmers as Pre-Columbian Landscape Engineers: Looking North from the San Jorge, Colombia," in *Prehistoric Intensive Agriculture in the Tropics*, 2 vols., ed. I. S. Farrington, International Series 232 (Oxford: British Archaeological Reports, 1985), 149–65.

16. Hodge, *Handbook of American Indians North of Mexico*, 481. For recent archaeological research regarding early Native American galena workings, see Bice, Davis, and Sundt, *Indian Mining of Lead for Use in Rio Grande Glaze Paint*.

17. William H. R. Lykins, "On the Mound-Builders' Knowledge of Metals," in *Kansas City Review of Science and Industry*, ed. Theo. S. Case, p. 535 (Kansas City, MO: Ramsey, Millet, and Hudson, 1882), 535; George I. Quimby Jr., "Indian Trade Objects in Michigan and Louisiana," *Michigan Academy of Science, Arts, and Letters* 27 (1941): 543–51; Gregory Perino, "The Krueger Site, Monroe County, Illinois," in *Mississippian Site Archaeology in Illinois I: Site Reports from the St. Louis and Chicago Areas*, Illinois Archaeological Survey, Bulletin no. 8 (Urbana: University of Illinois, 1971): 187–91; John A. Walthall, *Galena and Aboriginal Trade in Eastern North America* (Springfield: Illinois State Museum, 1981).

18. Cyrus Thomas, *Report on the Mound Explorations of the Bureau of Ethnology, Twelfth Annual Report of the Bureau of Ethnology 1890–91* (Washington, DC: Government Printing Office [GPO], 1894), 202–5.

19. Walthall, *Galena and Aboriginal Trade in Eastern North America*, 18–25.

20. W. R. Lethaby, *Leadwork: Old and Ornamental and for the Most Part English* (London: Macmillan, 1893), 29. A. R. N. Roberts, "The Life and Work of W. R. Lethaby," *Journal of the Royal Society of Arts* 105, no. 5000 (March 29, 1957): 355–71, https://www.jstor.org/stable/41366042. Harn, *Lead: The Precious Metal*, 44–45.

21. "Only the gold and the silver, the brass, the iron, the tin and the lead—everything that may abide the fire you shall make it go through the fire and it shall be clean." In addition, "They sank as lead in the mighty waters." Numbers 31: 22–23 and Exodus 15:10, respectively.

22. Harn, *Lead: The Precious Metal*; Rowe, *Lead Manufacturing in Britain*, chap. 21; Jerome. O. Nriagu, "Paleoenvironmental Research: 'Tales Told in Lead,'" *Science* 281, no. 5383 (September 11, 1998): 1622–23.

23. Mary Elizabeth Good, *Guebert Site: An 18th Century Historic Kaskaskia Indian Village in Randolph County, Illinois* (n.p.: Central States Archaeological Societies and Mary Elizabeth Good, 1972); White, *The Middle Ground*.

24. Rowland and Sanders, *Mississippi Provincial Archives, 1701–1729*, vol. 2; White, *The Middle Ground*.

25. Carl H. Chapman, "The Little Osage and Missouri Indian Village Sites," *Missouri Archaeologist* 21 (1959): 1–67; Carl H. Chapman, "A Preliminary Survey of Missouri Archaeology. Part I. Historic Indian Tribes," *Missouri Archaeologist* 10 (1946): 19–20; Margaret B. Brown, "The Zimmerman Site: Further Excavations at the Grand Village of Kaskaskia," *Illinois State Museum Reports of Investigations*, no 9 (1975): 116–20.

26. Agricola, *De Re Metallica*, Book VIII.

27. Agricola, *De Re Metallica*, Book I; Harn, *Lead: The Precious Metal*; Gordon and Malone, *The Texture of Industry*; Opie, *Nature's Nation*, 223–25.

28. Bern Dibner, *Agricola on Metals* (Norwalk, CT: Burndy Library, 1958), 5–22.

29. Laudan, *From Mineralogy to Geology*.

30. Plattes, G, et al.., *A Collection of Scarce and Valuable Treatises upon Metals, Mines and Minerals [...]* (London: Printed for J. Hodges, 1740), 38–40, 89–91; T. A. Rickard, *Man and Metals: A History of Mining in Relation to the Development of Civilization* (1932; Reprint, Arno Press, 1974); Rabbitt, *Minerals, Lands, and Geology for the Common Defense and General Welfare*; Stoll, *Larding the Lean Earth*; Spanagel, "Great Convulsions and Parallel Scratches," 179–82; Spanagel, *DeWitt Clinton and Amos Eaton*; George W. White, "Early Geological Observations in the American Midwest."

31. Dibner, *Agricola On Metals*, 5–9.

32. Hechenberger, "The Jesuits: History and Impact"; Thwaites, *The Jesuit Relations*, vol. 65, 103–5.

33. Thwaites, *The Jesuit Relations*, vol. 65, 173.

34. Rowland and Sanders, *Mississippi Provincial Archives, 1702–1729*, vol. 2, 162; See also "Journal of Diron D'Artaguiette," in Newton D. Mereness, ed., *Travels in the American Colonies, 1690–1783*, Edited Under the Auspices of the National Society of the Colonial Dames of America, (New York: Antiquarian Press, 1961), 15–94.

35. G. Malcolm Lewis, ed., *Cartographic Encounters: Perspectives on Native American Mapmaking and Map Use* (Chicago: University of Chicago Press, 1998); Conrad E. Heidenreich and Edward H. Dahl, "The French Mapping of North America, 1700–1760," *Map Collector* 19 (June 1982): 2–7; Conrad E. Heidenreich, "Mapping the Great Lakes: The Period of Imperial Rivalries, 1700–1760," *Cartographica* 18, no. 3 (1981): 74–109.

36. White, *The Middle Ground*, 67–93; Nellis M. Crouse, *Lemoyne d'Iberville: Soldier of New France* (Ithaca, NY: Cornell University Press, 1954).

37. Robert W. Karrow Jr. and David Buisseret, *Gardens of Delight: Maps and Travel Accounts of Illinois and the Great Lakes from the Collection of Hermon Dunlap Smith, An Exhibition at the Newberry Library, 29 October 1984–31 January 1985* (Chicago: Newberry

Library, 1984); Alan Gallay, *The Indian Slave Trade: The Rise of the English Empire in the American South, 1670–1717* (New Haven, CT: Yale University Press, 2001), ix–xi.

38. Des Ursin, "Relation of the Journey to the Mines in Illinois Country"; John E. Rothensteiner, "Earliest History of Mine La Motte," *The Missouri Historical Review*, vol. 10, (January 1926), 199–209; See also Clarence Walworth Alvord, *The Illinois Country, 1673–1818* (Springfield: Illinois Centennial Commission, 1920), 154. For a discussion of different Native American and Euro-American extraction practices, see John Mack Faragher, *Sugar Creek: Life on the Illinois Prairie* (New Haven, CT: Yale University Press, 1986); John Mack Faragher, *Women and Men on the Overland Trail* (New Haven, CT: Yale University Press, 2001); Ronald Trosper, "That Other Discipline: Economics and American Indian History," in *New Directions in American Indian History*, ed. Colin G. Calloway (Norman: University of Oklahoma Press, 1988), 208, 219; William Cronon, *Nature's Metropolis: Chicago and the Great West* (New York: W. W. Norton, 1992).

39. Usner, *Indians, Settlers, & Slaves in a Frontier Exchange Economy*.

40. Geoff Egan, *Lead Cloth Seals and Related Items in the British Museum* [with Mike Cowell and Hero Granger Taylor], Occasional Paper, no. 93 (London: Dept. of Medieval and Later Antiquities, British Museum, 1995); Timothy J. Kent, *Ft. Ponchartrain at Detroit: A Guide to the Daily Lives of Fur Trade and Military Personnel, Settlers, and Missionaries at French Posts*, 2 vols. (Ossineke, MI: Silver Fox Enterprises, 2001).

41. See account of the sale of widow Madame Gadobert's home and shot-making equipment to François Vallé in 1774, MSS 24, Ste. Geneviève Archives, Mine La Motte, Mines Collection, Missouri Historical Society, St. Louis.

42. Maubeuge is a former mining town in northern France where foundries, forges, and blast furnaces operated.

43. Elizabeth Shown Mills, "Parallel Lives: Philippe de La Renaudière and Philippe (de) Renault, Directors of the Mines, Company of the Indies," *The Natchitoches Genealogist* 22 (April 1998): 3–18.

44. La Renaudière "Relation of the Journey to the Mines in Illinois Country"; John E. Rothensteiner, "Earliest History of Mine La Motte," *The Missouri Historical Review*, vol. 10, (January 1926), 199–209;.

45. Galena is the main ore of lead and has been used since ancient times. Because of its low melting point, it was easy to obtain lead by smelting. In some deposits, galena contains about 1–2% silver, a source that outweighs the main lead ore in revenue. For designs similar to objects made by Native Americans, see Lykins, "On the Mound-Builders' Knowledge of Metals," 535; George E. Fay, "Lead-Silver Molds of the Osage Indians," *Transactions of the Kansas Academy of Science* 52, no. 2 (1949): 205–8.

46. La Renaudière, "Account of the Mines of M. de la Motte"; Des Ursin, "Relation of the Journey to the Mines in Illinois Country by des Ursin."

47. See note 45.

48. Native Americans and French miners worked with lead at the Guebert site, and traded lead products. Pottery decorated with glaze paints and catlinite were also recovered

near Kaskaskia Village. See John S. Sigstad, "A Field Test for Catlinite," *American Antiquity* 35, no. 3 (1970): 377–82; and George I. Quimby Jr., "Indian Trade Objects in Michigan and Louisiana," *Michigan Academy of Science, Arts, and Letters* 27 (1941): 543–51; Brown, "The Zimmerman Site: Further Excavations at the Grand Village of Kaskaskia," 80–91; J. T. Penman and J. N. Gundersen, "Pipestone Artifacts from Upper Mississippi Valley Sites" *Plains Anthropologist* 44, no. 167 (February 1999): 47–57, https://www.jstor.org/stable/25669585; Sarah Wisseman, Randall Hughes, Thomas Emerson, and Kenneth Farnsworth, "Refining the Identification of Native American Pipestone Quarries in the Midcontinental United States," *Journal of Archaeological Science* 39 (2012): 2496–505.

49. For comprehensive writings on early mining processes, see Alvaro Alonso Barba, *A Collection of Scarce and Valuable Treatises upon Metals*, 38–40, 89–91.

50. Des Ursin, "Relation of the Journey to the Mines in Illinois Country by des Ursin."

51. Robert M. Hazen, "The Founding of Geology in America: 1771 to 1818," *Geological Society of America Bulletin* 85 (1974): 1827–34.

52. Caleb Atwater was from Ohio, served as a commissioner at Prairie du Chien in 1829, and viewed the lead mines. See Caleb Atwater, *Writings of Caleb Atwater (Travel in America)* (Columbus, OH: Atwater, 1833), 340, http://books.google.com/books?id=H88i3y35liEC&pg=PP2&lpg=PP2&dq=Caleb+Atwater+writings; See also Walter Havighurst, *Wilderness for Sale: The Story of the First Western Land Rush* (New York: Hastings House, 1956), 140; Charles R. Birk, "Shortest Route to the Galena Lead Mines: The Lewistown Road," *Journal of the Illinois State Historical* Society 66, no. 2 (1973): 187–97.

53. John E. Weaver, *North American Prairie* (Lincoln: Johnsen, 1954), 69–70.

54. Esau Johnson, "Reminiscence," in *A Chronological History of Indian Lead Mining in the Upper Mississippi Valley from 1643 to 1848*, by Philip Millhouse (Unpublished Paper for History Special Projects, Galena Public Library, Galena, IL, 1993).

55. Des Ursin, "Relation of the Journey to the Mines in Illinois Country by des Ursin." See also Alvord, *The Illinois Country*, 154.

56. Thomas Houghton, *Royal institutions being proposals for articles to establish and confirm laws, liberties, & customs of silver & gold mines, to all the king's subjects, in such parts of Africa and America, which are now (or shall be) annexed to, and dependant on the crown of England : with rules, laws and methods of mining and getting precious stones, the working and making of salt-petre, and also, the digging and getting of lead, tin, copper, and quick-silver oars*[...] (London: Printed for Author, 1694), 169–70.

57. Des Ursin, "Relation of the Journey to the Mines in Illinois Country by des Ursin"; La Renaudière "Account of the Mines of M. de la Motte."

58. For comprehensive information on European and indigenous prospecting in New Spain, see Barba, *A Collection of Scarce and Valuable Treatises upon Metals*, 38–40, 89–91.

59. La Renaudière "Account of the Mines of M. de la Motte."

60. Bice, Davis, and Sundt, *Indian Mining of Lead for Use in Rio Grande Glaze Paint*.

61. Des Ursin, "Relation of the Journey to the Mines in Illinois Country by des Ursin." See also Alvord, *The Illinois Country*, 154.

62. Des Ursin, "Relation of the Journey to the Mines in Illinois Country by des Ursin"; La Renaudière, "Account of the Mines of M. de la Motte."

63. La Renaudière, "Account of the Mines of M. de la Motte."

64. Agricola, *De Re Metallica*.

65. Barba, *A Collection of Scarce and Valuable Treatises upon Metals*, 3, 18, 28.

66. Des Ursin, "Relation of the Journey to the Mines in Illinois Country by des Ursin"; J. Lyman Hayward, *The Los Cerrillos Mines [N. M.] and Their Mineral Resources: A Description of the Mines in the Los Cerrillos and Galisteo Mining Districts, Accompanied by a Map of the Same, Drawn from Actual Surveys* (South Framingham, MA: C. Clark, 1880), 26–28.

67. Agricola, *De Re Metallica*; See also Houghton, *Royal Institutions: Being Proposals for [. . .] the Digging and Getting of Lead*, 17.

68. Des Ursin, "Relation of the Journey to the Mines in Illinois Country by des Ursin"; J. Lyman Hayward, *The Los Cerrillos Mines [N. M.] and Their Mineral Resources*, 26–28; La Renaudière "Account of the Mines of M. de la Motte."

69. Barba, *A Collection of Scarce and Valuable Treatises upon Metals*, 74.

70. Barba, *A Collection of Scarce and Valuable Treatises upon Metals*, 75.

71. Benavides, Alonso de, *The Memorial of Fray Alonso de Benavides, 1630*, trans. Mrs. Edward Ayer (Chicago: privately printed, 1916), 217; *The Compleat Collier: Or, The Whole Art of Sinking, Getting, and Working, Coal-Mines [. . .] about Sunderland and New-Castle* (London: G. Conyers, 1708); See also Smith, *Mining America*, 27.

72. Perino, "The Krueger Site, Monroe County, Illinois."

73. Laudan, *From Mineralogy to Geology*; Agricola, *De Re Metallica*, Book VII.

74. Thwaites, *The Jesuit Relations*, vol. 65, 103–105; Des Ursin, "Relation of the Journey to the Mines in Illinois Country by des Ursin"; La Renaudière "Account of the Mines of M. de la Motte."

75. Thwaites, *The Jesuit Relations*, vol. 65, 101; Margry, *Discoveries and Settlements of the French*, 408; Swartzlow, "The Early History of Lead Mining in Missouri, Part I".

76. Des Ursin, "Relation of the Journey to the Mines in Illinois Country by des Ursin."

77. La Renaudière "Account of the Mines of M. de la Motte."

78. Usner, *Indians, Settlers, & Slaves in a Frontier Exchange Economy*; John E. Rothensteiner, "Earliest History of Mine La Motte," *Missouri Historical Review* 20, no. 2 (January 1926): 199–213; Swartzlow, "The Early History of Lead Mining in Missouri, Part I"; Ruby Swartzlow, "The Early History of Lead Mining in Missouri, Part II" *Missouri Historical Review* 28, no. 4 (July 1934): 287–95; Carl J. Ekberg, "Antoine Valentin de Gruy: Early Missouri Explorer," *Missouri Historical Review* 76 (January 1982): 136–50; Ekberg, *Colonial Ste. Geneviève*; Carl J. Ekberg, *French Roots in the Illinois Country: The Mississippi Frontier in Colonial Times* (Champaign: University of Illinois Press, 2000); Carl J. Ekberg, *François Vallé and His World: Upper Louisiana Before Lewis and Clark* (Columbia: University of Missouri Press, 2002).

79. Mills, "Parallel Lives: Philippe de La Renaudière and Philippe (de) Renault, 3–18.

80. Renault's report would surely be a fascinating document, but it has never been found. There is visual evidence of Renault's brick furnace. Louis Houck, *The Spanish Regime in Missouri: A Collection of Papers and Documents Relating to Upper Louisiana Principally within the Present Limits of Missouri During the Dominion of Spain* [...] (Chicago: R. R. Donnelley and Sons, 1909), 1:372, 392.

81. Contract for purchase, Kaskaskia Manuscripts, 41:12:29:2.

82. Contract for purchase, Kaskaskia Manuscripts, 42:3:28:1.

83. David Brion Davis, *The Problem of Slavery in Western Culture* (Ithaca, NY: Cornell University Press, 1966), 106–7; Ekberg, *Colonial Ste. Geneviève*, 196–201.

84. Olive Patricia Dickason, *Canada's First Nations: A History of Founding Peoples from Earliest Times* (Norman: University of Oklahoma Press, 1992); Alan Gallay, *The Indian Slave Trade*.

85. Ekberg, *Colonial Ste. Geneviève*, 196–201.

86. Des Ursin, "Relation of the Journey to the Mines in Illinois Country by des Ursin."

87. Des Ursin, "Relation of the Journey to the Mines in Illinois Country by des Ursin."

88. Contract for purchase, Kaskaskia Manuscripts, 42:3:28:1.

89. Antoine Valentine de Gruy documented his second expedition, on April 15, 1743, to the mines in "Memoir of Sieur de Guis Concerning Lead Mines in the Illinois Country." The original French document, translated by Carl J. Ekberg, is located in the Archives Nationales, Paris, series G1 465, folios 3–15, a photostat in the Illinois Historical Survey in Urbana, Illinois. For a complete translation see Ekberg, "Antoine Valentin de Gruy: Early Missouri Explorer," 136–50; Maxine Benson, Nicolas de Finiels, Carl J. Ekberg, and William E. Foley, "An Account of Upper Louisiana," *Journal of the Early Republic* 10, no. 3 (1990): 427.

90. R. F. Tylecote, *The Early History of Metallurgy*, 2nd ed. (London: Institute of Materials, 1992); Paul T. Craddock, *Early Metal Mining and Production* (London: Archetype Publications, 2010), 122–55; Martin, *Wonderful Power*, 113–38.

91. Robert H. Lamborn, *A Treatise on the Metallurgy of Silver and Lead*, 132.

92. La Renaudière "Account of the Mines of M. de la Motte." See also Alvord, *The Illinois Country*, 154.

93. Pioneers molded bullets, and from a very early period, pewterers mixed it with alloys for household goods. For sample inventory of a fur trader's personal effects including musket balls and musket manufacturing materials, see Reuben Gold Thwaites, "Notes on Early Lead Mining in the Fever (or Galena) River Region," *Collections of the State Historical Society of Wisconsin* 13 (1895): 278; Harn, *Lead: The Precious Metal*, 18, 44.

94. James Phinney Baxter, Jean François de La Roque Roberval, and Jean Alfonce, *A Memoir of Jacques Cartier, Sieur de Limoilou: His Voyages to the St. Lawrence* [...] (New York: Dodd, Mead, 1906), https://books.google.com/books?id=GrxiAAAAMAAJ; Samuel de Champlain, *Voyages of Samuel de Champlain: 1604–1610*, trans. Charles Pomeroy Otis (Boston: Prince Society, 1878), https://books.google.com/books?id=Q_kWAAAAYAAJ.

95. Frederick Overman, *A Treatise on Metallurgy: Comprising Mining, and General and Particular Metallurgical Operations, with a Description of Charcoal, Coke, and Anthracite Furnaces, Blast Machines, Hot Blast, Forge Hammers, Rolling Mills, Etc., Etc.* (New York: D. Appleton, 1887), 656–57; Craddock, *Early Metal Mining and Production*, 205.

96. Agricola, *De Re Metallica*, Book VIII.

97. Agricola, *De Re Metallica*, Book IV.

98. George R. Milner, "American Bottom Mississippian Cultures: Internal Development and External Relations," in *New Perspectives on Cahokia Archaeology: Views from the Periphery*, ed. James B. Stoltmann, Monographs in World Archaeology, no. 2 (Madison: Prehistory Press, 1991), 29–47.

99. Ekberg, "Antoine Valentin de Gruy: Early Missouri Explorer," 136–50.

100. Ekberg, *François Vallé and His World*, 22–41; Concerning travels, see Contract, Kaskaskia Manuscripts 44:5:5:1; Lucy Elizabeth Hanley, "Lead Mining in the Mississippi Valley during the Colonial Period" (unpublished master's thesis, St. Louis University, 1942), 30–37, 50–85.

101. Bice, Davis, and Sundt, *Indian Mining of Lead for Use in Rio Grande Glaze Paint*; Frank Hamilton Cushing, "Primitive Copper Working: An Experimental Study," *American Anthropologist* 7, no. 1 (January 1894): 93–117.

102. Agricola, *De Re Metallica*, Book VIII.

103. W. Y. Woods, "A Strange Pre-Historic Find," *The Wisconsin Naturalist* 1, no. 1 (1890), 25.

104. Agricola, *De Re Metallica*, Book VIII.

105. Thwaites, "Notes on Early Lead Mining in the Fever (or Galena) River Region," 276–77.

106. Ekberg, "Antoine Valentin de Gruy: Early Missouri Explorer," 136–50.

107. Ekberg, "Antoine Valentin de Gruy: Early Missouri Explorer," 136–50.

108. Ekberg, "Antoine Valentin de Gruy: Early Missouri Explorer," 136–50.

109. Henry Rowe Schoolcraft gives a similar description of this log-type furnace. He also compares it to the mill hopper furnace used during the early nineteenth century. Henry Rowe Schoolcraft, *A View of the Lead Mines of Missouri, Including Some Observations on the Mineralogy, Geology, Geography, Antiquities, Soil, Climate [...] of Missouri and Arkansas, and Other Sections of the Western Country.* (New York: Wiley, 1819) 45, 174, 195; Reuben Gold Thwaites, *How George Rogers Clark Won the Northwest: And Other Essays in Western History.* (Nabu Press, 2010), 314.

110. De Gruy does not offer a description, however, later manuals discuss the process of reheating to obtain a high yield of lead; see James Woodhouse, *The Young Chemist's Pocket Companion; Connected with a Portable Laboratory; Containing a Philosophical Apparatus, and a Great Number of Chemical Agents* [...] (Philadelphia: Printed by J. H. Oswald, 1797); Thomas Dobson, *A Compendious System of Mineralogy & Metallurgy; Extracted from the American Edition of the Encyclopaedia* [...] (Philadelphia: Printed by Thomas Dobson, 1794).

111. By 1855 *saumon* was described as leaden mass shaped as a *saumon*. These accounts are concerned with 1822 reports of Fox and Sac groups who trade in the Galena region. James H. Lockwood, *Early Times and Events in Wisconsin* (Madison: State Historical Society of Wisconsin, 1856), 131–32.

112. Thwaites, *How George Rogers Clark Won the Northwest*, 300–315.

113. Moses Meeker, *Early History of the Lead Region of Wisconsin* (Madison: State Historical Society of Wisconsin, 1872), 271–96.

114. There was a colloquial designation with a variety of spellings including *escourie, escouris,* and *recourie.* Jean-Antoine-Claude Chaptal, Comte De Chanteloup, *Elements of Chemistry*, 2nd American ed., trans. William Nicholson (Philadelphia: Printed by Lang & Ustick for M. Carey, 1796), 333–34.

115. Schoolcraft, *A View of the Lead Mines of Missouri*, chap. 3.

116. Mereness, "Journal of Diron D'Artaguiette," 77; Alvord, *The Illinois Country*, 159; Reuben Gold Thwaites, ed., "The French Regime in Wisconsin–I: 1634–1727," in *Collections of the State Historical Society of Wisconsin*, vol.16 (Madison: State Historical Society of Wisconsin, 1925). A note on page 400 explains that the prices in livres, sols, and deniers might be given the following values: 1 livre about 20 cents; 1 sol about 1 cent; 1 denier about 1/12 of a cent. Note these prices did not mention the source of the lead, so the computation is rough; See also Hanley, "Lead Mining in the Mississippi Valley during the Colonial Period," 10, 27.

117. Louis Houck, *A History of Missouri from the Earliest Explorations and Settlements Until the Admission of the State into the Union* (Chicago: R. R. Donnelley & Sons, 1908), 1:378.

118. Phone conversations on September 12, 2011, with tour guide and supervisor, Missouri Mines State Historic Site of the Missouri Department of Natural Resources.

119. Conevery Bolton Valenčius, *The Health of the Country: How American Settlers Understood Themselves and Their Land* (New York: Basic Books, 2002), 137–41. Gregg Mitman, "In Search of Health: Landscape and Disease in American Environmental History," *Environmental History* 10 (2005): 184–209; Hanley, "Lead Mining in the Mississippi Valley during the Colonial Period," 27–28.

120. Dibner, *Agricola on Metals*, 66–72.

121. Howard J. Cohen and Jeffrey S. Birkner, "Respiratory Protection," *Clinics in Chest Medicine* 33, no. 4 (December 2012): 783–93.

122. Agricola, *De Re Metallica*, 214–15, https://www.gutenberg.org/files/38015/38015-h/38015-h.htm.

123. Bernardino Ramazzini, *De morbis artificum diatriba: Diseases of Workers*, trans. Wilmer Cave Wright (Chicago: University of Chicago Press, 1940), chap. 5. See also A. Meiklejohn, "The Successful Prevention of Lead Poisoning in the Glazing of Earthenware in the North Staffordshire Potteries," *British Journal of Industrial Medicine* 20, no. 3 (1963): 169–80.

124. Samuel Stockhausen, *Traite des mauvais effets de la fumee de la litharge*, Nicolás Ruault ed. (Paris: Chez Ruault, 1776; Reprint Nabu Press, 2012).

125. S. A. D. Tissot, *Advice to the People in General, with Regard to Their Health* [...], trans. James Kirkpatrick (London: T. Becket, 1765; Reprint Philadelphia: J. Sparhawk, 1771); Frederick Gregory, *Natural Science in Western History* (Boston: Houghton Mifflin, 2008). Lawrence I. Conrad et al., *The Western Medical Tradition: 800 BC to AD 1800* (Cambridge: Cambridge University Press, 2011).

126. Tissot, *Advice to the People in General, with Regard to Their Health.*

127. Thomas Le Roux, *Le Laboratoire des Pollutions Industrielle: Paris, 1770–1830* (Paris: Editions Albin Michel, 2011) quoted in Christopher Sellers, "To Place or Not to Place: Toward an Environmental History of Modern Medicine," *Bulletin of the History of Medicine* 92, no. 1 (Spring 2018): 1–45.

128. Ekberg, *Colonial Ste. Geneviève*, 126–76; White, *The Middle Ground*; Daniel K. Richter, *Facing East from Indian Country: A Native History of Early America* (Cambridge, MA: Harvard University Press, 2001); Loretta Fowler, review of "One Vast Winter Count: The Native American West before Lewis and Clark," by Colin G. Calloway, *The Western Historical Quarterly* 36, no. 1 (April 1, 2005): 71, https://doi.org/10.2307/25443102; Cynthia J. Van Zandt, *Brothers among Nations: The Pursuit of Intercultural Alliances in Early America, 1580–1660* (Oxford: Oxford University Press, 2008).

129. Timothy Flint, *The History and Geography of the Mississippi Valley: To Which is Appended a Condensed Physical Geography of the Atlantic United States and the Whole American Continent.* 2nd Ed. 2 vols. (Cincinnati: E. H. Flint and L. R. Lincoln, 1832), 285–332.

130. Cronon, *Changes in the Land*, 49–51.

131. Carl H. Chapman, "Osage Village Locations and Hunting Territories to 1808," in *Osage Indians IV: A Preliminary Survey of Missouri Archaeology*, eds. Carl H. Chapman and D. R. Henning (New York: Garland, 1974), 17–30.

132. John Polemon, *Second Part of the Book of Battalias, Faught in Our Age* [...], (London: 1587; Reprint Amsterdam: Theatrum Orbis Terrarum / New York: Da Capo Press, 1972); Ekberg, *Colonial Ste. Geneviève*, 126–43.

133. Margaret Kimball Brown and Lawrie Cena Dean, *The Village of Chartres in Colonial Illinois, 1720–1765* (Baton Rouge, LA: Provincial Press, 2010); Ekberg, *Colonial Ste. Geneviève*, 126–57; Ekberg, *French Roots in the Illinois Country*, 99, 171.

134. Des Ursin, "Relation of the Journey to the Mines in Illinois Country by des Ursin"; La Renaudière "Account of the Mines of M. de la Motte"; "Memoir of Sieur de Guis Concerning Lead Mines in the Illinois Country."

135. Anthony Crozat's emphasis on the abundant forest in the region is noteworthy. Mereness, "Journal of Diron D'Artaguiette," 15–75; Moses Austin documented similar concerns. Moses Austin, *A Summary Description of the Lead Mines in Upper Louisiana: Also, an Estimate of Their Produce for Three Years Past* (City of Washington: A. and G. Way, 1804), 10, 21.

136. Benavides, *The Memorial of Fray Alonso de Benavides*, 217; *The Compleat Collier*; See also Smith, *Mining America*, 27.

137. "Memoir of Sieur de Guis Concerning Lead Mines in the Illinois Country."

138. For more on Jean-Frédéric Phélypeaux, Comte de Maurepas see John C. Rule, "Jean-Frédéric Phélypeaux, Comte de Pontchartrain et Maurepas: Reflections on His Life and His Papers," *Louisiana History: The Journal of the Louisiana Historical Association* 6, no. 4 (1965): 365–77. http://www.jstor.org/stable/4230863.

139. Maurepas to Bienville and Salmon, Versailles, October 6, 1741, Archives Nationales, Paris, Colonies, MSS B72: 476–77, (Missouri Historical Society, St. Louis).

140. The central government, notably the Ministry of the Marine, had a branch that engaged the scientific arm of France's colonial machine. The research sent to Louisiana supported French colonization. Their activities included cartography and mineralogy discoveries. Soon after Louis XV began to rule, Maurepas sponsored new letters patent to govern colonial commerce. McClellan, James E., III, and François Regourd. "The Colonial Machine: French Science and Colonization in the Ancien Regime." In "Nature and Empire: Science and the Colonial Enterprise," edited by Roy McCloud. *Osiris*, 2nd Series, vol. 15 (2000): 31–50.

141. Houghton, *Royal Institutions: Being Proposals for [. . .] the Digging and Getting of Lead*, 13, 15, 22.

142. Ekberg, "Antoine Valentin de Gruy: Early Missouri Explorer," 136–50.

143. Ekberg, "Antoine Valentin de Gruy: Early Missouri Explorer," 136–50.

144. Ekberg, "Antoine Valentin de Gruy: Early Missouri Explorer," 136–50.

145. Des Ursin, "Relation of the Journey to the Mines in Illinois Country by des Ursin"; La Renaudière "Account of the Mines of M. de la Motte."

146. Surrey, *Calendar of Manuscripts in Paris Archives*, 297; Joseph H. Schlarman, *From Quebec to New Orleans: The Story of the French in America, Illustrated; Fort de Chartres* (Belleville, IL: Buechler, 1929).

147. Ekberg, "Antoine Valentin de Gruy: Early Missouri Explorer," 136–50.

148. Mereness, "Journal of Diron D'Artaguiette," 67.

149. Hanley, "Lead Mining in the Mississippi Valley during the Colonial Period," 23; Lockwood, *Early Times and Events in Wisconsin*, 131–32. Lockwood does not use the word *saumon*, but he describes a leaden mass shaped like a *saumon* was probably shaped. His account is concerned with 1822 reports of indigenous trade in the Galena region.

150. Ekberg, "Antoine Valentin de Gruy: Early Missouri Explorer," 136–50.

151. Minister of the Marine to de Bertet, January 1, 1744, Paris, Archives Nationales, Paris, Colonies, MSS. B78, 443:443. (Missouri Historical Society, St. Louis).

152. The continuing significance of lead is noted in the Chouteau Family Papers, 1752–1946, which consist of correspondence, bills, accounts, inventories, contracts of engagement with various men, packing accounts, bills of lading, and other business papers related to the sale of lead from Auguste Chouteau, Pierre Chouteau Sr., Pierre Chouteau

Jr., and Rene Chouteau in their activities as fur traders, merchants, and financiers of Missouri. Includes 53 ledger account books of the American Fur Company's Western Division, Collection of Chouteau Family Papers, Missouri Historical Society, St. Louis, and Yale University Libraries, New Haven, CT.

153. See account of the sale of widow Madame Gadobert's home and shot-making equipment to François Vallé in 1774, MSS 24, Ste. Geneviève Archives.

154. Louise Kellogg, *The French Regime in Wisconsin and the Northwest* (Madison: State Historical Society of Wisconsin, 1925), 221, 360–63.

155. Bernard H. Schockel, "History of Development of Jo Daviess County," in *Geography of the Galena and Elizabeth Quadrangles*, eds. Arthur Trowbridge and Eugene Shaw, Illinois State Geological Survey, Bulletin no. 26 (Urbana: Illinois State Geological Survey and University of Illinois, 1916), 180.

156. Jean Arnold Valentine Bobe Desloseaux was a successor to Joseph Buchet as guardian of the King's warehouse in Illinois sometime after 1757, since Buchet served until that year. Little else is known of Desloseaux. Alvord, *The Illinois Country*, 96.

157. Archives Nationales, Paris, Colonies, MSS 38: 39–40. (Missouri Historical Society, St. Louis).

158. This was not only a problem at the mines, American colonial mines also lacked skilled workers. See Nicholas Roosevelt, Jacob Mark, Edward Livingston, and William Langworthy, *Papers, Relative to an Application to Congress, for an Exclusive Right of Searching for and Working Mines, in the North-West and South-West Territory* (Philadelphia: Printed by Samuel H. Smith, 1797).

159. De Gruy and Robineau to the Minister, December 6, 1744, Archives Nationales, Paris, Colonies, MSS B78: 443:443 (Missouri Historical Society, St. Louis).

160. Chouteau Family Papers, 1752–1946. American Fur Company's Western Division, Missouri Historical Society, St. Louis, and Yale University Libraries, New Haven, CT.

CHAPTER 2

1. Meinig, *The Shaping of America*. 193–202.
2. Account of La Rose and Gadobert, 1769–1771, MSS 29, Ste. Geneviève Archives, Mines Collection, Missouri Historical Society, St. Louis; Account of the expenses of Vallé regarding the amount of timber carted, MSS 29, Ste. Geneviève Archives, Mines Collection, Missouri Historical Society, St. Louis.
3. Austin, *A Summary Description of the Lead Mines*, 7–17.
4. Lewis, *Coal, Iron, and Slaves*.
5. Studnicki-Gizbert, "Exhausting the Sierra Madre: Mining Ecologies in Mexico over the Lougue Duree," 20–23; Gabriel B. Paquette, *Enlightenment, Governance and Reform in Spain and Its Empire, 1759–1808* (Basingstoke, UK: Palgrave Macmillan, 2011); Gabriel Paquette, ed., *Enlightened Reform in Southern Europe and Its Atlantic Colonies, c. 1750–1830* (Farnham, UK: Ashgate, 2009).

6. David A. Brading, "Mexican Silver Mining in the Eighteenth Century: The Revival of Zacatecas," in *Mines of Silver and Gold in the Americas*, ed. Peter J. Bakewell (Gower House–Brookfield, VT: Variorum, 1997) 303–19.

7. Dana Velasco Murillo, *Urban Indians in a Silver City: Zacatecas, Mexico, 1546–1810* (Stanford, CA: Stanford University Press, 2016), 161, 171.

8. Franco Rui to Don Antonio Ulloa in the Spanish Illinois Country, May 26, 1768, legajo (file) 1091, Archives of the Indies (Seville, Spain) Collection, Missouri Historical Society, St. Louis.

9. Charles (Don Carlos) DeHault Delassus (1764–1846), captain and commandant of the New Bourbon post was the second son of Pierre-Charles and Josepha, born in Bouchaine in April 1764. By 1799, under orders from Spain, Delassus was appointed lieutenant governor and commander in chief of Upper Louisiana. He was stationed in St. Louis before turning over control of the territory to the American agent Amos Stoddard when Louisiana was transferred to the United States. Trudeau was born in New Orleans, November 28, 1748. He was well educated and had a family of several sons. See Houck, *A History of Missouri from the Earliest Explorations*, vol. 1:57–59.

10. Ekberg, *Colonial Ste. Geneviève*, Appendix; Schroeder, *Opening the Ozarks*, 14–15, Appendix.

11. Sale of Madame Gadobert's home and shot making equipment to François Vallé in 1774, MSS 24, Ste. Geneviève Archives.

12. Ekberg, "Antoine Valentin de Gruy: Early Missouri Explorer," 136–50.

13. Kathleen DuVal, *The Native Ground: Indians and Colonists in the Heart of the Continent* (Philadelphia: University of Pennsylvania Press, 2006), 172; Daniel H. Usner. "Weaving Material Objects and Political Alliances: The Chitimacha Indian Pursuit of Federal Recognition" in *Native American and Indigenous Studies* 1, no. 1 (2014): 25–48; Janet Hoskins, *Biographical Objects: How Things Tell the Stories of People's Lives* (New York: Routledge, 1998), 193–97. Usner, *Indians, Settlers, and Slaves in a Frontier Exchange Economy*.

14. Inventory and deed of sale to Datchurut, December 17, 1766, and January 27, 1767, MSS 12, St. Louis Archives, Missouri Historical Society, St. Louis. The first twenty-five documents of the St. Louis Archives have been translated and typed by the WPA Project. Doc. 12 is in Vol. 1 of the typed translations.

15. Studnicki-Gizbert, "Exhausting the Sierra Madre: Mining Ecologies in Mexico over the Lougue Duree," 20–35; LeCain, *Mass Destruction*; Francaviglia, *Hard Places*.

16. Pedro Vial lived in Spanish Upper Louisiana from 1797 to 1801 at Mine à Breton, and exactly how much previous experience he had at the mines is difficult to record. Houck, *The Spanish Regime in Missouri*, vol. 1:350–58.

17. Christopher Clark, *The Roots of Rural Capitalism: Western Massachusetts, 1780–1860* (Ithaca, NY: Cornell University Press, 1990), 14. White, *The Middle Ground*.

18. François Vallé's items are included in the Missouri Historical Society Mines Collection among the bulk of the materials concerning official matters. This collection contains about ten items, which include copies from the Paris Archives Nationales on Mine La Motte. The collection includes information about early mining in Missouri, including Mine à Breton and Mine La Motte, as well as accounts of the history and

ownership status of the Mine La Motte (or Lamothe) lead mine area in Missouri. Mines Collection, Missouri Historical Society, St. Louis; The Western Historical Manuscripts Collection also contains the Vallé letter books. Also, see note on page 54 of Houck, *The Spanish Regime in Missouri*, vol. 1.

19. Mary Dalton, "Notes on the Genealogy of the Vallé Family," *Missouri Historical Society Collections* 2, no. 7 (1906), 54–82; John Stewart, "A Walking Tour in Old Ste. Geneviève," *Missouri Life* 6, no. 3 (July-August 1978), 50–57; Lawrence O. Christensen, William E. Foley, Gary R. Kremer, and Kenneth H. Winn, eds., *Dictionary of Missouri Biography* (Columbia: University of Missouri Press, 1999), 761–62.

20. Ekberg, "Antoine Valentin de Gruy: Early Missouri Explorer," 136–50.; François Vallé's account books show multiple lead exchanges for products outside of Ste. Geneviève. See François Vallé Papers, Missouri Historical Society, St. Louis; Vallé Mining Company Account Books, The Western Historical Manuscript Collection, Rolla, Missouri.

21. The Vallé family came from Rouen, Normandy, and settled near Quebec. See the François Vallé Papers, 1742–1846, and Mine La Motte activity, Mines Collection, 1715–1900, Missouri Historical Society, St. Louis. Houck, *The Spanish Regime in Missouri*, vol. 1:54; Hanley, "Lead Mining in the Mississippi Valley during the Colonial Period," 30–37; Ekberg, *François Vallé and His World*, 22–41.

22. Ekberg, *Colonial Ste. Geneviève*, 148–49.

23. Declaration of Nicolas Noel dit La Rose, August 19, 1773, MSS 27, Ste. Geneviève Archives, Mines Collection, Missouri Historical Society, St. Louis.

24. Inventory and deed of sale to Datchurut, December 17, 1766, and January 27, 1767, MSS 12, St. Louis Archives. See chapter 2, note 14.

25. Declaration of Nicolas Noel dit La Rose, August 19, 1773, MSS 27, Ste. Geneviève Archives; N. M. Miller Surrey, *The Commerce of Louisiana during the French Regime, 1699–1763* (New York: Columbia University, 1916).

26. For use of the term *deputy husband*, see Laurel Thatcher Ulrich, *Good Wives: Image and Reality in the Lives of Women in Northern New England 1650–1750* (New York: Vintage Books, 1991), 9; LeeAnn Whites, Mary Neth, and Gary R. Kremer, *Women in Missouri History: In Search of Power and Influence* (Columbia: University of Missouri Press, 2004), 25.

27. Agreement between Vallé and Madame Gadobert, November 9, 1773, MSS 25, Ste. Geneviève Archives, Mines Collection, Missouri Historical Society, St. Louis; Also see reference in Hanley, "Lead Mining in the Mississippi Valley during the Colonial Period," 75–77.

28. Account of Madame Gadobert, 1769–1771, MSS 29, Ste. Geneviève Archives, Mines Collection, Missouri Historical Society, St. Louis; Ekberg, *Colonial Ste. Geneviève*, 150; Hanley, "Lead Mining in the Mississippi Valley during the Colonial Period," 50–85.

29. Carl Ekberg, *Colonial Ste. Geneviève*, 207–10.

30. Ekberg, "Antoine Valentin de Gruy: Early Missouri Explorer," 136–50.

31. Plattes, *A Discovery of Subterraneal Treasure: of All Manner of Mines and Minerals* [...] (London, 1653), 304–8.

32. Declaration of Datchurut regarding the Castor Vein, October 28, 1773, MSS 20, Ste. Geneviève Archives, Mines Collection, Missouri Historical Society, St. Louis.

33. Datchurut's petition for accounting of La Rose's mineral, July 7, 1770, MSS 20, Ste. Geneviève Archives, Mines Collection, Missouri Historical Scoeity, St. Louis.

34. Carver, Jonathan, John Coakley Lettsom, and Isaiah Thomas. *Three Years' Travels throughout the Interior Parts of North America, for More than Five Thousand Miles [. . .]*. Walpole, NH: Isaiah Thomas, 1813, 63.

35. Plattes, G, et al.., *A Collection of Scarce and Valuable Treatises upon Metals, Mines and Minerals*.

36. Account of the expenses of Vallé in exploiting Mine La Motte sometime during the 1790s, MSS 29, Ste. Geneviève Archives, Mines Collection, Missouri Historical Society, St. Louis.

37. Austin, *A Summary Description of the Lead Mines*, 10–17.

38. Moses Austin's report outlines the working mines in 1797. This report also mentions the importance of trading lead products for goods made in the United States. Austin, *A Summary Description of the Lead Mines*, 17–22.

39. La Rose lead sale, May 22, 1770, MSS 26, Ste. Geneviève Archives, Mines Collection, Missouri Historical Society, St. Louis.

40. Account of the expenses of Vallé regarding the amount of timber carted, MSS 29, Ste. Geneviève Archives; Account of the expenses of Vallé in exploiting Mine La Motte sometime during the 1790s, MSS 29, Ste. Geneviève Archives.

41. Datchurut's petition for accounting of La Rose's mineral, July 7, 1770, MSS 20, Ste. Geneviève Archives.

42. Moses Austin, *A Summary Description of the Lead Mines*, 17–22.

43. Hartley, *Iron Works on The Saugus*; Bining, *Pennsylvania Iron Manufacture in the Eighteenth Century*; Dew, *Bond of Iron*; Alan Taylor, "'Wasty Ways': Stories of American Settlement," *Environmental History* 3 (July 1998): 291–310.

44. Houck, *The Spanish Regime in Missouri*, 1:55.

45. Application for competitive bidding for the conveyance of the mineral, January 28, 1775, MSS 25, Ste. Geneviève Archives, Mine La Motte, Mines Collection, Missouri Historical Society, St. Louis.

46. For significance of metals such as iron and lead to Europeans and Americans, see Harn, *Lead: The Precious Metal*; Gordon and Malone, *The Texture of Industry*; Opie, *Nature's Nation*, 223–25; Cowan, *A Social History of American Technology*, 57–63.

47. Inventory and deed of sale to Datchurut, December 17, 1766, and January 27, 1767, MSS 12, St. Louis Archives. See chapter 2, note 14.

48. Thomas A. Richard, *A History of American Mining* (New York: McGraw-Hill, 1932), 18; John R. Stilgoe, *Common Landscape of America, 1580 to 1845* (New Haven, CT: Yale University Press, 1982), 273–74.

49. John Stilgoe explains how Plattes's plan shaped early Spanish and English mining settlements with analysis of the technological transfer of mining practices from Europe; this suggests that European expertise made possible the construction of a mining settlement. See John R. Stilgoe, *Common Landscape of America*, 277–88; Plattes, G, et al.., *A Collection of Scarce and Valuable Treatises upon Metals, Mines and Minerals*; Charles Swift Riché Hildeburn, *A Century of Printing: The Issues of the Press in Pennsylvania, 1685–1784*,

Volume 2, American Culture Series, Library of American Civilization (Philadelphia: Press of Matlack & Harvey, 1886), 445.

50. For a more recent study of mining settlements as places of diversity, see Donald L. Hardesty, *Mining Archaeology in the American West: A View from the Silver State* (Lincoln: University of Nebraska Press and the Society for Historical Archaeology, 2010).

51. Thomas Houghton, *Rara avis in terris: or The Compleat Miner, in Two Books; the First Containing the Liberties, Laws and Customs, of the Lead-Mines, within [...] Wirksworth in Derbyshire, in Fifty Nine Articles [...]. The Second Teacheth the Art of Dialling and Levelling* (Derby, UK: Printed by Samuel Hodgkinson, 1729), 16.

52. Houck, *The Spanish Regime in Missouri*, 1:365–72; See also Henry Clay Thompson II, "A History of Madison County Missouri," *Democrat News* (Fredricktown, MO), serially published ca. 1940, 27–47.

53. Schroeder, *Opening the Ozarks*, 267–283.

54. La Renaudière "Account of the Mines of M. de la Motte."

55. Account of Vallé in exploiting Mine La Motte, MSS 29, Ste. Geneviève Archives; Work contract of Joseph Trudel, June 3, 1773, Mining File, 1770–1806, nos. 1–40. Ste. Geneviève Archives, Folder 22at; Schroeder, *Opening the Ozarks*, 267–68.

56. Zenón Trudeau Report, 1787, legajo (file) 194–96, Archives of the Indies (Seville, Spain) Collection, Missouri Historical Society, St. Louis.

57. Adam Smith, *An Inquiry into the Nature and Causes of the Wealth of Nations*, vol. 2 (London: W. Strahan and T. Cadell, 1776), retrieved October 17, 2018, Google Books.

58. Immanuel Kant, *Fundamental Principals of the Metaphysics of Morals* (Gloucester, UK: Dodo Press, 2005).

59. Account of Vallé in exploiting Mine La Motte, MSS 29, Ste. Geneviève Archives; Plattes, *A Discovery of Subterraneal Treasure: of All Manner of Mines and Minerals* [...], 276; Tristan Stubbs, *Masters of Violence: The Plantation Overseers of Eighteenth-Century Virginia, South Carolina, and Georgia* (Columbia: University of South Carolina Press, 2018).

60. Account of Vallé in exploiting Mine La Motte, MSS 29, Ste. Geneviève Archives.

61. "Memoir of Sieur de Guis Concerning Lead Mines in the Illinois Country."

62. Account of the expenses of François Vallé and La Rose to exploit half of the Vallé Castor Vein at Mine La Motte sometime during the 1790s, MSS 29, Ste. Geneviève Archives, Mines Collection, Missouri Historical Society, St. Louis.

63. For a concise history of the blasting methods, see Andrew Ure, *A Dictionary of Arts, Manufactures, and Mines* (London: Longman, Orme, Brown, Greene, & Longmans, 1839), 835–37. Native Americans and American miners used similar techniques at the Galena, Illinois, lead mines. See Philip Millhouse, *A Chronological History of Indian Lead Mining in the Upper Mississippi Valley from 1643 to 1848* (Unpublished Paper for History Special Projects, Galena Public Library, Galena, IL, 1993), 6–8.

64. Ure, *A Dictionary of Arts, Manufactures, and Mines*, 837.

65. See chapter 2, note 6; Houck, *A History of Missouri from the Earliest Explorations*, 1:57–59; Carl J. Ekberg, *A French Aristocrat in the American West: The Shattered Dreams of De Lassus de Luzières* (Columbia: University of Missouri, 2010).

66. See chapter 2, note 6; Houck, *A History of Missouri from the Earliest Explorations*, 1:57–59.

67. Houck, *The Spanish Regime in Missouri*, 1:373–89; Ekberg, *Colonial Ste. Geneviève*, 445. On Bourbon Reforms, see Studnicki-Gizbert, "Exhausting the Sierra Madre: Mining Ecologies in Mexico over the Lougue Duree;" and Antonio Avalos-Lozano and Miguel Aguilar-Robledo, "Reconstructing the Environmental History of Colonial Mining: The Real del Catorce Mining District, Northeast New Spain/Mexico, Eighteenth and Nineteenth Centuries," in *Mining North America: An Environmental History Since 1522*, eds. John Robert McNeill and George Vrtis (Oakland: University of California Press, 2017).

68. George P. Garrison, ed., "A Memorandum of M. Austin's Journey from the Lead mines in the County of Wythe in the State of Virginia to the Lead Mines in the Province of Louisiana West of the Mississippi, 1796–1797," *American Historical Review* 5 (1900): 539; See also David B. Gracy, "Moses Austin and the Development of the Missouri Lead Industry," *Gateway Heritage* 1, no. 4 (Spring 1981), 42–48.

69. Austin, *A Summary Description of the Lead Mines*, 10–17.

70. Austin, *A Summary Description of the Lead Mines*, 16.

71. Austin, *A Summary Description of the Lead Mines*, 14–17. Also, see Schoolcraft, *A View of the Lead Mines of Missouri*, 93–104. For illustrations of early furnaces see Agricola, *De Re Metallica*, Book IX.

72. Agricola, *De Re Metallica*, Book VIII.

73. Agricola, *De Re Metallica*, Book IX; Ure, *A Dictionary of Arts, Manufactures, and Mines*, 752.

74. Schoolcraft, *A View of the Lead Mines of Missouri*, 103–6.

75. Agricola, *De Re Metallica*, Book IV.

76. Thwaites, *How George Rogers Clark Won the Northwest*, 314.

77. Millhouse, *A Chronological History of Indian Lead Mining*, 38–40.

78. Schoolcraft, *A View of the Lead Mines of Missouri*, 94–106.

79. Giacomo Beltrami traveled up the Mississippi in 1823 to explore, to map, and to interact with the local Native Americans. Giacomo Constantino Beltrami, *A Pilgrimage in Europe and America*, 2 vols. (London: Hunt and Clarke, 1828), vol. 2. 161–64.

80. Schoolcraft, *A View of the Lead Mines of Missouri*, 94–106.

81. Agricola, *De Re Metallica*, Book VIII.

82. Lewis Mumford, *Technics and Civilization* (Reprint, Chicago: University of Chicago Press, 2010).

83. Agricola, *De Re Metallica*, 214–217. https://www.gutenberg.org/files/38015/38015-h/38015-h.htm.

84. Henry Marie Brackenridge, *Views of Louisiana: Containing Geographical, Statistical and Historical Notices of that Vast and Important Portion of America* (Baltimore: Printed by Schaeffer & Maund, 1817), 207.

85. Austin, *A Summary Description of the Lead Mines*.

86. Flint, *Condensed Geography*, 2:86.

87. Ibid.

88. Frederick Bates to Richard Bates, December 17, 1807, Frederick Bates Letter Book, April 1807 – July 1809. In Bates Family Papers at Missouri Historical Society, St. Louis.

89. Account of the expenses of Vallé regarding the amount of timber carted, MSS 29, Ste. Geneviève Archives.

90. Alexander Von Humboldt, *Political Essay on the Kingdom of New Spain* [. . .], vol. 3, trans. John Black (London: Longman, Hurst, Rees, Orme, and Brown, 1811).

91. Request of J. B. Vallé, S .J. Beauvais, François Vallé, and Pratte for two square leagues of land, October 15, 1800, MSS 17, Ste. Geneviève Archives, Mines Collection, Missouri Historical Society, St. Louis??. For land deal, see Walter Lowrie, ed., *American State Papers: Documents, Legislative and Executive, of the Congress of the United States in Relation to the Public Lands from the First Session of the First Congress to the First Session of the Twenty-Third Congress, March 4, 1789 to June 15, 1834* (Washington, DC: Printed by D. Green, 1834), vol. 1, 534–41.

92. Austin, *A Summary Description of the Lead Mines*; Schoolcraft, *A View of the Lead Mines of Missouri*, 129–30.

93. Houck, *The Spanish Regime in Missouri*, 1:358–60.

94. Carlos (Charles) DeHault Delassus's occupation section of the Spanish Census labels immigrants as cultivators and artisans who settled in New Bourbon, the village closest to the lead mines. De Luzières's Report and Census of the population in the Spanish Illinois Country, 1797, legajo (file) 2365, Archives of the Indies (Seville, Spain) Collection, Missouri Historical Society, St. Louis; Schroeder, *Opening the Ozarks*, 106–108.

95. Vial became famous after forging a trading route between Santa Fe, New Mexico, and St. Louis, Missouri, in 1792 to open communication between these two important posts of the Spanish colonies. See the documents concerning Vial's expedition in Houck, *The Spanish Regime in Missouri*, 1:350–58. Pedro Vial lived in Spanish Upper Louisiana from 1797 to 1801 at Mine à Breton. How much previous experience he had at the mines is difficult to determine.

96. Francis Cruzat, Padron General de Los Pueblos de Sn. Luis Y Ste. Geneoveva de Ilinueses, 1787, trans. Walter B. Douglas, (in the Census Collection of the Missouri Historical Society) p,28.

97. Cruzat, Padron General de Los Pueblos de Sn. Luis y Ste. Geneoveva de Ilinueses, p. 28.

98. Luna's natural history collections identify him as a cultivator artisan who shared these types of activities with the serious scientists. Francisco Luna reveals to Pedro Vial his collection of minerals, suggesting a link with European practices. See Protest by François Luna to François Vallé, March 29, 1799, MSS 8, Ste. Geneviève Archives, Mines Collection, Missouri Historical Society, St. Louis.

99. Herbert Eugene Bolton, *Guide to Materials for the History of the United States in the Principal Archives of Mexico* (Washington, DC: Carnegie Institution, 1913), p 308, lepajo (file) 1787–1807, Notebooks, #95, #96, #158; William E. Burns, *The Scientific Revolution in Global Perspective* (New York: Oxford University Press, 2016); Roy MacLeod, "Introduction," in "Nature and Empire: Science and the Colonial Enterprise," ed. Roy M. MacLeod, *Osiris*, 2nd Series, vol. 15 (2000): 1–13; Carlos Sempat Assadourian, "The Colonial Economy: The Transfer of the European System of Production to New Spain and Peru," *Journal of Latin American Studies* 24 (1992): 55–68.

100. Protest by François Luna to François Vallé, March 29, 1799, MSS 8, Ste. Geneviève Archives.

101. Bolton, *Guide to Materials for the History of the United States in the Principal Archives of Mexico*, p 308, lepajo (file) 1787–1807, Notebooks, #95, #96, #158.

102. For details of the initiation and structure of the Royal College Mines, see L. R. Caswell and R. W. S. Daley, "The Delhuyar Brothers, Tungsten, and Spanish Silver," *Bull. Hist. Chem.* 23 (1999): 11–19; Laudan, *From Mineralogy to Geology*.

103. Donation of lead ashes to Pedro Vial, September 27, 1798, Mines Collection, Missouri Historical Society, St. Louis.

104. Apparently when Austin formed the mining association with Delassus, in addition to receiving land about the mines, he was also entitled to the lead ashes left on the ground. Letter from Carlos Dehault Delassus to François Vallé, transmitted by him to Jean Baptiste Vallé, August 19, 1799, Vallé Papers, Missouri Historical Society, St. Louis.

105. Alexandra Oleson and Sanborn Conner Brown, *The Pursuit of Knowledge in the Early American Republic: American Scientific and Learned Societies from Colonial Times to the Civil War* (Baltimore : Johns Hopkins University Press, 1976).

106. Chaptal, *Elements of Chemistry*.

107. Cowan, *A Social History of American Technology*, 19.

108. For examples of Spanish concerns about entrusting the defense of Louisiana to Americans, see Georges-Henri-Victor Collot, *Journey in North America, Containing a Survey of the Countries Watered by the Mississippi, Ohio, Missouri, and Other Affluing Rivers [. . .]*, Volumes *1 and 2* (Paris: Printed for Arthus Bertrand, 1826), online facsimile edition at https://content.wisconsinhistory.org/digital/collection/aj/id/11319; Concerning Collot's activities in North America, see Houck, *The Spanish Regime in Missouri*, 1:133–138; V. Collot, "General Collot's Plan for a Reconnaissance of the Ohio and Mississippi Valleys, 1796," *The William and Mary Quarterly* 9, no. 4 (October 1952): 512–20.

109. De Luzières's Report and Census of the population in the Spanish Illinois Country, 1797, legajo (file) 2365, Archives of the Indes, Papeles de Cuba.

110. De Luzières's Report and Census of the population in the Spanish Illinois Country, 1797, legajo (file) 2365, Archives of the Indes, Papeles de Cuba.

111. Thompson, *A History of Madison County Missouri*, 27–47; Ekberg, *A French Aristocrat in the American West*, 150–53; "William C. Carr to Moses Austin," in Eugene C. Barker, ed., *The Austin Papers, Part I*, Annual Report of the American Historical Association for the Year 1919, in Two Volumes, vol. 2 (Washington, DC: GPO, 1924), 105–7.

112. Richard S. Chew, "'Certain Victims of an International Contagion: The Panic of 1797 and the Hard Times of the Late 1790s in Baltimore." *Journal of the Early Republic* 25, no. 4 (2005): 565–613; Bruce H. Mann, *Republic of Debtors: Bankruptcy in the Age of American Independence* (Cambridge, MA: Harvard University Press, 2002).

113. Schroeder, *Opening the Ozarks*, 97–101.

114. For early history of migration to the region, see Lawrence Kinnaird, "American Penetration of Louisiana," in *New Spain and the Anglo-American West*, vol. 1, ed. George P. Hammond (Lancaster, PA: Lancaster Press, 1932), 211–37; Gilbert Din, "Spain's Immigration Policy in Louisiana and the American Penetration, 1792–1803," *Southwestern Historical Quarterly* 76 (1973): 255–76.

115. Barker, *The Austin Papers, Part I*, 9–10; Ure, *A Dictionary of Arts, Manufactures, and Mines*, 761.

116. John Frederick Amelung, *Remarks on Manufactures: Principally on the New Established Glass-House, near Frederick-Town, in the State of Maryland* (Printed for Author, 1787); Dwight P. Lanmon, "The Baltimore Glass Trade, 1780 to 1820" *Winterthur Portfolio* 5 (1969): 15–48.

117. Harn, *Lead: The Precious Metal*; Gordon and Malone, *The Texture of Industry*.

118. For a detailed summary of the importance of lead discovery and use by settlers, see "Account of the Lead Mines in Derbyshire, England, with the Manner of Working the Mines," *The New York Magazine, or Literary Repository*, 1790–1797 (May 1797), 240.

119. Thompson, *A History of Madison County Missouri*, 30–33.

120. "Moses Austin to Lieutenant Governor Delassus," in Barker, *The Austin Papers, Part I*, 47–53.

121. Austin to Kendall and Bates, December 1797, in Barker, *The Austin Papers, Part I*, 32–39.

122. Roosevelt et al., *Papers, Relative to an Application to Congress, for an Exclusive Right of Searching for and Working Mines*, 3–28. See also Collamer M. Abbott, "Colonial Copper Mines," in *William and Mary Quarterly*, Third Series 27, no.2 (April 1970): 295–309.

123. Roosevelt et al., *Papers, Relative to an Application to Congress, for an Exclusive Right of Searching for and Working Mines*, 9–11.

124. See Benjamin Henfrey, *A Plan with Proposals for Forming a Company to Work Mines in the United States; and to Smelt and Refine the Ores Whether of Copper, Lead, Tin, Silver, or Gold* (Philadelphia: Snowden & M'Corkle, 1797), 18; For the manufacture of shot, see James Cutbush, *The American Artist's Manual, or Dictionary of Practical Knowledge in the Application of Philosophy to the Arts and Manufactures* [. . .] (Philadelphia: Johnson & Warner, and R. Fisher, 1814); John Redman Coxe, "Lead" in *Emporium of Arts & Sciences*, vol. 2 (August 1, 1814).

125. Adelman and Aron, "From Borderlands to Borders [. . .]," 814–41; White, *The Middle Ground*; Patricia Nelson Limerick, *The Legacy of Conquest: The Unbroken History of the American West* (New York: W. W. Norton, 1987).

126. Austin, *A Summary Description of the Lead Mines*, 19, 20.

127. Barbara M. Tucker, *Samuel Slater and the Origins of the American Textile Industry, 1790–1860* (Ithaca, NY: Cornell University Press, 1984); Charles Grier Sellers, *The Market Revolution: Jacksonian America, 1815–1846* (New York: Oxford University Press, 1994).

128. Garrison, "A Memorandum of M. Austin's Journey from the Lead Mines," 518–42; *Proposals For Establishing an Association for Working Mines and Manufacturing Metals in the United States*, (Philadelphia: Samuel H. Smith, 1796), 5.

129. Roosevelt et al., *Papers, Relative to an Application to Congress, for an Exclusive Right of Searching for and Working Mines*, 12–15.

130. Henfrey, *A Plan with Proposals for Forming a Company to Work Mines in the United States*, iii–v.

131. Edward G. Mason, ed., "Early Chicago and Illinois," *Chicago Historical Society's Collection*, vol. 4, (Chicago: Fergus Printing, 1890), 36, 230–51; Barker, *The Austin Papers, Part I*, 2–9.

132. Garrison, "A Memorandum of M. Austin's Journey from the Lead Mines," 523–24;

133. Austin, *A Summary Description of the Lead Mines*, 17–18; For early writings of travels along these water routes, see Zadok Cramer, *The Navigator: Containing Directions for*

Navigating the Monongahela, Allegheny, Ohio, and Mississippi Rivers [...]. *Sixth Edition–Improved and Enlarged* (Pittsburgh: Cramer & Spear, 1801), 334–36.

134. An account of Moses Austin lists the total number of slaves he owned as 16 men, 3 women, and 6 small children at the lead mines in Virginia. It is not known how many of these slaves he transported with him to Spanish Louisiana. See Barker, *The Austin Papers, Part I*, 48, 60.

135. In 1796, Moses Austin's brother Stephen Austin traveled to Great Britain to enlist the services of two lead miners: Timothy Mullins and John Storts. See Bankruptcy Records, United States District Court for the Eastern District of Pennsylvania, RG 21. The records of Stephen Austin's bankruptcy proceeding constitute Case #204 (Microfilm publication M-993, Roll 24); Contract between Austin and two employees, Timothy Mullins and John Storts, October 28, 1797, MSS 12, Ste. Geneviève Archives, Mines Collection, Missouri Historical Society, St. Louis.

136. Austin to Kendall and Bates, December 1797, in Barker, *The Austin Papers, Part I*, 38–39; Garrison, "A Memorandum of M. Austin's Journey from the Lead Mines," 518.

137. Austin to Kendall and Bates, December 1797, in Barker, *The Austin Papers, Part I*, 38, 48–51; Austin, *A Summary Description of the Lead Mines*, 19.

138. Austin, *A Summary Description of the Lead Mines*, 15–16; John Bradbury, *Travels in the Interior of America, in the Years 1809, 1810, and 1811* [...], (Liverpool: Smith and Galway, 1817).

139. Enclosure, Illinois Adventure To Stephen and Moses Austin, Dr., in Barker, *The Austin Papers, Part I*, 40–41; Moses Austin to Lieutenant Governor De Lassus, in Barker, *The Austin Papers, Part I*, 49–53; Schoolcraft, *A View of the Lead Mines of Missouri*, 104–6.

140. Garrison, "A Memorandum of M. Austin's Journey from the Lead Mines," 518–42.

141. Moses Austin to John S. Brickey, in Barker, *The Austin Papers, Part I*, 248.

142. Moses Austin to John S. Brickey, in Barker, *The Austin Papers, Part I*, 249.

143. Austin, *A Summary Description of the Lead Mines*, 8–9.

144. Austin to Kendall and Bates, December 1797, in Barker, *The Austin Papers, Part I*, 39; Austin, *A Summary Description of the Lead Mines*, 7–10.

145. Numerous mining and smelting manuals were reproduced in the United States. Alvaro Alonso Barba, Gabriel Plattes, and Thomas Green, *A Collection of Scarce and Valuable Treatises upon Metals, Mines and Minerals: In Four Parts. Part I. and II. Containing the Art of Metals, Written Originally in Spanish* [...], *Translated by Edward Montagu Sandwich* [...] (London: Printed by C. Jephson for O. Payne, 1738), 304–8; *Proposals For Establishing an Association for Working Mines*, 14–16.

146. For descriptions of Austin's mining activity, see Brackenridge, "Sketches of the Territory of Louisiana" and "Lead Mines in the District of Ste. Geneviève," *Missouri Gazette*, June 20, 1811.

147. Schoolcraft, *A View of the Lead Mines of Missouri*, 90–93.

148. Moses Austin to James Bryan, Durham Hall, November 25, 1814, in Barker, *The Austin Papers, Part I*, 243–44.

149. S. Terry Childs and David Killick, "Indigenous African Metallurgy: Nature and Culture," *Annual Review of Anthropology* 22, no. 1 (October 1993): 317–37.

150. Miners followed instructions from manuals on the various stages of prospecting, assaying, cleaning, and smelting ores. Woodhouse, *The Young Chemist's Pocket Companion*; Cutbush, *The American Artist's Manual*, "Lead" section; Schoolcraft, *A View of the Lead Mines of Missouri*, 98–103.

151. For the aspects of production, see Cutbush, *The American Artist's Manual*, "Mine" section.

152. Most of the literature refers to the ash furnace as a "blast furnace." It is hard to know why the term is interchanged. Chaptal, *Elements of Chemistry*, 333–34.

153. Schoolcraft, *A View of the Lead Mines of Missouri*, 102–103.

154. Schoolcraft, *A View of the Lead Mines of Missouri*, 103–106.

155. Chaptal, *Elements of Chemistry*, 338; Schoolcraft, *A View of the Lead Mines of Missouri*, 103–5.

156. Schoolcraft, *A View of the Lead Mines of Missouri*, 22.

157. Schoolcraft, *A View of the Lead Mines of Missouri*, 102–3.

158. Amelung, *Remarks on Manufactures*.

159. Dorothy Daniel, "The First Glasshouse West of the Alleghenies" *The Western Pennsylvania Historical Magazine* 32 (September-December 1949): 97–113; Dorothy Daniel, *Cut and Engraved Glass, 1771–1905* (New York: M. Barrows, 1950).

160. Barker, *The Austin Papers, Part I*, 9–10; Ure, *A Dictionary of Arts, Manufactures, and Mines*, 761.

161. Barker, *The Austin Papers, Part I*, 10.

162. Chaptal, *Elements of Chemistry*, 338.

163. William Bingley, *Useful Knowledge of a Familiar and Explanatory Account of the Various Productions of Nature: Mineral, Vegetable, and Animal* (Philadelphia: A. Small, 1818), 179.

164. Schoolcraft, *A View of the Lead Mines of Missouri*, 2.

165. For descriptions of *saumon*, "the oval-shaped masses of lead that were two feet long, six to eight inches wide, two to four inches thick and weighing 30 to 40 pounds," see Datchurut's petition for a survey of Mine La Motte, July 7, 1770, MSS 23, Ste. Geneviève Archives, Mines Collection, Missouri Historical Society, St. Louis; Lockwood, *Early Times and Events in Wisconsin*, 131–32.

166. Phone interview with Gwendolyn Midlo Hall by Mark M. Chambers (January 27, 2007).

167. Cramer, *The Navigator*, 34–40.

168. Cramer, *The Navigator*, 18.

169. Cramer, *The Navigator*, 34.

170. Barker, *The Austin Papers, Part I*, 69–74.

171. Barker, *The Austin Papers, Part I*, 75–76.

172. Austin, *A Summary Description of the Lead Mines*, 20–21.

173. Complaint against Amable Partenay dit Mason, April 21, 1804, MSS 40, Ste. Geneviève Archives, Mines Collection, Missouri Historical Society, St. Louis.

CHAPTER 3

1. *Amos* Stoddard to Thomas Jefferson, June 16, 1804, in United States and Thomas Jefferson, *Message from the President of the United States to Both Houses of Congress, 8th November, 1804*; Amos Stoddard, *Sketches, Historical and Descriptive, of Louisiana*, [...] (Philadelphia: M. Carey, 1812); Gracy, *Moses Austin: His Life*, 95–118.

2. Jefferson, *Notes on the State of Virginia*; Edmund H. Fulling, "Thomas Jefferson. His Interest in Plant Life as Revealed in His Writings—II." *Bulletin of the Torrey Botanical Club* 72, no. 3 (1945): 248–70.

3. Moses Austin, *A Summary Description of the Lead Mines*, 10; Houck, *A History of Missouri From the Earliest Explorations*, vol. 2:355–70.

4. "To Thomas Jefferson from Amos Stoddard, 16 June 1804," Founders Online, National Archives, https://founders.archives.gov/documents/Jefferson/01-43-02-0504. Original source: *The Papers of Thomas Jefferson*, vol. 43, 11 March–30 June 1804, ed. James P. McClure (Princeton, NJ: Princeton University Press, 2017), 609–10.

5. Thomas Jefferson to the Senate and House of Representatives, November 8, 1804, *American State Papers, Public Lands*, 1:13.

6. For information regarding Jefferson's instructions to Meriwether Lewis concerning the Lewis and Clark Expedition's purposes for paving the way for the development of American commerce, see James P. Ronda, *Jefferson's West: A Journey with Lewis and Clark* (Charlottesville, VA: Thomas Jefferson Foundation, 2000); James P. Ronda, *Lewis and Clark Among the Indians* (Lincoln: University of Nebraska Press, 1984).

7. William E. Burns, *The Scientific Revolution in Global Perspective*, 39–56; Recently, scholarship portrays the founding intellects in spirited discussions, for example, Chaplin, *The First Scientific American: Benjamin Franklin and the Pursuit of Genius*; Shachtman, *Gentlemen Scientists and Revolutionaries*; Cohen, *Benjamin Franklin's Science*; and Cohen, *Science and The Founding Fathers*.

8. Carroll D. Wright and William C. Hunt, History and Growth of the United States Census: 1790–1890, Prepared for Senate Committee on the Census, 56th Congress, 1st Session, February 24, 1900, Doc. 194, Washington: U.S. Govt. Print. Off, 1900, 17–20.

9. Meinig, *The Shaping of America*, 4–22; Kastor, *The Nation's Crucible*.

10. The facsimile of the 1814 edition of the *American Mineralogical Journal* provides interesting reading for historians of geology and mineralogy. George W. White, ed., *The American Mineralogical Journal. Archibald Bruce, M.D. Contributions to the History of Geology*. vol. 1; (New York: Hafner, 1968) 7–9.

11. Edward and Dorn, *Science and Technology in World History: An Introduction*.

12. Gracy, *Moses Austin: His Life*.

13. Description of manufacturing procedures at the Christian Wilt red lead mill derives from Ure, *A Dictionary of Arts, Manufactures, and Mines*, 744–46; Schoolcraft Papers, Container 82.

14. Schoolcraft Papers, Container 82; Cutbush, *American Artist's Manual*, "Lead" section. See also Rowe, *Lead Manufacturing in Britain: A History*.

15. The Wetherill White Lead Works offers examples of trade connections between merchants and western settlers. Hussey and Wharton School Industrial Research Unit, The Wetherill Papers, 1762–1899; John L. Larson, *Internal Improvement: National Public Works and the Promise of Popular Government in the Early United States* (Chapel Hill: University of North Carolina Press, 2001).

16. Richard White, "American Environmental History: The Development of a New Historical Field," *Pacific Historical Review* 54 (1985): 297–335.

17. Austin, *A Summary Description of the Lead Mines*, 8, 11,17–19.

18. For early nineteenth-century immigrant prospecting, mining, and smelting methods, see Charlotte Erickson, *Invisible Immigrants: The Adaptation of English and Scottish Immigrants in Nineteenth-Century America* (Florida: University of Miami Press, 1972), 411–20.

19. Austin, *A Summary Description of the Lead Mines*, 8, 10, 13, 15.

20. Austin, *A Summary Description of the Lead Mines*, 8.

21. Austin, *A Summary Description of the Lead Mines*, 8.

22. Austin, *A Summary Description of the Lead Mines*, 19; Schoolcraft, *A View of the Lead Mines of Missouri*, 90–92.

23. Austin, *A Summary Description of the Lead Mines*, 19; See also Schoolcraft, *A View of the Lead Mines of Missouri*, 90–92.

24. For more on mining technologies transfer, see Isenberg, *Mining California*, 24.

25. James X. Corgan, ed., *Geological Sciences in the Antebellum South* (Tuscaloosa, AL: University of Alabama Press, 1982), 9–25; Benjamin R. Cohen, "Surveying Nature: Environmental Dimensions of Virginia's First Scientific Survey, 1835–1842," *Environmental History* 11, no. 1 (2006): 37–69; Isenberg, *Mining California*, 24.

26. For a historical description of the scientific scene in the early American republic and development of mineralogical studies in the United States, see John C. Greene and John G. Burke, "The Science of Minerals in the Age of Jefferson," in *Transactions of the American Philosophical Society, New Series*, vol. 68, no. 4 (1978): 1–113; Laudan, *From Mineralogy to Geology*, 87–88

27. Austin, A *Summary Description of the Lead Mines*, 7–8.

28. Laudan, *From Mineralogy to Geology, 101–2*

29. Roosevelt et al., *Papers, Relative to an Application to Congress, for an Exclusive Right of Searching for and Working Mines*, 17–18; Cutbush, *The American Artist's Manual*, "Ores" section.

30. For comprehensive explanation of European terms and mining techniques, see W. Pryce, *Mineralogia Cornubiensis: A Treatise on Minerals, Mines, and Mining [...] to Which is Added, an Explanation of the Terms and Idioms of Miners* (London: James Phillips, 1778), 112, 127–128; Cutbush, *The American Artist's Manual*, "Mine" section.

31. Chaptal, *Elements of Chemistry*, 330–39.

32. Austin, *A Summary Description of the Lead Mines*, 18–19.

33. William Maclure and others wrote essays on the mineral productions of the United States: William Maclure, "Observations on the Geology of the United States, Explanatory of a Geological Map," *Transactions of the American Philosophical Society* 6 (1809): 411–28; Benjamin De Witt and Sylvain Godon wrote essays on the mineral productions of New York and Maryland, respectively. Benjamin De Witt, "Mineral Productions of the State of New York," *Memoirs of the American Academy of Arts and Sciences* 2, no. 2 (1804): 73–81: Sylvain Godon, "Observations to Serve for the Mineralogical Map of the State of Maryland," *Transactions of the American Philosophical Society* 6 (1809), 319–23.

34. Abraham G. Werner, *On the External Characters of Minerals*, trans. Albert V. Carozzi (Urbana: University of Illinois Press, 1962); William Maclure made a number of references to Werner's processes in "Observations on the Geology of the United States," 415–17.

35. Early mineralogists' and chemists' manuals included mining instructions and methods to construct furnaces, to conduct tests, and for melting lead ore. Pryce, *Mineralogia Cornubiensis*, 131–41; Chaptal, *Elements of Chemistry*, 333.

36. See Richard Kirwan, *Elements of Mineralogy in Two Volumes* (London: Printed by J. Nichols for P. Elmsly, 1794–96).

37. Samuel Miller, *A Brief Retrospect of the Eighteenth Century, Part First in Two Volumes Containing A Sketch of the Revolutions and Improvements In Science, Arts, and Literature During That Period* (Philadelphia: T. and J. Swords, 1803), 145–56; George W. White, "The History of Geology and Mineralogy as Seen by American Writers, 1803–1835: A Bibliographic Essay," *Isis* 64, no. 2 (1973): 197–214.

38. Moses Austin's list of assets includes "70 VI Cyclopedia and cost" on October 7, 1820. In author's conversation on October 20, 2008, with Evan Hocker, archivist at the Barker Texas History Center at University of Texas at Austin, Mr. Hocker advised that no list of Austin's Cyclopedias has been located. Barker, *The Austin Papers, Part I*, 360–62.

39. Partnership agreement between Austin, Vallé, Delassus, January 26, 1797 in Barker, *The Austin Papers, Part I*, 29–31.

40. Austin, *A Summary Description of the Lead Mines*, 10, 17.

41. Schoolcraft, *A View of the Lead Mines of Missouri*, 38–40, chap. 7.

42. De Witt, Benjamin. "Mineral Productions of the State of New York." Memoirs of the American Academy of Arts and Sciences 2, no. 2 (1804): 73–81.

43. Austin, *A Summary Description of the Lead Mines*, 18–20.

44. For travel experiences, see Lewis C. Beck, *A Gazetteer of the State of Illinois and Missouri* (Albany, NY: Charles R. and George Webster, 1823), 252; "Journal of Voyage Down the Mississippi," in Barker, *The Austin Papers, Part I*, 69–75;

45. Austin, *A Summary Description of the Lead Mines*, 10.

46. Austin, *A Summary Description of the Lead Mines*. 7, 17.

47. Austin, *A Summary Description of the Lead Mines*, 20. For the nature of European descriptions of native practices in the African and Asian context, see Michael Adas,

Machines as the Measure of Men: Science, Technology, and Ideologies of Western Dominance (Ithaca, NY: Cornell University Press, 1989), 1–3.

48. Austin, *A Summary Description of the Lead Mines*, 7.

49. Hunt's Minutes, Recorder's Office, St. Louis, Missouri, 1825, typed. Hunt Family Papers, Missouri Historical Society, St. Louis; Schoolcraft, *View of the Lead mines of Missouri*, 18; Houck, *The Spanish Regime in Missouri*, 1:74, 283; Thomas Maitland Marshall, *The Life and Papers of Frederick Bates*, vol. 1 (St. Louis: Missouri Historical Society, 1926), 188, 275.

50. Austin, *A Summary Description of the Lead Mines*, 18, 19.

51. Austin, *A Summary Description of the Lead Mines*, 8.

52. Austin, *A Summary Description of the Lead Mines*, 8, 15.

53. Agricola, *De Re Metallica*, Book I.

54. The English word *vein* corresponds to the French term *filon*, which indicates all the deposits of this ore. Kirwan follows Werner's methods to describe the variety of veins, see Kirwan, *Elements of Mineralogy*, 202–23.

55. Pryce, *Mineralogia Cornubiensis*, 179.

56. Pryce, *Mineralogia Cornubiensis*, 92–94.

57. Agricola, *De Re Metallica*, Book I; Ure, *Dictionary of Arts, Manufactures, and Mines*, 745–48.

58. Austin, *A Summary Description of the Lead Mines*, 15.

59. Broxton W. Bird et al., "Pre-Columbian Lead Pollution from Native American Galena Processing and Land Use in the Midcontinental United States," *Geology* 47, no.12 (2019): 1193–97; Walthall, *Galena and Aboriginal Trade in Eastern North America*, 37–41.

60. Kirwan, *Elements of Mineralogy*, 49, 179.

61. Chaptal, *Elements of Chemistry*, 331; Schoolcraft, *A View of the Lead Mines of Missouri*, 77–80, 103–106.

62. Mine La Motte and Derbyshire, England, lead mines note limestone. See Cutbush, *The American Artist's Manual*, "Lead" section; Ure, *Dictionary of Arts, Manufactures, and Mines*, 837.

63. Schoolcraft refers to many European encyclopedias and experiments, and he conducted experiments while traveling through the lead district in 1818. Schoolcraft Papers, Container 82.

64. Schoolcraft discusses types of galena used to manufacture flint glass. Schoolcraft Papers, Container 82.

65. Chaptal, *Elements of Chemistry*, 335.

66. Ure, *Dictionary of Arts, Manufactures, and Mines*, 541; Schoolcraft, *A View of the Lead Mines of Missouri*, 77–80.

67. Schoolcraft Papers, Container 82.

68. Chaptal, *Elements of Chemistry*, 1–3.

69. Scientific or experimental chemists required a place as well as specific tools for conducting their operations, which were designed to discover how to effectively smelt and manufacture minerals. Cutbush, *The American Artist's Manual*, "Laboratory" section.

70. Agricola, *De Re Metallica*, Book VII.

71. Ure, *Dictionary of Arts, Manufactures, and Mines*, 837.

72. It is important to examine influences of scientific and technological development on European views of non-Western peoples' mining cultures. This reveals how Europeans perceived traditional practices, which affected their view of African and Asian technological acumen into the industrial era. See Adas, *Machines as the Measure of Men*.

73. Austin to Kendall and Bates, December 1797, in Barker, *The Austin Papers, Part I*, 32–39.

74. Dobson, *A Compendious System of Mineralogy & Metallurgy*, 30–40.

75. Early mineralogists' and chemists' manuals included mining instructions and methods to construct furnaces. They carried pocket-size books instructing them in the various stages of prospecting, assaying, and smelting metals. Chaptal, *Elements of Chemistry*, 2–8, 330–39.

76. Woodhouse, *The Young Chemist's Pocket Companion*; Dobson, *A Compendious System of Mineralogy & Metallurgy*, 420–23.

77. Miller, *Brief Retrospect of the Eighteenth Century*, 145–55.

78. Woodhouse, *The Young Chemist's Pocket Companion*, 26–27.

79. Woodhouse, *The Young Chemist's Pocket Companion*, 24–25.

80. Dobson, *A Compendious System of Mineralogy & Metallurgy*, 30–40.

81. Thomas Thomson, *A System of Chemistry in Four Volumes: Volume 1* (Edinburgh: Bell & Bradfute, 1802), 150–60.

82. Woodhouse, *The Young Chemist's Pocket Companion*, 24.

83. Schoolcraft, *A View of the Lead Mines of Missouri*, 90–94.

84. Woodhouse, *The Young Chemist's Pocket Companion*, 26.

85. Woodhouse, *The Young Chemist's Pocket Companion*, 14–20.

86. Austin, *A Summary Description of the Lead Mines*, 16.

87. Schoolcraft, *A View of the Lead Mines of Missouri*, 80.

88. Schoolcraft, *A View of the Lead Mines of Missouri*, 81.

89. Oleson and Brown, *Pursuit of Knowledge*, 39–46.

90. Greene and Burke, "The Science of Minerals in the Age of Jefferson," 21–22.

91. Thomas Jefferson to the Senate and House of Representatives, November 8, 1804, *American State Papers: Foreign Relations* 1:63.

92. Joseph Charless, "The Subscribers Having Erected at the New Diggings District at St. Geneviève, an Assay Furnace," *Missouri Gazette*, April 12, 1809, (St. Louis, Missouri).

93. Reviewing the amount of Missouri pig lead received by the Wetherills from the buying agent in New Orleans between 1810 and 1826, it is clear they had increased the production of red and white lead. Wetherill Production Records, Item 17, 18, and 25, [Wetherill Collection, University of Pennsylvania]; Also, see Walter Renton Ingalls, *Lead and Zinc in the United States: Comprising an Economic History of the Mining and Smelting of the Metals and the Conditions, Which Have Affected the Development of the Industries* (New York: Hill, 1908).

94. Similar reports of rich minerals deposits in the American Midwest by explorers Peter Kalm in 1753, Major Robert Rogers in 1765, and Thomas Hutchins in 1778 stimulated interest in the growing nation's mineral wealth. The science of mineralogy acquired a solid footing in the United States only gradually, remaining on the descriptive level well into the nineteenth century. Miller, *A Brief Retrospect of the Eighteenth Century*, 145–55.

95. Tench Coxe was appointed by President George Washington as Commissioner of the Revenue of the United States, to oversee the "Report on Manufactures." See Tench Coxe, *A View of the United States of America, In a Series of Papers Written at Various Times, between 1787 and 1794* (Philadelphia: Printed for William Hall, Wrigley & Berriman, 1794), viii.

96. Federic Louis Billon, *Annals of St. Louis in Its Territorial Days, from 1804 to 1821: Being a Continuation of the Author's Previous Work, the Annals of the French and Spanish Period* (St. Louis: Printed for the Author, 1888), 19, 115, 229.

97. Joseph Charless, "At Herculaneum a Shot Manufactory is Now Erecting by an Active Enterprising Citizen of Our Territory," *Missouri Gazette*, March 8, 1809.

98. Cramer, *The Navigator*, 87–90.

99. Joseph Charless, "The Erection of a Patent Shot Manufactory At Herculaneum," *Missouri Gazette*, November 16, 1809.

100. Brackenridge, "Sketches of the Territory of Louisiana," *Missouri Gazette*, June 20, 1811.

101. Thompson, *A History of Madison County Missouri*, 20.

102. Joseph Charless, "Shipment Received at the General Store of Aaron Elliot & Son," *Missouri Gazette*, November 7, 1810.

103. Wetherill Collection, Letter book, Item #1, July 8, 1812, to Benjamin Morgan, New Orleans.

104. Wetherill Collection, Letter book, Item #1, June 3, 1789, to Brandram, Templeman, & Jacques, London.

105. Ure, *A Dictionary of Arts, Manufactures, and Mines*, 747–48, 572; Schoolcraft Papers, Container 82.

106. Ure, *A Dictionary of Arts, Manufactures, and Mines*, 573; Schoolcraft Papers, Container 82.

107. Description of manufacturing procedures at the Christian Wilt red lead mill derives from William Nicholson, *A Dictionary of Chemistry, Exhibiting the Present State of the Theory and Practice of That Science, Its Application to Natural Philosophy, the Processes of Manufactures, Metallurgy* [...], 2 vols. (London: G. G. and J. Robinson, 1795), 616–33.

108. Joseph Charless, "The Proper Country to Establish Manufactures of Red Lead, White Lead," *Missouri Gazette*, October 26, 1811.

109. Joseph Hertzog to Christian Wilt, April 30, 1811, Christian Wilt Letter book, Missouri Historical Society, St. Louis. For a detailed review of the number of glass factories and potters located along the Ohio River, see Cramer, *The Navigator*, 18.

110. Hertzog to Wilt, May 13, 1811.

111. In Pittsburgh, Bakewell & Company manufactured red lead from lead acquired from Wilt. Joseph Hertzog to Z. Musiana, June 10, 1811, Christian Wilt Papers, Letterbook, Missouri Historical Society, St. Louis; Benjamin G. Bakewell, *The Family Book of Bakewell* (Pittsburgh: W. G. Johnston, 1896), 70.

112. Hertzog to Wilt, May 13, 1811; and May 23, 1811.

113. Hertzog to Wilt, May 25, 1811; and see the discussion between Sparke and Hertzog. Apparently, Hertzog was considering hiring a Mr. Donahue, a former employee of the Wetherills', and Hertzog wanted a reference from Sparke, Hertzog to Wilt, September 26, 1811.

114. From the letters, it is difficult to know the precise time when they realized Henderson's experiment would not produce what he suggested. Wilt to Hertzog, December 3, 1814.

115. Hertzog to Wilt, June 27, 1811.

116. François Victor Merat's work was published in Matthieu Joseph Bonaventure Orfila, *General System of Toxicology, or, A Treatise on Poisons, Found in the Mineral, Vegetable, and Animal Kingdoms* [. . .], trans. John A. Waller and Joseph G. Nancrede (Philadelphia: M. Carey & Son, 1817), 184–204, https://archive.org/details/generalsystemoftooorfi.

117. Complaint against Amble Partenay dit Mason, April 21, 1804, MSS 40, Ste. Geneviève Archives.

118. Complaint against Amble Partenay dit Mason, April 21, 1804, MSS 40, Ste. Geneviève Archives.

119. Orfila, *A General System of Toxicology*, 184–204.

120. L. Tanquerel Des Planches, *Lead Diseases: A Treatise from the French, with Notes and Additions on the Use of Lead Pipe and Its Substitutes*, trans. Samuel Luther Dana (Lowell, MA: D. Bixby, 1848).

121. The letter is silent about Hertzog's purchasing of slaves from Maryland. Wilt to Hertzog, September 6, 1812.

122. Wilt to Hertzog, December 31, 1814. Letter provides a detailed description of the factory's operation as melted lead moves from one stage to the next. In addition, the Cutbush manual is used to provide additional detail where necessary. Cutbush, *The American Artist's Manual*, "Lead" section.

123. Stephen T. Whitman, *The Price of Freedom: Slavery and Manumission in Baltimore and Early National Maryland*, (Louisville: University of Kentucky Press, 1997).

124. Slaves figured prominently in late eighteenth- and early nineteenth-century production of iron, another capital-intensive, heat-powered industry that subjected workers to extremely hot working conditions. See Lewis, *Coal, Iron, and Slaves*, chap. 1.

125. The men working the furnaces refused to work at night and on Sundays, except in cases when Wilt paid them additional wages, December 31, 1814, Christian Wilt Papers, Missouri Historical Society, St. Louis.

126. Massicot or massicoc, a yellow oxide of lead before it turns red, is used as yellow pigment. See Cutbush, *The American Artist's Manual*.

127. Calcinations are the reduction of lead by fire to powder or ashes. Calcinations are the actual process of oxidation, changing the lead into red lead. See Cutbush, *The American Artist's Manual*.

128. George Baker, *An Essay Concerning the Cause of the Endemical Colic of Devonshire* [...] (London: Printed by J. Hughs, 1767), reprinted in *Medical Transactions* 1 (1768): 1–64.

129. Joseph Charless, "Arrival of Individual Gifted in the Two Sciences," *Missouri Gazette*, February 22, 1810.

130. Brackenridge, *Views of Louisiana*, 154.

131. Austin, *Summary Description of the Lead Mines*, 10; Austin Report 1816, *American State Papers, Public Lands*, 3:609–13; Brackenridge, *Views of Louisiana*, 154; Schoolcraft, *A View of the Lead Mines of Missouri*, 126–27.

132. Schoolcraft, *A View of the Lead Mines of Missouri*, 90–93.

133. Miller, *A Brief Retrospect of the Eighteenth Century*, 145–55.

CHAPTER 4

1. Schoolcraft, *A View of the Lead Mines of Missouri*, 4–7, 26–27.

2. See Alexis de Tocqueville, "A Fortnight in the Wilds," in *Journey to America*, ed. J. P. Mayer, trans. George Lawrence (New Haven: Yale University Press, 1959), http://www.iwu.edu/~matthews/journey1.html. Also, see David Nye, *America as Second Creation: Technology and Narratives of New Beginnings* (Cambridge, MA: MIT Press, 2003), 9.

3. Schoolcraft, *A View of the Lead Mines of Missouri*. 4, 113–33.

4. Austin, *A Summary Description of the Lead Mines*, 8, 15; Schoolcraft, *A View of the Lead Mines of Missouri*, 4–7.

5. Brackenridge, *Views of Louisiana*, 125, 256.

6. Linda Nash notes how Americans understood new landscapes not only in terms of impending resource possibility, but also in terms of health. Linda Nash, "Finishing Nature: Harmonizing Bodies and Environments in Late-Nineteenth-Century California," *Environmental History* 8, no. 1 (January 2003): 25.

7. Recently, environmental historians have begun to examine the scale of material flows among various landscapes to understand more fully how the past has shaped modern environments. Valenčius, *The Health of the Country*, 137–41.; Gregg Mitman, Michelle Murphy, and Christopher Sellers, eds., "Landscapes of Exposure: Knowledge and Illness in Modern Environments," *Osiris*, 2nd Series, vol. 19, (2004).

8. Brackenridge, *Views of Louisiana*, 125, 256.

9. Schoolcraft, *A View of the Lead Mines of Missouri*, 4, 90–112.

10. For example, Nathan Haley, an English immigrant, described in letters to his parents how he could easily apply a "pick and wooden shovel to discovered ore," like earlier miners. Erickson, *Invisible Immigrants*, 411–20.

11. Schoolcraft, *A View of the Lead Mines of Missouri*, 4.

12. Miller, *A Brief Retrospect of the Eighteenth Century*, vii-xvi.

13. Brackenridge, *Views of Louisiana*, 150–151.
14. Brackenridge, *Views of Louisiana*, 20.
15. Brackenridge, *Views of Louisiana*, 4, 268.
16. Brackenridge, *Views of Louisiana*, 269.
17. Brackenridge, *Views of Louisiana*, 236.
18. Austin, *A Summary Description of the Lead Mines*, 10.
19. Schoolcraft, *A View of the Lead Mines of Missouri*, 133; Austin, *A Summary Description of the Lead Mines*, 16, 19.
20. Lewis Fields Linn Papers, Box 2, Folder 4, Missouri Historical Society, St. Louis.
21. Thompson, *A History of Madison County Missouri*, 40.
22. Thompson, *A History of Madison County Missouri*, 77.
23. Thompson, *A History of Madison County Missouri*, 78–79.
24. Rozier Family Papers, Box 3, Folder 2, Unpublished Journal No. 8, June 28, 1832, to July 20, 1838, Missouri Historical Society, St. Louis.
25. For personal travel and lead shipments between New Orleans and Ste. Geneviève, see Henry Rozier Ledger. Henry Rozier Ledger, Missouri Historical Society, St. Louis.
26. Rozier Family Papers, Box 3, Folder 2, Unpublished Journal, 1. Henry Rozier, "Old Ste. Genevieve," manuscript, Missouri Historical Society, St. Louis
27. Rozier Family Papers, Box 3, Folder 2, Unpublished Journal, 3.
28. Rozier Family Papers, Box 3, Folder 2, Unpublished Journal, 17.
29. Rozier Family Papers, Box 3, Folder 2, Unpublished Journal. 10.
30. Ure, *A Dictionary of Arts, Manufactures, and Mines*, 830–31.
31. Rozier Family Papers, Box 3, Folder 2, Unpublished Journal, 14.
32. Rozier Family Papers, Box 3, Folder 2, Unpublished Journal, 17.
33. Rozier Family Papers, Box 3, Folder 2, Unpublished Journal, 22.
34. Rozier Family Papers, Box 3, Folder 2, Unpublished Journal, 23.
35. George William Featherstonhaugh and United States Army Corps of Engineers, *Report of a Geological Reconnoissance Made in 1835* [...] (Washington, DC: Gales and Seaton, 1836), 74, 158, 168.
36. George William Featherstonhaugh and United States Army Corps of Engineers, *Geological Report of an Examination Made in 1834 of the Elevated Country between the Missouri and Red Rivers* (Washington, DC: Gales and Seaton, 1835), 5–51, Lynn Morrow Personal Collection, Missouri Department of Natural Resources, Missouri State Archives, Jefferson City.
37. Cohen, "Surveying Nature," 37–69; Corgan, *Geological Sciences in the Antebellum South*, 9–25.
38. Featherstonhaugh, *Geological Report*, 47–51.
39. Thomas G. Clemson, "From a Discourse Delivered before the Citizens of Fredericktown [...] Governing Mine La Motte," in *Observations of the La Motte Mines and Domain in the State of Missouri* [...] by Lewis Fields Linn (Baltimore: Royston &

Brown, 1839), 22–32, Lynn Morrow Personal Collection, Missouri Department of Natural Resources, Missouri State Archives, Jefferson City.

40. Charles Gregoire, *Observations of the La Motte Mines and Domain in the State of Missouri* [...] by Lewis Fields Linn (Baltimore: Royston & Brown, 1839), 3–6, Lynn Morrow Personal Collection, Missouri Department of Natural Resources, Missouri State Archives, Jefferson City.

41. Featherstonhaugh, *Geological Report*, 45–46.

42. Clemson, "From a Discourse Delivered before the Citizens of Fredericktown," in *Observations of the La Motte Mines*, 27.

43. Clemson, "From a Discourse Delivered before the Citizens of Fredericktown," in *Observations of the La Motte Mines*, 29.

44. Clemson, "From a Discourse Delivered before the Citizens of Fredericktown," in *Observations of the La Motte Mines*, 28–30.

45. Featherstonhaugh, *Geological Report*, 48–49.

46. Featherstonhaugh, *Geological Report*, 42–51.

47. For a visual of Mine La Motte furnaces and machines, see 1839 "Map of the La Motte Copper & Lead Mines, Containing 24,010 acres," in *Observations of the La Motte Mines and Domain in the State of Missouri* [...] by Lewis Fields Linn (Baltimore: Royston & Brown, 1839), 1–6, Lynn Morrow Personal Collection, Missouri Department of Natural Resources, Missouri State Archives, Jefferson City.

48. Ingalls, *Lead and Zinc in the United States*, 38, 106; Thompson, *A History of Madison County Missouri*, 84.

49. Ingalls, *Lead and Zinc in the United States*, 63–64, 107.

50. Arrell Morgan Gibson, *Wilderness Bonanza: The Tri-State District of Missouri, Kansas, and Oklahoma* (Norman: University of Oklahoma Press, 1972), 24, 119.

51. Clemson, "From a Discourse Delivered before the Citizens of Fredericktown," in *Observations of the La Motte Mines*, 22–32.

52. Rozier Family Papers, Box 3, Folder 2, Unpublished Journal, 18.

53. Gregoire, *Observations of the La Motte Mines*, 4–5; For lead shipments between New Orleans and Ste. Geneviève and Herculaneum, see Henry Rozier Ledger. Henry Rozier Ledger, Missouri Historical Society, St. Louis.

54. Clemson, "From a Discourse Delivered before the Citizens of Fredericktown," in *Observations of the La Motte Mines*, 30–32.

55. Featherstonhaugh, *Geological Report*, 48–49.

56. For personal travel between New Orleans and Ste. Geneviève and Herculaneum, see Henry Rozier Ledger. Henry Rozier Ledger, Missouri Historical Society, St. Louis.

57. Ruby Swartzlow, "The Early History of Lead Mining in Missouri, Part IV," *Missouri Historical Review* 29, no. 2 (January 1935): 109–14; Ruby Swartzlow, "The Early History of Lead Mining in Missouri, Part V," *Missouri Historical Review* 29, no. 3 (April 1935): 195–205; Thompson, *A History of Madison County Missouri*, 74–75.

58. Gregoire, *Observations of the La Motte Mines*, 20–22

59. The report included a map of Mine La Motte, which truly depicts the mines growing into an industrial complex. See "Map of the La Motte Copper & Lead Mines, Containing 24,010 acres," in *Observations of the La Motte Mines.*

60. Rozier Family Papers, Box 3, Folder 2, Unpublished Journal, 1–4.

61. Clemson, "From a Discourse Delivered before the Citizens of Fredericktown," in *Observations of the La Motte Mines*, 32.

62. Charles-Alexandre Lesueur, *Drawings and Sketches of the Places We Passed on the Way from Philadelphia to Pittsburgh and from Pittsburgh to New Harmony During our Voyage on the Keelboat while Descending the Ohio from November 27, 1825 to January 26, 1826*, Reprinted in Adrien Loir, "Charles-Alexandre Lesueur Artiste Et Savant Français; En Amerique De 1816 a 1839" (PhD diss., Universite De Caen, 1920), 1–72.

63. Lesueur, *Drawings and Sketches*, Plate XVII, [reprinted] in Loir 1920.

64. Lesueur, *Drawings and Sketches*, Plate XIX, [reprinted] in Loir 1920.

65. Lesueur, *Drawings and Sketches*, Plate XIX, [reprinted] in Loir 1920.

66. Quoted in William E. Foley, *The Genesis of Missouri: From Wilderness Outpost to Statehood* (Columbia: University of Missouri Press, 1989), 120.

67. Ekberg, *Colonial Ste. Geneviève*, 116.

68. Foley, *The Genesis of Missouri*, 120–121; K.W. Townsend, *First Americans: A History of Native Peoples (2nd ed.)* (New York: Routledge, 2019), 229–230.

69. Andrew Jackson, Fifth Annual Message, December 3, 1833, in James D. Richardson, ed., *A Compilation of Messages and Papers of the Presidents*, 11 vols. (New York: Published by authority of Congress, 1896–1899), 3:1021–22.

70. Delegate Scott to the Secretary of War, September 21, 1820, *TP* 15: 645–46; Clarence E. Carter, vol 13, Louisiana-Missouri, 1803–1806 (1948.), Washington, DC: GPO, 1950; Charles J. Kappler, *Indian Affairs: Laws and Treaties*, vol. 2, *Treaties* (Washington, DC: GPO, 1904), 262–64, 304–5, 370–72; *Statutes at Large of USA* 7 (1832): 397–99; Schroeder, *Opening the Ozarks.*

71. Austin, *A Summary Description of the Lead Mines*, 17–22.

72. Schoolcraft, *A View of the Lead Mines of Missouri*, 134–49.

73. Wetherill Collection, Letter book, Item # 1: June 3, 1789–December 5, 1796, Item # 2: 1815–1829.

74. Miriam Hussey, *From Merchants to Colour Men: Five Generations of Samuel Wetherill's White Lead Business* (Philadelphia: University of Pennsylvania Press, 1956), xiv & 4.

75. Wetherill Collection, Letter book, Item #1, October 29, 1789, to Dr. Young Eagle; Letter book, Item #1, January 10, 1810, to Brandram, Templeman & Co., London.

76. One example in the Wetherill letters shows that William Nicholson spent a great deal of time outlining the correct procedure to manufacture red lead, see Nicholson, *A Dictionary of Chemistry.*

77. Wetherill Collection, Letter book, Item #1, June 3, 1798, to Brandram, Templeman, & Jacques, London. This letter confirms the interest of the Wetherill brothers in manufacturing lead products.

78. Hussey and Wharton School Industrial Research Unit, The Wetherill Papers, 1762–1899, 12–23.

79. Wetherill Collection, Letter book, Item #1, July 8, 1808.

80. Wetherill Daybook 21, September 19, 1809, 30. The only James Lyle listed in the city directories of 1809 and 1810 is a merchant. A description of the early nineteenth-century factory appears on June 14, 1845, billhead. Ten sheds were on the property at 12th and Cherry Streets in the 1820s. Sheds contained logwood, straw, linseed oil, set white lead bed, vinegar cistern, white lead barrels, and bricks; there were also horse stables. Wetherill Daybook 70, June 14, 1845.

81. Cutbush, *American Artist's Manual*, "Laboratory" section.

82. Hussey, *From Merchants to Colour Men*, 76–78.

83. Cutbush, *American Artist's Manual*, "Lead" section.

84. Ure, *A Dictionary of Arts, Manufactures and Mines*, 572–73, 744–45; Henry R. Schoolcraft researched and documented the manufacture of white and red lead, and glass. See Schoolcraft Papers, Container 82.

85. Cutbush, *American Artist's Manual*, "Lead" section; Also, Thomas Andrews used the term *workscapes* for the interplay between human labor and the environment. See Thomas G. Andrews, *Killing for Coal: America's Deadliest Labor War* (Cambridge, MA: Harvard University Press, 2008), 125.

86. Wetherill Collection, Letter book, Item #1, July 8, 1812, to Benjamin Morgan, New Orleans.

87. Wetherill Collection, Letter book, Item #1, January 26, 1813, Henry Thompson, Baltimore.

88. Wetherill Collection, Letter book, Item #1, February 12, 1813, John Kipp, Baltimore.

89. Wetherill Collection, Letter book, Item #1, June 13, 1815, Benjamin Morgan, New Orleans.

90. Wetherill Collection, Letter book, Item #4, March 30, 1835, John P. Wetherill.

91. Beck, *A Gazetteer of the State of Illinois and Missouri*, 256–58.

92. Gregoire made introductions, *Observations of the La Motte Mines*, 5–6.

93. Leland Dewitt Baldwin, *The Keelboat Age on Western Waters*, (Pittsburgh: University of Pittsburgh Press, 1941).

94. Frederick J. Dobney, *River Engineers on the Middle Mississippi: A History of the St. Louis District, U.S. Army Corps of Engineers* (Washington, DC: GPO, 1978), 13–14

95. Wetherill outlines the water route in Wetherill Production Records, Item #25, Routing of Pig Lead, July 10, 1835.

96. Wetherill Miscellaneous, Item #22, "Factory Hauling," 1838.

97. Wetherill Letter book, Item #1, January 2, 1813.

98. Brackenridge, *Views of Louisiana*, 262–63.

99. Brackenridge, *Views of Louisiana*, 207.

100. See Bernardino Ramazzini, *A Treatise of the Diseases of Tradesmen, Shewing the Various Influence of Particular Trades upon the State of Health* [...] (London: 1705); Thomas Percival, *Observations and Experiments on the Poison of Lead* (London: Printed for J. Johnson, 1774).

101. Schroeder, *Opening the Ozarks*, 51.

102. Brackenridge, *Views of Louisiana*, 262.

103. Brackenridge, *Views of Louisiana*, 73.

104. Anthony Fothergill, "General Effects of the Poison of Lead," in *Cautions to the Heads of Families, In Three Essays* (Bath, UK: Printed by R. Cruttwell, 1790), http://books.google.com/books?id=vT4HFtF6HZMC&pg=PA33&dq.

105. Schoolcraft, *A View of the Lead Mines of Missouri*, 30–31.

106. Valenčius, *The Health of the Country*.

107. Clemson, "From a Discourse Delivered before the Citizens of Fredericktown," in *Observations of the La Motte Mines*, 24.

108. Clemson, "From a Discourse Delivered before the Citizens of Fredericktown," in *Observations of the La Motte Mines*, 26.

109. Clemson, "From a Discourse Delivered before the Citizens of Fredericktown," in *Observations of the La Motte Mines*, 30.

110. Clemson, "From a Discourse Delivered before the Citizens of Fredericktown," in *Observations of the La Motte Mines*, 22–32.

111. Hardage Lane, in Hardage Lane Scrapbook, Missouri Historical Society, St. Louis; Biographical information, Max A. Goldstein, *One Hundred Years of Medicine and Surgery in Missouri* [...] (St. Louis, MO: St. Louis Star, 1900); See also Carey P. McCord, "Lead and Lead Poisoning in Early America: Lead Mines and Lead Poisoning," *Industrial Medicine and Surgery* 22, no. 11 (1953): 534–39.

112. Lane, in Hardage Lane Scrapbook, Missouri Historical Society, St. Louis.

113. Hardage Lane, "Case of Poison from Lead," *St. Louis Medical and Surgical Journal* 1, no. 2, (May 15, 1843): 32.

114. Lane, "Case of Poison from Lead," 33.

115. Lane, in Hardage Lane Scrapbook, Missouri Historical Society, St. Louis.

116. Terri L. Snyder, "Suicide, Slavery, and Memory in North America," *Journal of American History* 97, no. 1 (2010): 39–62, http://www.jstor.org/stable/40662817; Herbert G. Gutman, *Work, Culture, and Society in Industrializing America: Essays in American Working-Class and Social History* (New York: Alfred A. Knopf, 1976), 15–32. Similar accounts in Alan Dawley and Paul Faler, "Working-Class Culture and Politics in the Industrial Revolution: Sources of Loyalism and Rebellion," *Journal of Social History* 9 (June 1976): 466–68; and in Bruce Laurie, *Artisans into Workers: Labor in Nineteenth-Century America* (New York: University of Illinois, 1997), 37–46.

117. Fothergill, "General Effects of the Poison of Lead," 73; Ramazzini, *Diseases of Workers*; George Baker, "An Inquiry Concerning the Cause of the Endemical Colic of Devonshire," *Medical Transactions of the Royal College of Physicians* 2 (1785):175–256; Jerome O. Nriagu, *Lead and Lead Poisoning in Antiquity* (New York: John Wiley and Sons, 1983), 437.

118. Fothergill, "General Effects of the Poisson of Lead," 61.

119. For references to suicide in slave narratives, see for example, Olaudah Equiano, *Interesting Narrative of the Life of Olaudah Equiano, or Gustavas Vasa, the African, Written by Himself*, vol. 1. (London: Published for the Author, 1789), 57; Harriet Jacobs, *Incidents in the Life of a Slave Girl, Written by Herself*, ed. Maria Child (Boston: Published for the Author, 1861), 61, 77–78, esp. 122, 162; Charles Ball, *Slavery in the United States: A Narrative of the Life and Adventures of Charles Ball, a Black Man, Who Lived Forty Years in Maryland, South Carolina and Georgia, as a Slave Under Various Masters, and was One Year in the Navy with Commodore Barney, During the Late War* (New York: John S. Taylor, 1837), 329, 69.

120. Harriet C. Frazier, *Slavery and Crime in Missouri, 1773–1865*, (Jefferson, NC: McFarland, 2001), 109–11.

121. Michael Angelo Gomez, *Exchanging Our Country Marks: The Transformation of African Identities in the Colonial and Antebellum South* (Chapel Hill: University of North Carolina Press, 2001), 133–34; Huey P. Newton, *Revolutionary Suicide* (New York: Harcourt Brace Jovanovich, 1973), 1–6.

122. Stephen Skeel, "Lead Colic, or Mine Sickness," *St. Louis Medical and Surgical Journal* 6, no. 2, (1848): 125–29. Copies located at the New York Academy of Medicine Library, NY; McCord, "Lead and Lead Poisoning in Early America," 534–39.

123. Fothergill, "General Effects of the Poison of Lead," 46.

124. Skeel, "Lead Colic, or Mine Sickness," 126–27.

125. Skeel, "Lead Colic, or Mine Sickness," 127–28.

126. Skeel, "Lead Colic, or Mine Sickness," 128.

127. Rozier Family Papers, Box 3, Folder 2, Unpublished Journal, 25.

128. Rozier Family Papers, Box 3, Folder 2, Unpublished Journal, 26.

129. Rozier Family Papers, Box 3, Folder 2, Unpublished Journal, 27.

130. Thompson, *A History of Madison County Missouri*, 40.

131. The New York State Medical Society proposed as a subject the influences of trades and occupations in the United States on the production of disease. Benjamin W. McCready, *On the Influence of Trades, Professions, and Occupations in the United States, in the Production of Disease* (Originally printed Albany: E. W. and C. Skinner, 1837; reprinted Baltimore: Johns Hopkins Press, 1943), introductory essay by Genevieve Miller. The topic was most likely influenced by a book on occupational diseases. See C. Turner Thackrah, *The Effects of Arts, Trades, and Professions and of Civic States and Habits of Living on Health and Longevity: with Suggestions for the Removal of Many of the Agents Which Produce Disease, and Shorten the Duration of Life* (London: Longman, 1832).

132. This two-volume survey includes information on diseases, geography, climate, and mineral resources. Daniel Drake, *A Systematic Treatise, Historical, Etiological, and Practical, on the Principal Diseases of the Interior Valley of North America* (Cincinnati: Winthrop B. Smith, 1850).

133. Valenčius, *The Health of the Country*.

134. Brackenridge, *Views of Louisiana*, 150; Skeel, "Lead Colic, or Mine Sickness," 125–29.

CONCLUSION

1. DeKaury, Spoon. "Narrative of Spoon Decorah," ed. Reuben Gold Thwaites, *Collections of the State Historical Society of Wisconsin*, vol. 13 (Madison: Wisconsin Historical Society, 1895): 448–62.

2. DeKaury, "Narrative of Spoon Decorah," 448–62.

3. Phone conversations on September 12, 2011, with tour guide and supervisor, Missouri Mines State Historic Site of the Missouri Department of Natural Resources.

4. Phone conversation on January 17, 2007, with the State Historical Society of Missouri Research Center, Rolla, about the St. Joseph Lead Company film.

BIBLIOGRAPHY

PRIMARY SOURCE COLLECTIONS AND PAPERS

Library of Congress, Washington, DC
  Henry Rowe Schoolcraft Papers
Missouri Department of Natural Resources, Missouri State Archives,
    Jefferson City, Missouri
  Hunt, Theodore, compiler. Testimony Before the Recorder of Land Titles.
    3 vols. St. Louis, 1825.
  Lynn Morrow Personal Collection
Missouri Historical Society, St. Louis, Missouri
  Alexander Craighead Papers
  Amos Stoddard Papers
  Amoureux-Bolduc Papers
  Archives of the Indies (Seville, Spain) Collection, Missouri Historical Society,
    St. Louis
  Bates Family Papers—Frederick Bates Letter Book
  Billon Papers
  Blow Family Papers
  Census Collection: Seven items arranged chronologically. Census reports of
    Missouri only.
  Charles Gratiot, Letter Book
  Chouteau Collections
  Christian Wilt Papers
  Daniel Bissell Papers
  Delassus–St. Vrain Family Collection
  François Vallé Papers
  George Oliver Carpenter Papers
  Henry Rozier Ledger

John James Audubon Collection
Joseph Desloge Collection
Journals and Diaries Collection
Kaskaskia, Illinois, Collection
Kennett Family Papers
Lewis Fields Linn Papers
Lewis Vital Bogy Family Papers
Mines Collection
Moses Austin Collection
Pierre Chouteau, Letter Book
Rose Mary Bogy Collection
Rozier Family Papers
Ste. Genevieve Papers
Vallé Mining Company Account Books

Lane, Hardage. *Hardage Lane Scrapbook*. Missouri Historical Society, St. Louis, 1843.

National Archives, New York City
Letters Received by the Secretary of War Relative to Indian Affairs
Letters Sent by the Secretary of War Relative to Indian Affairs

Newberry Library Cartographic Collection

New York Public Library
Print Collection
Rare Books

The State Historical Society of Missouri
Western Historical Manuscripts Collection
Lead Mining Co., Washington County, Missouri, Records, 1809–1954
Mine La Motte Estate, Madison County, Missouri, Records, 1875–1897
Ste. Genevieve, Missouri, Archives, 1756–1930

University of Pennsylvania
Kislak Center for Special Collections, Rare Books and Manuscripts
Wetherill Collection

Washington University Bernard Becker Medical Library

Yale University Library
Papers of the St. Louis Fur Trade, 1725–1925
Chouteau Collection
Fur Company Ledgers & Account Books, 1802–1871

INTERVIEWS AND COMMUNICATION

Hocker, Evan. Archivist, Barker Texas History Center, University of Texas at Austin. Conversation with author. October 20, 2008.

NEWSPAPERS

Literary Repository
The New York Magazine
St. Louis Enquirer
St. Louis Missouri Gazette

UNITED STATES GOVERNMENT DOCUMENTS

American State Papers: Finance, Foreign Relations, Indian Affairs, Military Affairs, Public Lands, and Miscellaneous. 38 vols. Washington, DC: Gales and Seaton, 1832–1861.

Carter, Clarence E., ed., *The Territorial Papers of the United States.* Vol. 13, *Louisiana-Missouri, 1803–1806* (1948); Vol. 14, *Louisiana-Missouri, 1806–1814* (1949); Vol. 15, *Louisiana-Missouri, 1815–1821* (1954). Washington, DC: US Government Printing Office (GPO), 1934–1962.

Fay, Albert Hill. *A Glossary of the Mining and Mineral Industry* (Washington, DC: GPO, 1920).

Jackson, Andrew. Fifth Annual Message, December 3, 1833, in James D. Richardson, ed., *A Compilation of Messages and Papers of the Presidents*, 11 vols. (New York: Published by authority of Congress, 1896–1899), 3:1021–22.

Laws for the Territory of Louisiana. St. Louis, 1808.

Rabbitt, Mary C. *Minerals, Lands, and Geology for the Common Defense and General Welfare*, vol. 1, *Before 1879: A history of public lands, federal science and mapping policy, and development of mineral resources in the United States.* Washington, DC: U.S. Government Printing Office, 1982.

Statutes at Large of USA 7 (1832).

Wright, Carroll D., and William C. Hunt. *History and Growth of the United States Census: 1790–1890.* Prepared for Senate Committee on the Census, 56th Congress, 1st Session, February 24, 1900. Doc. 194, 17–20.

PRIMARY SOURCES

Agricola, Georgius. *De Re Metallica.* Translated by Herbert Clark Hoover and Lou Henry Hoover. New York: Dover, 1950. https://www.gutenberg.org/files/38015/38015-h/38015-h.htm

Amelung, John Frederick. *Remarks on Manufactures: Principally on the New Established Glass-House, near Frederick-Town, in the State of Maryland.* Fredricktown, MD: Printed by Matthias Bartgis for the Author, 1787.

Atwater, Caleb. *Writings of Caleb Atwater (Travel in America).* Columbus, OH: Atwater, 1833.

Austin, Moses. *A Summary Description of the Lead Mines in Upper Louisiana. Also, An Estimate of their Produce for Three Years Past.* City of Washington: A. and G. Way Printers, 1804.

Baker, George. *An Essay Concerning the Cause of the Endemical Colic of Devonshire [...]* London: Printed by J. Hughs, 1767. Reprinted in *Medical Transactions* 1 1768.

Bakewell, Benjamin G. *The Family Book of Bakewell.* Pittsburgh: W. G. Johnston, 1896.

Ball, Charles. *Slavery in the United States: A Narrative of the Life and Adventures of Charles Ball, a Black Man, Who Lived Forty Years in Maryland, South Carolina and Georgia, as a Slave Under Various Masters, and was One Year in the Navy with Commodore Barney, During the Late War.* New York: John S. Taylor, 1837.

Barba, Alvaro Alonso. *A Collection of Scarce and Valuable Treatises upon Metals, Mines and Minerals* [...] London: Printed for J. Hodges, 1684.

Barker, Eugene C., ed. *Annual Report of the American Historical Association for the Year 1919: In Two Volumes, Volume II, The Austin Papers, Part I.* Washington, DC: GPO, 1924.

Baxter, James Phinney, Jean François de La Roque Roberval, and Jean Alfonce. *A Memoir of Jacques Cartier: Sieur de Limoilou, His Voyages to the St. Lawrence.* [...] New York: Dodd, Mead, 1906.

Beck, Lewis C. *A Gazetteer of the State of Illinois and Missouri.* Albany: Charles R. and George Webster, 1823.

Beltrami, Giacomo Costantino. *A Pilgrimage in Europe and America: Leading to the Discovery of the Sources of the Mississippi and Bloody River: with a Description of the Whole Course of the Former and of the Ohio, in Two Volumes,* vol. 2. London: Hunt and Clarke, 1828.

Bingley, William. *Useful Knowledge of a Familiar and Explanatory Account of the Various Productions of Nature: Mineral, Vegetable, and Animal.* Philadelphia: A. Small, 1818.

Bolton, Herbert Eugene. *Guide to Materials for the History of the United States in the Principal Archives of Mexico.* Washington, DC: Carnegie Institution, 1913.

Boyle, Robert. *The Works of the Honorable Robert Boyle in Five Volumes.* London: Thomas Birch, 1772.

Brackenridge, Henry Marie. "Sketches of the Territory of Louisiana," Missouri Gazette, June 20, 1811.

———. *Views of Louisiana: Containing Geographical, Statistical and Historical Notices of that Vast and Important Portion of America.* Baltimore: Printed by Schaeffer & Maund, 1817.

Bradbury, John. *Travels in the Interior of America, in the Years 1809, 1810, and 1811* [...]. Liverpool: Smith and Galway, 1817.

Broadhead, Garland C. *Report of the Geological Survey of the State of Missouri, Including Field Work of 1873–1874, Missouri Geological Survey.* Jefferson City, MO: Regan & Carter, 1874.

Carver, Jonathan, John Coakley Lettsom, and Isaiah Thomas. *Three Years' Travels throughout the Interior Parts of North America, for More than Five Thousand Miles* [...]. Walpole, NH: Isaiah Thomas, 1813.

Champlain, Samuel de. *Voyages of Samuel de Champlain: 1604–1610, Volumes 11–13 of Publications of the Prince Society*, vol. 2. Translated by Charles Pomeroy Otis. Boston: Prince Society, 1878.

Chaptal, Jean-Antoine-Claude. Comte De Chanteloup. *Elements of Chemistry*, 2nd American ed., trans. William Nicholson (Philadelphia: Printed by Lang & Ustick for M. Carey, 1796.

Charless, Joseph. "At Herculaneum a Shot Manufactory is Now Erecting by an Active Enterprising Citizen of Our Territory," *Missouri Gazette*, March 8, 1809.

———. "The Subscribers Having Erected at the New Diggings District at St. Geneviève, an Assay Furnace," *Missouri Gazette*, April 12, 1809, (St. Louis, Missouri).

———. "The Erection of a Patent Shot Manufactory At Herculaneum," *Missouri Gazette*, November 16, 1809.

———. "Shipment Received at the General Store of Aaron Elliot & Son," *Missouri Gazette*, November 7, 1810.

———. "The Proper Country to Establish Manufactures of Red Lead, White Lead," *Missouri Gazette*, October 26, 1811.

———. "Arrival of Individual Gifted in the Two Sciences," *Missouri Gazette*, February 22, 1810.

Clemson, Thomas G. "From a Discourse Delivered before the Citizens of Fredericktown [. . .] Governing Mine La Motte." In *Observations of the La Motte Mines and Domain in the State of Missouri with Some Accounts of the Advantages and Inducements* [. . .] by Lewis Fields Linn, proprietor of Mine La Motte. Baltimore: Royston & Brown, 1839.

Collot, Georges-Henri-Victor. *Journey in North America, Containing a Survey of the Countries Watered by the Mississippi, Ohio, Missouri, and Other Affluing Rivers* [. . .]. 2 vols. Paris: Printed for Arthus Bertrand, 1826.

The Compleat Collier: Or, The Whole Art of Sinking, Getting, and Working, Coal-Mines [. . .] *about Sunderland and New-Castle*. London: G. Conyers, 1708.

Coxe, John Redman. "Lead" in Emporium of Arts & Sciences, vol. 2 August 1, 1814.

Coxe, Tench. *A View of the United States of America, In a Series of Papers Written at Various Times, between 1787 and 1794*. Philadelphia: Printed for William Hall, Wrigley & Berriman, 1794).

Cramer, Zadok. *The Navigator: Containing Directions for Navigating the Monongahela, Allegheny, Ohio, and Mississippi Rivers* [. . .]. Sixth Edition, Improved and Enlarged. Pittsburgh: Cramer & Spear, 1801.

Cruzat, Francis. Padron General de Los Pueblos de Sn. Luis Y Ste. Geneoveva de Ilinueses, 1787, trans. Walter B. Douglas. in the Census Collection of the Missouri Historical Society.

Cutbush, James. *The American Artist's Manual: Dictionary of Practical Knowledge in the Application of Philosophy to the Arts and Manufactures* [. . .]. Philadelphia: Johnson & Warner, and R. Fisher, 1814.

Dalton, Mary. "Notes on the Genealogy of the Vallé Family." *Missouri Historical Society Collections* 2, no. 7 (1906), 54–82.

De Witt, Benjamin. "Mineral Productions of the State of New York." *Memoirs of the American Academy of Arts and Sciences* 2, no. 2 (1804): 73–81.

DeKaury, Spoon. "Narrative of Spoon Decorah." Edited by Reuben Gold Thwaites. *Collections of the State Historical Society of Wisconsin* 13 (1895):448–62. Online facsimile, http://www.wisconsinhistory.org/turningpoints/search.asp?id=34.

Des Ursin, "Relation of the Journey to the Mines in Illinois Country; John E. Rothensteiner, "Earliest History of Mine La Motte," The Missouri Historical Review, vol. 10, January 1926, 199-209.

Dobson, Thomas. *A Compendious System of Mineralogy & Metallurgy: Extracted from the American Edition of the Encyclopaedia* [...]. Philadelphia: Printed by Thomas Dobson, 1794.

Drake, Daniel. *A Systematic Treatise, Historical, Etiological, and Practical, on the Principal Diseases of the Interior Valley of North America*. Cincinnati, OH: Winthrop B. Smith, 1850.

Eissler, M. *The Metallurgy of Silver: A Practical Treatise on the Amalgamation, Roasting, and Lixiviation of Silver Ores: Including the Assaying, Melting, and Refining of Silver Bullion*. London: Crosby Lockwood and Son, 1889.

Equiano, Olaudah. *Interesting Narrative of the Life of Olaudah Equiano, or Gustavas Vasa, the African, Written by Himself, vol. 1*. London: Published for the Author, 1789.

Featherstonhaugh, George William and United States Army Corps of Engineers. *Geological Report of an Examination Made in 1834 of the Elevated Country between the Missouri and Red Rivers*. Washington, DC: Gales and Seaton, 1835.

———. *Report of a Geological Reconnoissance Made in 1835, from the Seat of Government, by Way of Green Bay and the Wisconsin Territory to the Coteau de Prairie, an Elevated Ridge Dividing the Missouri from the St. Peter's River*. Washington, DC: Gales and Seaton, 1836.

Flint, Timothy. *The History and Geography of the Mississippi Valley, to Which is Appended a Condensed Physical Geography of the Atlantic United States and the Whole American Continent*. 2nd ed., 2 vols. Cincinnati: E. H. Flint and L. R. Lincoln, 1832.

Fothergill, Anthony. "General Effects of the Poison of Lead." In *Cautions to the Heads of Families, In Three Essays*. Bath, UK: Printed by R. Cruttwell, 1790. http://books.google.com/books?id=vT4HFtF6HZMC&pg=PA33&dq.

Founders Online, National Archives, https://founders.archives.gov/documents/Jefferson/01-43-02-0504. [Original source: *The Papers of Thomas Jefferson*, vol. 43, 11 March–30 June 1804, ed. James P. McClure, Princeton, NJ: Princeton University Press, 2017, 609–10.

"The French Regime in Wisconsin–I: 1634–1727," in *Collections of the State Historical Society of Wisconsin*, vol.16, Madison: State Historical Society of Wisconsin, 1925.

Godon, Sylvain. "Observations to Serve for the Mineralogical Map of the State of Maryland." *American Philosophical Society Transactions* 6 (1809): 319–23.

Haricot, Thomas. *A Brief and True Report of the New Found Land of Virginia, of the Commodities and of the Nature and Manners of the Natural Inhabitants*. Translated by Richard Hakluyt. New York: J. Sabin & Sons, 1871. http://docsouth.unc.edu/nc/hariot/hariot.html.

Henfrey, Benjamin. *A Plan with Proposals for Forming a Company to Work Mines in the United States; and to Smelt and Refine the Ores Whether of Copper, Lead, Tin, Silver, or Gold*. Philadelphia: Snowden & M'Corkle, 1797.

Houghton, Thomas. *Rara avis in terris, or, The Compleat Miner, In Two Books; the First Containing the Liberties, Laws and Customs, of the Lead-Mines, within [...] Wirksworth in Derbyshire in Fifty Nine Articles [...]. The Second Teacheth the Art of Dialling and Levelling*. Derby, UK: Printed by Samuel Hodgkinson, 1729.

———. *Royal Institutions: Being Proposals for Articles to Establish and Confirm Laws, Liberties, & Customs of Silver & Gold Mines [...], the Digging and Getting of Lead [...]*. London: Printed for Author, 1694.

Humboldt, Alexander von. *Political Essay on the Kingdom of New Spain [...]; with Physical Sections and Maps [...]*, vol. 3. Translated by John Black. London: Longman, 1811.

Hussey, Miriam, and Wharton School Industrial Research Unit. *The Wetherill Papers, 1762–1899; Being the Collection of Business Records of the Store and White Lead Works Founded by Samuel Wetherill in the Late Eighteenth Century [...]*. Philadelphia: Industrial Research Dept., Wharton School of Finance and Commerce, University of Pennsylvania, 1942.

Jacobs, Harriet. *Incidents in the Life of a Slave Girl, Written by Herself*, ed. Maria Child, Boston: Published for the Author, 1861.

Jefferson, Thomas. *Notes on the State of Virginia*. London: J. Stockdale, 1787.

Johnson, Esau Johnson. "Reminiscence," in *A Chronological History of Indian Lead Mining in the Upper Mississippi Valley from 1643 to 1848*, by Philip Millhouse. Unpublished Paper for History Special Projects, Galena Public Library, Galena, IL, 1993.

Kirwan, Richard. *Elements of Mineralogy in Two Volumes*. London: Printed by J. Nichols for P. Elmsly, 1794–96.

La Renaudière "Relation of the Journey to the Mines in Illinois Country; John E. Rothensteiner, "Earliest History of Mine La Motte," *The Missouri Historical Review*, vol. 10. January 1926, 199-209.

Lahontan, Louis Armand Baron. *New Voyages to North America Containing an Account of the Several Nations of That Vast Continent; Their Customs, Commerce, and the Way of Navigation upon Lakes and Rivers [...]*. 2 vols. London: Printed for H. Bonwicke and others, 1703.

Lamborn, Robert H. *A Rudimentary Treatise on the Metallurgy of Silver and Lead: Containing a Description of the Argentiferous and Plumbiferous Minerals [...]*. London: John Weale, 1861.

Lesueur, Charles-Alexandre. *Drawings and Sketches of the Places We Passed on the Way from Philadelphia to Pittsburgh and from Pittsburgh to New Harmony during our Voyage on the Keelboat While Descending the Ohio from November 27, 1825 to January 26, 1826.* The Natural History Museum of Havre, France.

Linn, Lewis Fields. *Observations of the La Motte Mines and Domain in the State of Missouri with Some Accounts of the Advantages and Inducements* [...]. Baltimore: Royston & Brown, 1839.

Lockwood, James H. *Early Times and Events in Wisconsin.* Madison: Wisconsin State Historical Society, 1856.

Loir, Adrien. "Charles-Alexandre Lesueur Artiste Et Savant Français; En Amerique De 1816 a 1839," (PhD diss., Universite De Caen, 1920).

Lowrie, Walter, ed. *American State Papers: Documents, Legislative and Executive, of the Congress of the United States in Relation to the Public Lands from the First Session of the First Congress to the First Session of the Twenty-Third Congress, March 4, 1789 to June 15, 1834.* Washington, DC: Printed by D. Green, 1834.

Maclure, William. "Observations on the Geology of the United States, Explanatory of a Geological Map." *Transactions of the American Philosophical Society* 6 (1809): 411–28.

Margry, Pierre. *Discoveries and Settlements of the French in the West and in the South of North America (1614–1754): Memories and Original Documents Collected and Published.* Paris: Imprimerie D. Jouaust, 1876–86.

Merat, François Victor. in Matthieu Joseph Bonaventure Orfila, *General System of Toxicology, or, A Treatise on Poisons, Found in the Mineral, Vegetable, and Animal Kingdoms* [...], trans. John A. Waller and Joseph G. Nancrede, Philadelphia: M. Carey & Son, 1817. https://archive.org/details/generalsystemoftooorfi.

Miller, Samuel. *A Brief Retrospect of the Eighteenth Century, Part First in Two Volumes Containing a Sketch of the Revolutions and Improvements in Science, Arts, and Literature During That Period.* Philadelphia: T. and J. Swords, 1803.

Nicholson, William. *A Dictionary of Chemistry, Exhibiting the Present State of the Theory and Practice of That Science, its Application to Natural Philosophy, the Processes of Manufactures, Metallurgy, and Numerous others Arts* [...]. 2 vols. London: G. G. and J. Robinson, 1795.

Percival, Thomas. *Observations and Experiments on the Poison of Lead.* London: Printed for J. Johnson, 1774.

Plattes, Gabriel. *A Discovery of Subterraneal Treasure: of All Manner of Mines and Minerals* [...]. London: Printed for J. E.; sold by Humphrey Moseley, 1653.

Proposals for Establishing an Association for Working Mines and Manufacturing Metals in the United States. Philadelphia: Printed by Samuel H. Smith, 1796.

Pryce, W. *Mineralogia Cornubiensis: A Treatise on Minerals, Mines, and Mining* [...] *to Which is Added, an Explanation of the Terms and Idioms of Miners.* London: James Phillips, 1778.

Ramazzini, Bernardino. *De morbis artificum diatriba: Diseases of Workers.* Translated and annotated by Wilmer Cave Wright. Chicago: University of Chicago Press, 1940.

———. *A Treatise of the Diseases of Tradesmen, Shewing the Various Influence of*

Particular Trades upon the State of Health [. . .]. London: Printed for Andrew Bell and others, 1705.

Roosevelt, Nicholas, Jacob Mark, Edward Livingston, and William Langworthy. *Papers, Relative to an Application to Congress, for an Exclusive Right of Searching for and Working Mines, in the North-West and South-West Territory.* Philadelphia: Printed by Samuel H. Smith, 1797.

Schoolcraft, Henry Rowe. *A View of the Lead Mines of Missouri, Including Some Observations on the Mineralogy, Geology, Geography, Antiquities, Soil, Climate* [. . .] *of Missouri and Arkansas, and Other Sections of the Western Country.* New York: Wiley, 1819.

Skeel, Stephen. "Lead Colic, or Mine Sickness," *St. Louis Medical and Surgical Journal* 6, no. 2, (1848): 125–29.

Smith, Adam. *An Inquiry into the Nature and Causes of the Wealth of Nations*, vol. 2. London: W. Strahan and T. Cadell, 1776.

Stockhausen, Samuel. *Traite des mauvais effets de la fumee de la litharge.* Edited by Nicolás Ruault. Paris: Chez Ruault, 1776; Reprint Nabu Press, 2012.

Stoddard, Amos. *Sketches, Historical and Descriptive, of Louisiana* [. . .]. Philadelphia: M. Carey, 1812.

Tanquerel Des Planches, L. *Lead Diseases: A Treatise from the French. With Notes and Additions on the Use of Lead Pipe and Its Substitutes.* Translated by Samuel L. Dana. Boston: Lowell, MA: D. Bixby, 1848.

Thackrah, C. Turner. *The Effects of Arts, Trades, and Professions and of Civic States and Habits of Living on Health and Longevity: with Suggestions for the Removal of Many of the Agents Which Produce Disease, and Shorten the Duration of Life.* London: Longman, 1832.

Thomson, Thomas. *A System of Chemistry in Four Volumes*, vol. 1. Edinburgh: Bell & Bradfute; sold by J. Murray, 1802.

Thwaites, Reuben Gold. *How George Rogers Clark Won the Northwest: And Other Essays in Western History.* Nabu Press, 2010.

———. *The Jesuit Relations and Allied Documents: Travels and Explorations of the Jesuit Missionaries in New France, 1610–1791: The Original French, Latin, and Italian Texts with English Translations and Notes* [. . .]. Vol. 63, *Lower Canada, Iroquois: 1667–1687.* New York: Pageant Book, 1959.

———. *The Jesuit Relations and Allied Documents: Travels and Explorations of the Jesuit Missionaries in New France, 1610–1791: The Original French, Latin, and Italian Texts with English Translations and Notes* [. . .]. Vol. 65, *Lower Canada, Mississippi Valley: 1696–1702.* Cleveland, OH: Burrows Brothers, 1900. http://moses.creighton.edu/kripke/jesuitrelations/relations_65.html.

———. "Notes on Early Lead Mining in the Fever (or Galena) River Region." *Collections of the State Historical Society of Wisconsin* 13 (1895): 271–92.

Tissot, Samuel-Auguste David. *Advice to the People in General, with Regard to Their Health* [. . .]. Translated by James Kirkpatrick. London: T. Becket, 1765; Reprint Philadelphia: J. Sparhawk, 1771.

Tylecote. R. F. *The Early History of Metallurgy*, 2nd ed., London: Institute of Materials, 1992.

Ure, Andrew. *A Dictionary of Arts, Manufactures, and Mines: Containing a Clear Exposition of Their Principles and Practice.* London: Longman, Orme, Brown, Greene, & Longmans, 1839.

Woodhouse, James. *The Young Chemist's Pocket Companion Connected with a Portable Laboratory; Containing a Philosophical Apparatus, and a Great Number of Chemical Agents* [. . .]. Philadelphia: J. H. Oswald, 1797.

SECONDARY SOURCE MATERIAL

Abbott, Collamer M. "Colonial Copper Mines." *The William and Mary Quarterly*, 3rd Series, vol. 27, no. 2 (April 1970): 295–309.

Adams, Sean P. *Old Dominion, Industrial Commonwealth: Coal Politics, and Economy in Antebellum America.* Baltimore: Johns Hopkins University Press, 2004.

Adas, Michael. *Machines as the Measure of Men: Science, Technology, and Ideologies of Western Dominance.* Ithaca, NY: Cornell University Press, 1989.

Adelman, Jeremy and Aron, Stephen. "From Borderlands to Borders: Empires, Nation-States, and the Peoples in Between in North American History." *American Historical Review* 104, no. 3 (June 1999): 814–41.

Agricola, Georgius. *De Re Metallica, Translated from the First Latin Edition of 1556, with Biographical Introduction, Annotations and Appendices upon the Development of Mining Methods, Metallurgical Processes, Geology, Mineralogy & Mining Law from the Earliest Times to the 16th Century.* Translated by Herbert Clark Hoover and Lou Henry Hoover. Reprint, New York: Dover, 1950.

Aitchison, Leslie. *A History of Metals.* London: MacDonald & Evans, 1960.

Alanen, Arnold R. "Documenting the Physical and Social Characteristics of Mining and Resource-Based Communities." *APT Bulletin* 11 (1979): 49–68.

Ali, Saleem H. *Mining, the Environment, and Indigenous Development Conflicts.* Tucson: University of Arizona Press, 2003.

Allen, Michael. "The Lower Mississippi in 1803: The Travelers' View." *Missouri Historical Review* 77 (1983): 253–71.

Allgor, Catherine. *Parlor Politics: In Which the Ladies of Washington Help to Build a City and a Government.* Charlotte: University Press of Virginia, 2000.

Alvord, Clarence Walworth. *The Illinois Country, 1673–1818.* Springfield: Illinois Centennial Commission, 1920.

Anderson, Hattie M. "Missouri, 1804-1828: Peopling a Frontier State." *Missouri Historical Review* 31 (1937): 150–80.

Andrews, Thomas G. *Killing for Coal: America's Deadliest Labor War.* Cambridge, MA: Harvard University Press, 2008.

Appadurai, Arjun, ed. *The Social Life of Things: Commodities in Cultural Perspective.* New York: Cambridge University Press, 1986.
Appleby, Joyce. *Capitalism and a New Social Order: The Republican Vision of the 1790s.* New York: New York University Press, 1984.
———. *Inheriting the Revolution: The First Generation of Americans.* Cambridge, MA: Belknap Press, 2000.
Armitage, David, and Braddick, Michael J., eds. *The British Atlantic World, 1500–1800.* New York: Palgrave Macmillan, 2002.
Aron, Stephen. *American Confluence: The Missouri Frontier from Borderland to Border States.* Bloomington: Indiana University Press, 2006.
Ashby, Leroy. *With Amusement for All: A History of American Popular Culture Since 1830.* Lexington: University Press of Kentucky, 2006.
Assadourian, Carlos Sempat. "The Colonial Economy: The Transfer of the European System of Production to New Spain and Peru." *Journal of Latin American Studies* 24 (1992): 55–68.
Avalos-Lozano, Antonio and Aguilar-Robledo, Miguel. "Reconstructing the Environmental History of Colonial Mining: The Real del Catorce Mining District, Northeast New Spain/Mexico, Eighteenth and Nineteenth Centuries." in *Mining North America: An Environmental History Since 1522,* eds. John Robert McNeill and George Vrtis. Oakland: University of California Press, 2017.
Axtell, James. *The Invasion Within: The Contest of Cultures in Colonial North America.* New York: Oxford University Press, 1985.
Baird, Robert. *View of the Valley of the Mississippi, or the Emigrant's and Traveller's Guide to the West* [. . .], 2nd ed. Philadelphia: H. S. Tanner, 1834.
Bakewell, Peter J. *Mines of Silver and Gold in the Americas.* Gower House–Brookfield, VT: Variorum, 1997.
Baldwin, Leland Dewitt. *The Keelboat Age on Western Waters.* Pittsburgh: University of Pittsburgh Press, 1941.
Bancroft, Hubert Howe. *The Native Races of the Pacific States,* vol. 4. New York: A. L. Bancroft, 1875.
Barger, Harold, and Sam H. Schurr. *The Mining Industries, 1899–1939: A Study of Output, Employment, and Productivity.* Originally published, New York: National Bureau of Economic Research, 1944; Reprint, Arno Press, 1975.
Bauxar, J. Joseph. "The Historic Period." *Illinois Archaeology,* Bulletin no.1 (1973): 40–58.
———. "History of the Illinois Area," in *Handbook of North American Indians,* vol. 15, *Northeast,* ed. Bruce G. Trigger. Washington, DC: Smithsonian Institution, 1978. 594–601
Beesley, David. *Crow's Range: An Environmental History of the Sierra Nevada.* Reno: University of Nevada Press, 2004.
Bek, William G. "George Engelman, Man of Science." Parts 1–4. *Missouri Historical Review* 23 (1929): 167–206, 427–46, 514–35; 24 (1929): 66–86.

Belting, Natalia Maree. *Kaskaskia under the French Regime*. Urbana: University of Illinois Press, 1948.

Benavides, Alonso de. *The Memorial of Fray Alonso de Benavides, 1630*. trans. Mrs. Edward Ayer. Chicago: privately printed, 1916.

Benson, Maxine, Nicolas de Finiels, Carl J. Ekberg, and William E. Foley. "An Account of Upper Louisiana." *Journal of the Early Republic* 10, no. 3 (1990): 427-428.

Berkeley, Edmund, and Dorothy Smith Berkeley. *George William Featherstonhaugh: The First U.S. Government Geologist*. History of American Science and Technology Series, ed. Lester D. Stephens. Tuscaloosa: University of Alabama Press, 1988.

Berlin, Ira. *Many Thousands Gone: The First Two Centuries of Slavery in North America*. Cambridge, MA: Harvard University Press, 1998.

Bice, Richard A., Phyllis S. Davis, and William M. Sundt. *Indian Mining of Lead for Use in Rio Grande Glaze Paint: Report of the AS-5 Bethsheba Project Near Cerrillos, New Mexico*. Albuquerque, NM: Archaeological Society, 2003.

Billon, Federic Louis. *Annals of St. Louis in Its Territorial Days, from 1804 to 1821: Being a Continuation of the Author's Previous Work, the Annals of the French and Spanish Period*. St. Louis: Printed for the Author, 1888.

Bining, Arthur C. *Pennsylvania Iron Manufacture in the Eighteenth Century*. Harrisburg: Pennsylvania Historical and Museum Commission, 1973.

Bird, Broxton W. et al. "Pre-Columbian Lead Pollution from Native American Galena Processing and Land Use in the Midcontinental United States." *Geology* 47. no.12 2019.

Birk, Charles R. "Shortest Route to the Galena Lead Mines: The Lewistown Road." *Journal of the Illinois State Historical* Society 66, no. 2 (1973): 187–97.

Blumin, Stuart M. *The Emergence of the Middle Class: Social Experience in the American City, 1760–1900*. Cambridge: Cambridge University Press, 1989.

Boydston, Jeanne. *Home & Work: Housework, Wages, and the Ideology of Labor in the Early Republic*. New York: Oxford University Press, 1990.

Brading, David A. "Mexican Silver Mining in the Eighteenth Century: The Revival of Zacatecas." in *Mines of Silver and Gold in the Americas*. ed. Peter J. Bakewell. Gower House–Brookfield, VT: Variorum, 1997.

Bray, Robert T. "The Missouri Indian Tribe in Archaeology and History." *Missouri Historical Review* 55 (April 1961): 213–25.

Brooks, James F. *Captives and Cousins: Slavery, Kinship, and Community in the Southwest Borderlands*. Chapel Hill: University of North Carolina Press, 2002.

Brosnan, Kathleen. *Uniting Mountain and Plain: Cities, Law and Environmental Change Along the Front Range*. Albuquerque, NM: University of New Mexico Press, 2002.

Brown, Margaret B. "The Zimmerman Site: Further Excavations at the Grand Village of Kaskaskia." *Illinois State Museum Reports of Investigations* 9 (1975), 116–20.

Brown, Margaret Kimball, and Lawrie Cena Dean, eds. *The Village of Chartres in Colonial Illinois 1720–1765*. New Orleans: Polyanthos, Published for La Compagnie des Amis de Fort de Chartres, 1977.

———. *The Village of Chartres in Colonial Illinois, 1720–1765.* Baton Rouge, LA: Provincial Press, 2010.

Brown, Ronald C. *Hard Rock Miners: The Intermountain West, 1860–1920.* College Station: Texas A&M University Press, 1979.

Browner, Tara. *Heartbeat of the People: Music and Dance of the Northern Pow-Wow.* Urbana: University of Illinois Press, 2002.

Buechler, Jeff, ed. *Proceedings of the Workshop on Historic Mining Resources, Defining the Research Questions for Evaluation and Preservation.* Vermillion, SD: State Historic Preservation Center, 1988.

Burkett, Paul. *Marx and Nature: A Red and Green Perspective.* New York: St. Martin's Press, 1999.

Burns, William E. *The Scientific Revolution in Global Perspective.* New York: Oxford University Press, 2016.

Bushman, Richard. *The Refinement of America: Persons, Houses, Cities.* New York: Vintage Books, 1993.

———. "Markets and Composite Farms in Early America" *William and Mary Quarterly,* 3rd Ser., 55 (July 1998): 351–74.

Cadigan, Sean. "The Moral Economy of the Commons: Ecology and Equity in the Newfoundland Cod Fishery, 1815–1855." *Labour/Le Travail* 43 (1999): 9–42.

Calloway, Colin G. *New Worlds for All: Indians, Europeans, and the Remaking of Early America.* Baltimore: Johns Hopkins University Press, 1997.

Canny, Nicholas, and Anthony Pagden, eds. *Colonial Identity in the Atlantic World, 1500–1800.* Princeton, NJ: Princeton University Press, 1987.

Carney, Judith Ann. *Black Rice: The African Origins of Rice Cultivation in the Americas.* Cambridge, MA: Harvard University Press, 2002.

Carr, Lois Green, Russell R. Menard, and Lorena S. Walsh. *Robert Cole's World: Agriculture and Society in Early Maryland.* Chapel Hill: Published for the Institute of Early American History and Culture by the University of North Carolina Press, 1991.

Caswell, L. R., and R. W. S. Daley. "The Delhuyar Brothers, Tungsten, and Spanish Silver." *Bull. Hist. Chem.* 23 (1999): 11–19.

Cayton, Andrew R. L., and Fredrika J. Teute, eds. *Contact Points: American Frontiers from the Mohawk Valley to the Mississippi, 1750–1830.* Chapel Hill: University of North Carolina Press, 1998.

Chandler, Alfred D., Jr. *The Visible Hand: The Managerial Revolution in American Business.* Cambridge, MA: Belknap Press, 1977.

Chaplin, Joyce. *The First Scientific American: Benjamin Franklin and the Pursuit of Genius.* New York: Basic Books, 2006.

Chapman, Carl H. "A Preliminary Survey of Missouri Archaeology. Part I. Historic Indian Tribes." *Missouri Archaeologist* 10 (1946): 19–20.

———. "The Little Osage and Missouri Indian Village Sites." *Missouri Archaeologist* 21 (1959): 1–67.

———. "Osage Village Locations and Hunting Territories to 1808." In *Osage Indians IV: A Preliminary Survey of Missouri Archaeology*, edited by Carl H. Chapman and D. R. Henning, 17–30. New York: Garland Publishing, 1974.

Chapman, Carl H. and Eleanor F. Chapman. *Indians and Archaeology of Missouri*. Columbia: University of Missouri Press, 1983.

Chew, Richard S. "Certain Victims of an International Contagion: The Panic of 1797 and the Hard Times of the Late 1790s in Baltimore." *Journal of the Early Republic* 25, no. 4. 2005: 565–613.

Childs, S. Terry, and David Killick. "Indigenous African Metallurgy: Nature and Culture." *Annual Review of Anthropology* 22, no. 1 (October 1993): 317–37.

Christensen, Lawrence O. William E. Foley, Gary R. Kremer, and Kenneth H. Winn, eds. *Dictionary of Missouri Biography*. Columbia: University of Missouri Press, 1999, 761–62.

Clark, Christopher. *The Roots of Rural Capitalism: Western Massachusetts, 1780–1860*. Ithaca, NY: Cornell University Press, 1990.

Clark, Claudia. *Radium Girls: Women and Industrial Health Reform, 1910–1935*. Chapel Hill: University of North Carolina Press, 1997.

Cleary, Patricia. "Contested Terrain: Environmental Agendas and Settlement Choices in Colonial St. Louis." In *Common Fields: An Environmental History of St. Louis*, edited by Andrew Hurley, 58–72. St. Louis: Missouri Historical Society Press, 1997.

Cochra, Thomas C. *Frontiers of Change: Early Industrialism in America*. Oxford: Oxford University Press, 1981.

Cohen, Benjamin R. "Surveying Nature: Environmental Dimensions of Virginia's First Scientific Survey, 1835–1842." *Environmental History* 11, no. 1 (2006): 37–69.

Cohen, Howard J., and Jeffrey S. Birkner. "Respiratory Protection." *Clinics in Chest Medicine* 33, no. 4 (December 2012): 783–93. https://doi.org/10.1016/j.ccm.2012.09.005.

Cohen, I. Bernard. *Benjamin Franklin's Experiments: A New Edition of Franklin's Experiments and Observations on Electricity*. Cambridge, MA: Harvard University Press, 1941.

———. *Benjamin Franklin's Science*. Cambridge, MA: Harvard University Press, 1990.

———. *Science and The Founding Fathers: Science in the Political Thought of Jefferson, Franklin, Adams, and Madison*. New York: W.W. Norton & Company Publisher, 1997.

Collot, V. "General Collot's Plan for a Reconnaissance of the Ohio and Mississippi Valleys, 1796." Translated by Durand Echeverria. *The William and Mary Quarterly* 9, no. 4 (October 1952): 512. https://doi.org/10.2307/1923755.

———. "General Collot's Plan for a Reconnaissance of the Ohio and Mississippi Valleys, 1796." *The William and Mary Quarterly* 9, no. 4. October 1952: 512–20.

Conrad, Lawrence I., Michael Neve, Vivian Nutton, Roy Porter, and Andrew Wear. *The Western Medical Tradition: 800 BC to AD 1800*. Cambridge: Cambridge University Press, 2011.

Corgan, James X., ed. *Geological Sciences in the Antebellum South*. Tucsaloosa: University of Alabama Press, 1982.

Cornell, Saul. *The Other Founders: Anti-Federalism and the Dissenting Tradition in America, 1788–1828*. Chapel Hill: University of North Carolina Press, 1999.

Cowan, Ruth Schwartz. *A Social History of American Technology*. New York: Oxford University Press, 1997.

Cowie, Jefferson. *Capital Moves: RCA's Seventy-Year Quest for Cheap Labor*. New York: New Press, 1999.

Craddock, Paul T. *Early Metal Mining and Production*. London: Archetype Publications, 2010.

Cronon, William. *Changes in the Land: Indians, Colonists, and the Ecology of New England*. New York: Hill and Wang, 1983.

———. "Kennecott Journey: The Paths out of Town." In *Under an Open Sky: Rethinking America's Western Past*, edited by William Cronon, George Miles, and Jay Gitlin, eds. New York: W. W. Norton, 1993. 28–51.

———. "Modes of Prophecy and Production: Placing Nature in History." *Journal of American History* 76 (March 1990), 1122–31.

———. *Nature's Metropolis: Chicago and the Great West*. London: W. W. Norton, 1991.

———. "A Place for Stories: Nature, History, and Narrative." *Journal of American History* 78 (March 1992): 1347–76.

———. ed. *Uncommon Ground: Rethinking the Human Place in Nature*. New York: W. W. Norton, 1996.

Cronon, William, George Miles, and Jay Gitlin. "Becoming West: Toward a New Meaning for Western History." in *Under an Open Sky: Rethinking America's Western Past*, eds. William Cronon, George Miles, and Jay Gitlin. New York: W. W. Norton, 1992, 3–27.

Crosby, Alfred W. "The Past and Present of Environmental History." *American Historical Review* 100, no. 4 (Oct. 1995): 1177–89.

Crouse, Nellis M. *Lemoyne d'Iberville: Soldier of New France*. Ithaca, NY: Cornell University Press, 1954.

Cumbler, John T. *Reasonable Use: The People, the Environment, and the State in New England, 1790–1930*. New York: Oxford University Press, 2001.

Curtis, Kent. "Greening Anaconda: EPA, ARCO, and the Politics of Space in Postindustrial Montana." In *Beyond the Ruins: The Meanings of Deindustrialization*, edited by Jefferson Cowie and Joseph Heathcott. Ithaca, NY: Cornell University Press, 2003. 91–111.

Cushing, Frank Hamilton. "Primitive Copper Working: An Experimental Study." *American Anthropologist* 7, no. 1 (1894): 93–117.

Dalzell, Robert. *Enterprising Elite: The Boston Associates and the World They Made*. Cambridge, MA: Harvard University Press, 1987.

Daniel, Dorothy. "The First Glasshouse West of the Alleghenies" *The Western Pennsylvania Historical Magazine 32 (September-December 1949): 97–113*.

———. *Cut and Engraved Glass, 1771–1905.* New York: M. Barrows, 1950.

Daniels, C., and M. Kennedy, eds. *Negotiated Empires: Centers and Peripheries in the Americas, 1500–1820.* New York: Routledge, 2002.

Dann, Kevin, and Gregg Mitman. "Essay Review: Exploring the Borders of Environmental History and the History of Ecology." *Journal of the History of Biology* 30 (1997): 291–302.

Davidson, Hugh. "The George Cresswell Lead Plantation." *Material Culture* 23, no. 2 (1991): 1–23.

Davis, David Brion. *The Problem of Slavery in Western Culture.* Ithaca, NY: Cornell University Press, 1966.

Davis, James E. *Frontier Illinois.* Bloomington: Indiana University Press, 1998.

Davis, Pearce. *The Development of the American Glass Industry.* Cambridge, MA: Harvard University Press, 1949.

Dawley, Alan. *Class and Community: The Industrial Revolution in Lynn.* Cambridge, MA: Harvard University Press, 1976.

Dawley, Alan, and Paul Faler. "Working-Class Culture and Politics in the Industrial Revolution: Sources of Loyalism and Rebellion," *Journal of Social History* 9. June 1976: 466–68.

Del Mar, Alexander. *A History of the Precious Metals: From the Earliest Times to the Present.* New York: Cambridge Encyclopedia Company, 1902.

Denevan, William M. "The Pristine Myth: The Landscape of the Americas in 1492," in *Annals of the Association of American Geographers* 82, no. 3, September 1992: 369–85. https://www.jstor.org/stable/2563351.

Deutsch, Sarah. *No Separate Refuge: Culture, Class, and Gender on an Anglo-Hispanic Frontier in the American Southwest, 1880–1940.* New York: Oxford University Press, 1987.

Dew, Charles B. *Bond of Iron: Master and Slave at Buffalo Forge.* New York: W. W. Norton, 1994.

Dibner, Bern. *Agricola on Metals.* Norwalk, CT: Burndy Library, 1958.

Dickason, Olive Patricia. *Canada's First Nations: A History of Founding Peoples from Earliest Times.* Norman: University of Oklahoma Press, 1992.

Din, Gilbert. "Spain's Immigration Policy in Louisiana and the American Penetration, 1792–1803." *Southwestern Historical Quarterly* 76 (1973): 255–76.

Dobney, Frederick J. *River Engineers on the Middle Mississippi: A History of the St. Louis District, U.S. Army Corps of Engineers.* Washington, DC: GPO, 1978.

Doerflinger, Thomas. *A Vigorous Spirit of Enterprise: Merchants and Economic Development in Revolutionary Philadelphia.* Chapel Hill: University of North Carolina Press, 2001.

Du Val, Kathleen. *The Native Ground: Indians and Colonists in the Heart of the Continent.* Philadelphia: University of Pennsylvania, 2006.

Dublin, Thomas. *Women at Work: The Transformation of Work and Community in Lowell, Massachusetts, 1826–1860.* New York: Columbia University Press, 1979.

Dupree, A. Hunter. "Comment: The Role of Technology in Society and the Need for Historical Perspective." *Technology and Culture* 10 (October 1969): 528–34.

Edelson, S. Max, Gwendolyn Midlo Hall, Walter Hawthorne, David Eltis, Philip Morgan, and David Richardson. "AHR Exchange: The Question of 'Black Rice'" *American Historical Review* 115, no. 1 (February 2010): 123–71.

Edward, James, and Harold Dorn. *Science and Technology in World History: An Introduction.* Baltimore: Johns Hopkins University Press, 2015.

Egan, Geoff. *Lead Cloth Seals and Related Items in the British Museum* [with Mike Cowell and Hero Granger Taylor], Occasional Paper, no. 93. London: Dept. of Medieval and Later Antiquities, British Museum, 1995.

Ekberg, Carl J. "Antoine Valentin de Gruy: Early Missouri Explorer." *Missouri Historical Review* 76 (January 1982): 136–50.

———. *Colonial Ste. Genevieve: An Adventure on the Mississippi Frontier.* Tucson: Patrice Press, 1985.

———. *François Vallé and His World: Upper Louisiana Before Lewis and Clark.* Columbia: University of Missouri Press, 2002.

———. *A French Aristocrat in the American West: The Shattered Dreams of De Lassus de Luzières.* Columbia: University of Missouri, 2010.

———. *French Roots in the Illinois Country: The Mississippi Frontier in Colonial Times.* Champaign: University of Illinois Press, 2000.

Ellis, Joseph J. *American Sphinx: The Character of Thomas Jefferson.* New York: Alfred A. Knopf, 1997.

Eltis, David. *The Rise of African Slavery in the Americas.* Cambridge: Cambridge University Press, 2000.

Emerson, Ralph Waldo. "Nature." In *Works by Ralph Waldo Emerson: Volume One, Nature, Addresses, and Lectures.* Boston: Phillips, Sampson, 1909.

Emerson, Thomas E., and R. Barry Lewis, eds. *Cahokia and the Hinterlands: Middle Mississippian Cultures of the Midwest.* Urbana: University of Illinois Press, 1991.

English, Peter. *Old Paint: A Medical History of Childhood Lead-Paint Poisoning in the United States to 1980.* New Brunswick, NJ: Rutgers University Press, 2001.

Erickson, Charlotte. *Invisible Immigrants: The Adaptation of English and Scottish Immigrants in Nineteenth-Century America.* Florida: University of Miami Press, 1972.

Faragher, John Mack. *Sugar Creek: Life on the Illinois Prairie.* New Haven, CT: Yale University Press, 1986.

———, ed. *Rereading Frederick Jackson Turner: The Significance of the Frontier in American History and Other Essays.* New York: Henry Holt, 1994.

———. *Women and Men on the Overland Trail.* New Haven, CT: Yale University Press, 2001.

Faue, Elizabeth. "Retooling the Class Factory." *Labour History*, no. 82 (May 2002): 109–119.

Fay, George E. "Lead-Silver Molds of the Osage Indians." *Transactions of the Kansas Academy of Science* 52, no. 2 (1949): 205–8.

Figueirôa, Silvia, and Clarete da Silva. "Enlightened Mineralogists: Mining Knowledge in Colonial Brazil, 1750–1825." In "Nature and Empire: Science and the Colonial Enterprise," edited by Roy MacLeod. *Osiris*, 2nd Series, vol. 15 (2000): 174–89.

Fisher, Marvin. *Workshops in the Wilderness: The European Response to American Industrialization, 1830–1860*. New York: Oxford University Press, 1967.

Fishkin, Shelley Fisher. "Crossroads of Cultures: The Transnational Turn in American Studies—Presidential Address to the American Studies Association, November 12, 2004." *American Quarterly* 57, no. 1. (March 2005): 17–57.

Foley, William E. *The Genesis of Missouri: From Wilderness Outpost to Statehood*. Columbia: University of Missouri Press, 1989.

Forbes, Jack D. "Frontiers in American History and the Role of the Frontier Historian." *Ethnohistory* 15, no.2 (Spring 1968): 203–35.

Foster, John Bellamy. *Marx's Ecology: Materialism and Nature*. New York: Monthly Review Press, 2000.

Fowler, Loretta. Review of "One Vast Winter Count: The Native American West before Lewis and Clark," by Colin G. Calloway. *The Western Historical Quarterly* 36, no. 1 (April 1, 2005): 71. https://doi.org/10.2307/25443102.

Fox-Genovese, Elizabeth. *Within the Plantation Household: Black and White Women of the Old South*. Chapel Hill: University of North Carolina Press Books, 1988.

Francaviglia, Richard V. *Hard Places: Reading the Landscapes of America's Historic Mining Districts*. Iowa City: University of Iowa Press, 1992.

Frazier, Harriet C. *Slavery and Crime in Missouri, 1773–1865*, Jefferson, NC: McFarland, 2001, 109–11.

Fulling, Edmund H. "Thomas Jefferson. His Interest in Plant Life as Revealed in His Writings—II." *Bulletin of the Torrey Botanical Club* 72, no. 3 (May 1945): 248.

Gallay, Alan. *The Indian Slave Trade: The Rise of the English Empire in the American South, 1670–1717*. New Haven, CT: Yale University Press, 2001.

Garrison, George P., ed. "A Memorandum of M. Austin's Journey from the Lead mines in the County of Wythe in the State of Virginia to the Lead Mines in the Province of Louisiana West of the Mississippi, 1796–1797." *American Historical Review* 5 (1900): 518–42.

Gibson, Arrell Morgan. *Wilderness Bonanza: The Tri-State District of Missouri, Kansas, and Oklahoma*. Norman: University of Oklahoma Press, 1972.

Gilje, Paul A., ed. *Wages of Independence: Capitalism in the Early American Republic*. Madison: Madison House, 1997.

Gilroy, Paul. *The Black Atlantic: Modernity and Double Consciousness*. Cambridge, MA: Harvard University Press, 1993.

Glacken, Clarence. *Traces on the Rhodian Shore: Nature and Culture in Western Thought From Ancient Times to the End of the Nineteenth Century*. Berkeley: University of California Press, 1967.

Goldstein, Max A. *One Hundred Years of Medicine and Surgery in Missouri [. . .]*. St. Louis, MO: St. Louis Star, 1900.

Gomez, Michael A. *Exchanging Our Country Marks: The Transformation of African Identities in the Colonial and Antebellum South.* Chapel Hill: University of North Carolina Press, 2001.

Good, Mary Elizabeth. *Guebert Site: An 18th Century Historic Kaskaskia Indian Village in Randolph County, Illinois.* n.p.: Central States Archaeological Societies and Mary Elizabeth Good, 1972.

Gordon, Robert B., and Patrick M. Malone. *The Texture of Industry: An Archaeological View of the Industrialization of North America.* New York: Oxford University Press, 1994.

Gracy, David B. "Moses Austin and the Development of the Missouri Lead Industry," *Gateway Heritage* 1, no. 4 Spring 1981, 42–48.

———. *Moses Austin: His Life.* San Antonio, TX: *Trinity University Press*, 1987.

Green, Michael D. "The Expansion of European Colonization to the Mississippi Valley, 1780–1880." In *North America: vol 1 . part 1*, edited by Bruce G. Trigger and Wilcomb E. Washburn, *The Cambridge History of the Native Peoples of North America*, 460-538. New York: Cambridge University Press, 1996.

Greene, John C. Greene and John G. Burke. "The Science of Minerals in the Age of Jefferson," in *Transactions of the American Philosophical Society, New Series*, vol. 68, no. 4 (1978): 1–113.

Gregory, Cedric Errol. *A Concise History of Mining.* Oxford: Pergamon Press, 1980.

Gregory, Frederick. *Natural Science in Western History.* Boston: Houghton Mifflin, 2008.

Griffiths, Tom and Libby Robin, eds. *Ecology and Empire: The Environmental History of Settler Societies.* Edinburgh: Edinburgh University Press, 1998.

Grove, Richard H. *Green Imperialism: Colonial Expansion, Tropical Islands Edens and the Origins of Environmentalism, 1600–1860.* Studies in Environment and History, edited by Donald Worster and Alfred W. Crosby. Cambridge: Cambridge University Press, 1995.

Gruesz, Kristen Silba. *Ambassadors of Culture: The Transamerica Origins of Latino Writing*, Princeton, NJ: Princeton University Press, 2002.

Gutman, Herbert G. *Work, Culture, and Society in Industrializing America: Essays in American Working-Class and Social History.* New York: Alfred A. Knopf, 1976.

Gutman, Herbert. *Power and Culture: Essays on the American Working Class.* Edited by Ira Berlin. New York: Pantheon Books, 1987.

———. *The Black Family in Slavery and Freedom 1750–1925*, New York: Pantheon Books, 1976.

Habakkuk, H. J. *American and British Technology in the Nineteenth Century.* Cambridge: Cambridge University Press, 1962.

Hall, Jacquelyn Dowd, James Leloudis, Robert Korstad, Mary Murphy, Lu Ann Jones, and Christopher B. Daly. *Like A Family: The Making of a Southern Mill World.* Chapel Hill: University of North Carolina Press, 1987.

Halpern, Joel Martin. "Thomas Jefferson and the Geological Sciences." *Rocks & Minerals* 26, no. 11–12 (November 1951): 601–2. https://doi.org/10.1080/00357529.1951.11768253.

Halttunen, Karen. *Confidence Men and Painted Women: A Study of Middle-Class Culture in America, 1830–1870*. New Haven, CT: Yale University Press, 1982.

Hammond, George P., and Agapito Rey, eds. *Narratives of the Coronado Expedition, 1540–1542: Don Hernando Alvarado, An Account of Don Hernando Alvarado's Travels among the Pueblos in 1540*. New York: AMS Press, 1977.

Handlin, Oscar, and Mary Flug Handlin. *Commonwealth; a Study of the Role of Government in the American Economy: Massachusetts, 1774–1861*. Cambridge, MA: Belknap Press, 1969.

Hanley, Lucy Elizabeth. "Lead Mining in the Mississippi Valley during the Colonial Period." Master's thesis, St. Louis University, 1942.

Hardesty, Donald L. *The Archeology of Mining and Miners: A View from the Silver State*. Special Publication Series, no. 6. Ann Arbor, MI: Society for Historical Archaeology, 1988.

———. *Mining Archaeology in the American West: A View from the Silver State*. Lincoln: University of Nebraska Press and Society for Historical Archaeology, 2010.

Hardin, Garrett. "The Tragedy of the Commons." *Science* 162 (1968): 1243–48.

Harley, J. B. *The New Nature of Maps: Essays in the History of Cartography*. Baltimore: John Hopkins University Press, 2002.

———. "Rereading the Maps of the Columbian Encounter." *Annals of the Association of American Geographers* 82, no. 3 (1992): 522–36.

Harn, Orlando C. *Lead: The Precious Metal*. New York: Century, 1924.

Hartley, E. N. *Iron Works on the Saugus: The Lynn and Braintree Ventures of the Company of Undertakers of the Ironworks in New England*. Tulsa: University of Oklahoma Press, 1957.

Haskell, Thomas, and Richard F. Teichgraeber, eds. *The Culture of the Market: Historical Essays*. Cambridge: Cambridge University Press, 1993.

Hauser, Raymond E. "The Illinois Indian Tribe: From Autonomy and Self-Sufficiency to Dependency and Depopulation," *Journal of the Illinois State Historical Society* 69, May 1976, 127–38;

Havighurst, Walter. *Wilderness for Sale: The Story of the First Western Land Rush*. New York: Hastings House, 1956.

Hays, Samuel P., and Barbara D. Hays. *Beauty, Health, and Permanence: Environmental Politics in the United States, 1955–1985*. Cambridge: Cambridge University Press, 1987.

Hayward, J. Lyman. *The Los Cerrillos Mines [N. M.] and Their Mineral Resources: A Description of the Mines in the Los Cerrillos and Galisteo Mining Districts, Accompanied by a Map of the Same, Drawn from Actual Surveys*. South Framingham, MA: J. C. Clark, 1880.

Hazen, Robert M. "The Founding of Geology in America: 1771 to 1818." *Geological Society of American Bulletin* 85 (1974): 1827–34.

Hechenberger, Daniel. "The Jesuits: History and Impact: From Their Origins Prior to the Baroque Crisis to Their Role in the Illinois Country," in *Journal of the Illinois State Historical Society (1998–)* 100, no. 2 (Summer 2007), 85–109. https://www.jstor.org/stable/40204675;

Heidenreich, Conrad E., and Edward H. Dahl. "The French Mapping of North America, 1700–1760." *Map Collector* 19 (June 1982): 2–7.

———. "Mapping the Great Lakes: The Period of Imperial Rivalries, 1700–1760." *Cartographica* 18, no. 3 (1981), 74–109.

Herod, Andrew. *Labor Geographies*. New York: Guilford Press, 2001.

Heyl, A. V., M. H. Delevaux, R. E. Zartman, and M. R. Brock. "Isotopic Study of Galenas from the Upper Mississippi Valley, the Illinois-Kentucky, and Some Appalachian Valley Mineral Districts." Economic Geology (1966) 61: 933–61.

Hildeburn, Charles Swift Riché. *A Century of Printing: The Issues of the Press in Pennsylvania, 1685–1784, Volume 2, American Culture Series, Library of American Civilization*. Philadelphia: Press of Matlack & Harvey, 1886.

Hildreth, Richard. *The History of the United States of America: Colonial, 1497–1688*, vol. 1. New York: Harper, 1863.

Hill, Christopher V. *South Asia: An Environmental History: Nature and Human Societies*. Santa Barbara: ABC-CLIO, 2008.

Hodge, Frederick W., ed. *Handbook of American Indians North of Mexico*, vol. 2. New York: Pageant Books, 1960.

Hodges, Henry. *Technology in the Ancient World*. New York: Barnes & Noble Books, 1992.

Hogan, Richard. *Class and Community in Frontier Colorado*. Lawrence: University Press of Kansas, 1990.

Horn, James, Jan Lewis, and Peter Onuf, eds. *The Revolution of 1800: Democracy, Race, and the New Republic*. Charlottesville: University of Virginia, 2002.

Hornbeck Tanner, Helen, ed. *Atlas of Great Lakes Indian History*. Norman: University of Oklahoma Press, 1987.

Horwitz, Morton J. *The Transformation of American Law 1780–1860*. Cambridge, MA: Harvard University Press, 1977.

Hoskins, Janet. *Biographical Objects: How Things Tell the Stories of People's Lives*. New York: Routledge, 1998.

Houck, Louis. *A History of Missouri from the Earliest Explorations and Settlements Until the Admission of the State into the Union*. 3 vols. Chicago: R. R. Donnelley & Sons, 1908.

———. *The Spanish Regime in Missouri: a Collection of Papers and Documents Relating to Upper Louisiana Principally within the Present Limits of Missouri during the Dominion of Spain* [...]. 2 vols. Chicago: R. R. Donnelley and Sons, 1909.

Hounshell, David A. *From the American System to Mass Production, 1800–1932: The Development of Manufacturing Technology in the United States*. Baltimore: Johns Hopkins University Press, 1984.

Hughes, Donald J. *Pan's Travail: Environmental Problems of the Ancient Greeks and Romans*. Baltimore: Johns Hopkins University Press, 1994.

Hurley, Andrew. *Environmental Inequalities: Class, Race, and Industrial Pollution in Gary, Indiana, 1945–1980*. Chapel Hill: University of North Carolina Press, 1995.

———, ed. *Common Fields: An Environmental History of St. Louis*. St. Louis: Missouri Historical Society Press, 1997.

Hussey, Miriam and Wharton School Industrial Research Unit, *The Wetherill Papers, 1762–1899; Being the Collection of Business Records of the Store and White Lead Works Founded by Samuel Wetherill in the Late Eighteenth Century* [. . .]. Philadelphia: Industrial Research Dept., Wharton School of Finance and Commerce, University of Pennsylvania, 1942.

Igler, David. *Industrial Cowboys: Miller and Lux and the Transformation of the Far West, 1850–1920*. Berkeley: University of California Press, 2005.

Immanuel Kant. *Fundamental Principals of the Metaphysics of Morals*. Gloucester, UK: Dodo Press, 2005.

Ingalls, Walter Renton. *Lead and Zinc in the United States: Comprising an Economic History of the Mining and Smelting of the Metals and the Conditions, Which have Affected the Development of the Industries*. New York: Hill, 1908.

Ingham, John N. *Making Iron and Steel: Independent Mills in Pittsburgh, 1820–1920*. Columbus: Ohio State University Press, 1991.

Innes, Stephen. *Labor in a New Land: Economy and Society in Seventeenth-Century Springfield*. Princeton, NJ: Princeton University Press, 1983.

Iseminger, William R. "Relationships Between Climate Change and Culture Change in Prehistory," *Illinois Antiquity* 24, Spring 1990.

———. "Culture and Environment in the American Bottom: The Rise and Fall of Cahokia Mounds." In *Common Fields: An Environmental History of St. Louis*, edited by Andrew Hurley, 38–57. St. Louis: Missouri Historical Society Press, 1997.

Isenberg, Andrew C. *Mining California: An Ecological History*. New York: Hill & Wang, 2005.

Jackson, Donald Dean, and James P. Ronda. *Thomas Jefferson & the Stony Mountains: Exploring the West from Monticello*. Tulsa: University of Oklahoma Press, 1993.

Jacobs, Nancy. *Environment, Power, and Injustice: A South African History*. Studies in Environment and History. Cambridge: Cambridge University Press, 2003.

Jacoby, Karl. *Crimes Against Nature: Squatters, Poachers, Thieves, and the Hidden History of American Conservation*. Berkeley: University of California Press, 2001.

Jaffee, David. "Peddlers of Progress and the Transformation of the Rural North, 1760–1850." *Journal of American History* 78 (September 1991): 511–35.

Jamieson, Duncan R. "American Environmental History." *Choice* 32 (Sept 1994): 49–61.

Jefferson, Thomas. *Notes on the State of Virginia*. Edited by Frank Shuffelton. New York: Penguin Books, 1999.

Jefferson, Thomas, and William Dunbar. *Documents Relating to the Purchase and Exploration of Louisiana*. New York: Houghton Mifflin, 1904

Jennings, Francis. *The Invasion of America: Indians, Empires, and Republics in the Great Lakes Region, 1650–1815*. Chapel Hill: University of North Carolina Press, 1991.

Jennings, Francis, and George Irving Quimby. "Indian Culture and European Trade Goods: The Archaeology of the Historic Period in the Western Great Lakes Region." *Ethnohistory* 18, no. 1 (1971): 71.

Jeremy, David. *Transatlantic Industrial Revolution: The Diffusion of Textile Technologies Between Britain and America, 1790–1830s*. Cambridge, MA: MIT Press, 1981.

John Robert McNeill, and George Vrtis. *Mining North America: An Environmental History since 1522*. Oakland, California: University of California Press, 2017.

Johnson, Esau. "Reminiscence," in *A Chronological History of Indian Lead Mining in the Upper Mississippi Valley from 1643 to 1848*, by Philip Millhouse (Unpublished Paper for History Special Projects, Galena Public Library, Galena, IL, 1993).

Johnson, Walter. *Soul by Soul: Life Inside the Antebellum Slave Market*. Cambridge, MA: Harvard University Press, 1999.

Jones, Gareth Steadman. *Languages of Class: Studies in English Working Class History, 1832–1982*. Cambridge: Cambridge University Press, 1983.

Jones, Jacqueline. *Labor of Love, Labor of Sorrow: Black Women, Work, and the Family from Slavery to the Present*. New York: Basic Books, 1986.

Jordan, Winthrop D. *White Over Black: American Attitudes Toward the Negro, 1550–1812*. Chapel Hill: University of North Carolina Press, 1968.

Joyce, Patrick. *Visions of the People: Industrial England and the Question of Class, 1848–1914*. New York: Cambridge University Press, 1991.

Judd, Richard W. *The Untilled Garden: Natural History and the Spirit of Conservation in America, 1740–1840*. Cambridge: Cambridge University Press, 2009.

Kappler, Charles J. *Indian Affairs: Laws and Treaties*, vol. 2, *Treaties*. Washington, DC: GPO, 1904.

Karrow, Robert W., and David Buisseret. *Gardens of Delight: Maps and Travel Accounts of Illinois and the Great Lakes from the Collection of Hermon Dunlap Smith, An Exhibition at The Newberry Library 29 October 1984–31 January 1985*. Chicago: Newberry Library, 1984.

Kasson, John F. *Civilizing the Machine: Technology and Republican Values in America, 1776–1900*. New York: Macmillan, 1976.

Kastor, Peter J. *Nation's Crucible: The Louisiana Purchase and the Creation of America*. New Haven, CT: Yale University Press, 2012.

Kelley, Robert L. *Gold vs. Grain: The Hydraulic Mining Controversy in California's Sacramento Valley*. Glendale: Arthur H. Clark, 1959.

Kellogg, Louise. *The French Regime in Wisconsin and the Northwest*. Madison: State Historical Society of Wisconsin, 1925.

Kelman, Ari. *A River and Its City: The Nature of Landscape in New Orleans*. Berkeley: University of California Press, 2003.

Kent, Timothy J. *Ft. Ponchartrain at Detroit: A Guide to the Daily Lives of Fur Trade and Military Personnel, Settlers, and Missionaries at French Posts*, 2 vols. Ossineke, MI: Silver Fox Enterprises, 2001.

Kinnaird, Lawrence. "American Penetration of Louisiana." In *New Spain and the Anglo-American West*, vol. 1., edited by George P. Hammond. Lancaster, PA: Lancaster Press, 1932.

Knoblauch, Frieda. *The Culture of Wilderness: Agriculture as Colonization in the American West*. Chapel Hill: University of North Carolina Press, 1996.

Kolodny, Annette. *The Land Before Her: Fantasy and Experience of the American Frontiers, 1630–1860*. Chapel Hill: University of North Carolina Press, 1984.

Kulikoff, Allan. "The Transition to Capitalism in Rural America." *William and Mary Quarterly*, 3rd Series, vol. 46 (January 1989): 121–44.

Kupperman, Karen Ordahl. *America in European Consciousness*. Chapel Hill: University of North Carolina, 1995.

———. *Indians and English: Facing Off in Early America*. New York: Cornell University Press, 2000.

———. *Providence Island: The Other Puritan Colony, 1630–1641*. Cambridge, MA: Harvard University Press, 1993.

LaDow, Beth. *The Medicine Line: Life and Death on the North American Borderland*. New York: Routledge, 2001.

Lamar, Howard, and Leonard Thompson. "Comparative Frontier History." In *The Frontier in History: North America and Southern Africa Compared*, edited by Howard Lamar and Leonard Thompson, 3–13. New Haven: Yale University Press, 1981.

Lamborn, Robert H. *A Rudimentary Treatise on the Metallurgy of Silver and Lead*. London: J. Weale, 1861.

Landau, Ralph, and Nathan Rosenberg. "Successful Commercialization in the Chemical Process Industries." In *Technology and the Wealth of Nations*, edited by Nathan Rosenberg, Ralph Landau, and David C. Mowery, 73 – 119. Stanford, CA: Stanford University Press, 1992.

Lander, Ernest M., Jr. *The Textile Industry in Antebellum South Carolina*. Baton Rouge: Louisiana State University Press, 1969.

Langston, Nancy. *Where Land and Water Meet: A Western Landscape Transformed*. Seattle: University of Washington Press, 2003.

Lankton, Larry D. *Cradle to Grave: Life, Work, and Death at the Lake Superior Copper Mines*. New York: Oxford University Press, 1991.

Lanmon, Dwight P. "The Baltimore Glass Trade, 1780 to 1820" *Winterthur Portfolio* 5 (1969): 15–48.

Lapp, Rudolph M. *Blacks in Gold Rush California*. New Haven, CT: Yale University Press, 1977.

Larson, Brooke. *Cochabamba: 1550–1900, Colonialism and Agrarian Transformation in Bolivia*. Durham, NC: Duke University Press, 1998.

Larson, John L. *Internal Improvement: National Public Works and the Promise of Popular Government in the Early United States*. Chapel Hill: University of North Carolina Press, 2001.

Laudan, Rachel. *From Mineralogy to Geology: The Foundations of a Science, 1650–1830*. Chicago: University of Chicago Press, 1993.

Laurie, Bruce. *Artisans into Workers: Labor in Nineteenth-Century America*. Urbana: University of Illinois Press, 1997.

———. *Working People of Philadelphia, 1800–1850*. Philadelphia: Temple University Press, 1980.

Layton, Edwin. "Mirror-Image Twins: The Communities of Science and Technology in 19th-Century America." *Technology and Culture* 12, no. 4 (1971): 562–80.

Le Page Du Pratz, Antoine Simon. *The History of Louisiana, or of the Western Parts of Virginia and Carolina* [. . .]. Translated by Stanley Arthur. London: 1764.

LeCain, Timothy J. *Mass Destruction: The Men and Giant Mines That Wired America and Scarred the Planet.* New Brunswick, NJ: Rutgers University Press, 2009.

———. "Moving Mountains: Technology and the Environment in Western Copper Mining" (PhD diss., University of Delaware, 1998).

Lepore, Jill. *The Name of War: King Philip's War and the Origins of American Identity.* New York: Vintage, 1998.

Lethaby, W. R. *Leadwork: Old and Ornamental and for the Most Part English.* London: Macmillan, 1893.

Lewis, G. Malcolm, ed. *Cartographic Encounters: Perspectives on Native American Mapmaking and Map Use.* Chicago: University of Chicago Press, 1998.

Lewis, Ronald L. *Coal, Iron, and Slaves: Industrial Slavery in Maryland and Virginia, 1715–1865.* Westport, CT: Greenwood Press, 1979.

Lewis, W. David. *Sloss Furnaces and the Rise of the Birmingham District: An Industrial Epic.* Tuscaloosa: University of Alabama Press, 1994.

Licht, Walter. *Industrializing America: The Nineteenth Century.* Baltimore: Johns Hopkins University Press, 1995.

Limerick, Patricia Nelson. *The Legacy of Conquest: The Unbroken Past of the American West.* New York: W. W. Norton, 1988.

Lingenfelter, Richard. *The Hardrock Miners: A History of the Mining Labor Movement in the American West, 1863–1893.* Berkeley: University of California Press, 1974.

Long, Priscilla. *Where the Sun Never Shines: A History of America's Bloody Coal Industry.* New York: Paragon House, 1989.

Lykins, William H. R. "On the Mound-Builders' Knowledge of Metals." *Kansas City Review of Science and Industry,* edited by Theo. S. Case, vol. VII, 535. Kansas City, MO: Ramsey, Millet, and Hudson, 1882.

MacLeod, Roy M., ed. "Nature and Empire: Science and the Colonial Enterprise." *Osiris,* 2nd Series, vol. 15 (2000).

Mancall, Peter C., and James H. Merrell, eds. *American Encounters: Natives and Newcomers from European Contact to Indian Removal, 1500–1850.* New York: Routledge, 2000.

Mann, Bruce H. *Republic of Debtors: Bankruptcy in the Age of American Independence.* Cambridge, MA: Harvard University Press, 2002.

Maree, Natalia. *Kaskaskia under the French Regime.* Carbondale: Southern Illinois University Press, 2003.

Markowitz, Gerald, and David Rosner. *Deceit and Denial: The Deadly Politics of Industrial Pollution.* Berkeley: University of California Press, 2002.

Marks, Paula Mitchell. *Precious Dust: The American Gold Rush Era, 1848–1900.* New York: William Morrow, 1994.

Marsh, George Perkins. *Man and Nature*. Edited by David Lowenthal. Cambridge, MA: Belknap Press, 1965.

Marshall, Thomas Maitland. *The Life and Papers of Frederick Bates*, vol. 1. St. Louis: Missouri Historical Society, 1926.

Martin, Calvin. *Keepers of the Game: Indian-Animal Relations and the Fur Trade*. Berkley: University of California Press, 1978.

Martin, Susan R. *Wonderful Power: The Story of Ancient Copper Working in the Lake Superior Basin*. Detroit: Wayne State University Press, 1999.

Marx, Jennifer. *The Magic of Gold*. New York: Doubleday, 1978.

Marx, Leo. *The Machine in the Garden: Technology and the Pastoral Ideal in America*. New York: Oxford University Press, 2000.

Mason, Edward G., ed. "Early Chicago and Illinois," *Chicago Historical Society's Collection*, vol. 4 Chicago: Fergus Printing, 1890, 36, 230–51.

Massey, Doreen. *Spatial Divisions of Labor: Social Structures and the Geography of Production*. London: Macmillan, 1984.

Matson, Cathy. *The Economy of Early America: Historical Perspectives & New Directions*, University Park: Pennsylvania State University Press, 2006.

McClellan, James E., III, and François Regourd. "The Colonial Machine: French Science and Colonization in the Ancien Regime." In "Nature and Empire: Science and the Colonial Enterprise," edited by Roy McCloud. *Osiris*, 2nd Series, vol. 15 (2000): 31–50.

McCord, Carey P. "Lead and Lead Poisoning in Early America: Lead Mines and Lead Poisoning." *Industrial Medicine and Surgery* 22, no. 11 (1953): 534–39.

McCoy, Drew R. *The Elusive Republic: Political Economy in Jeffersonian America*. Chapel Hill: University of North Carolina Press, 1980.

McCready, Benjamin W. *On the Influence of Trades, Professions, and Occupations in the United States, in the Production of Disease*. Originally printed Albany: E. W. and C. Skinner, 1837. Reprinted Baltimore: Johns Hopkins University Press, 1943. Introductory essay by Genevieve Miller.

McEvoy, Arthur F. *The Fisherman's Problem: Ecology and Law in the California Fisheries, 1850–1980*. New York: Cambridge University Press, 1986.

———. "Working Environments: An Ecological Approach to Industrial Health and Safety." *Technology and Culture*, supplement to vol. 36, no. 2 (April 1995): 145–73.

McGaw, Judith. *Early American Technology: Making and Doing Things from the Colonial Era to 1850*. Chapel Hill: University of North Carolina Press, 1994.

———. *Most Wonderful Machine: Mechanization and Social Change in Berkshire Paper Making, 1801–1885*. Princeton, NJ: Princeton University Press, 1987.

McNeil, John. *Something New Under The Sun: An Environmental History of the Twentieth-Century World*. New York: W. W. Norton, 2000.

McNeill, J. R. "The Nature of Environmental History: Observations on the Nature and Culture of Environmental History." *History and Theory, Theme Issue* 42 (2003): 5–43.

Meeker, Moses. *Early History of the Lead Region of Wisconsin.* Madison: Wisconsin. State Historical Society of Wisconsin, 1872.

Meiklejohn, A. "The Successful Prevention of Lead Poisoning in the Glazing of Earthenware in the North Staffordshire Potteries," *British Journal of Industrial Medicine* 20, no. 3 (1963): 169–80.

Meinig, Donald. W. *The Shaping of America: A Geographical Perspective on 500 Years of History,* vol. 2, *Continental America, 1800–1867.* New Haven, CT: Yale University Press, 1993.

Merchant, Carolyn. *Ecological Revolutions: Nature, Gender, and Science in New England.* Chapel Hill: University of North Carolina Press, 1989.

———. "Gender and Environmental History." *Journal of American History* 76 (March 1990): 1117–21.

Mereness, Newton D., ed. *Travels in the American Colonies, 1690–1783.* Edited under the Auspices of the National Society of the Colonial Dames of America. New York: Antiquarian Press, 1961.

Mihesuah, Devon A. *Natives and Academics: Researching and Writing about American Indians.* Lincoln: University of Nebraska Press, 1998.

Miller, Char, ed. *American Forests: Nature, Culture, Politics.* Lawrence: University Press of Kansas, 1997.

Miller, Randall M. *The Cotton Mill Movement in Antebellum Alabama.* New York: Arno Press, 1978.

Miller Surrey, Nancy Maria. *The Commerce of Louisiana during the French Regime, 1699–1763.* New York: Columbia University, 1916.

Mills, Elizabeth Shown. "Parallel Lives: Philippe de La Renaudière and Philippe (de) Renault, Directors of the Mines, Company of the Indies," *The Natchitoches Genealogist* 22 (April 1998): 3–18.

Milner, George R. "American Bottom Mississippian Cultures: Internal Development and External Relations," in *New Perspectives on Cahokia Archaeology: Views from the Periphery,* ed. James B. Stoltmann, Monographs in World Archaeology, no.2, Madison: Prehistory Press, 1991, 29–47.

Misa, Thomas J. *A Nation of Steel: The Making of Modern America, 1865–1925.* Baltimore: Johns Hopkins University Press, 1995.

Mitman, Gregg. "In Search of Health: Landscape and Disease in American Environmental History." *Environmental History* 10 (2005): 184–209.

Mitman, Gregg, Michelle Murphy, and Christopher Sellers, eds. "Landscapes of Exposure: Knowledge and Illness in Modern Environments." *Osiris,* 2nd Series, vol. 19 (2004).

Molloy, Peter M. *The History of Metal Mining and Metallurgy: An Annotated Bibliography.* New York: Garland, 1986.

Montgomery, David. *Fall of the House of Labor: The Workplace, the State, and American Labor Activism, 1865–1925.* Cambridge: Cambridge University Press, 1987.

———. *Workers in Control of America: Studies in the History of Work, Technology, and Labor Struggles*. Cambridge: Cambridge University Press, 1979.

Morgan, Edmund S. *American Slavery, American Freedom: The Ordeal of Colonial Virginia*. New York: W. W. Norton, 1975.

Morgan, Philip D. *Slave Counterpoint: Black Culture in the Eighteenth-Century Chesapeake and Lowcountry*. Chapel Hill: University of North Carolina Press, 1998.

Morlot, A. "On the Date of the Copper Age in the United States." *Proceedings of the American Philosophical Society* 9, no. 68 (1862): 111–14.

Morrissey, Katherine. *Mental Territories: Mapping the Inland Empire*. Ithaca, NY: Cornell University Press, 1997.

Morrow, Lynn. "New Madrid and Its Hinterland: 1783–1826." *Bulletin of the Missouri Historical Society* 34, no. 4, part 2 (1980): 241–50.

Morse, Kathryn. *The Nature of Gold: An Environmental History of the Klondike Gold Rush*. Seattle: University of Washington Press, 2003.

Motten, Clement G. *Mexican Silver and the Enlightenment*. Philadelphia: Octagon Books, 1950.

Mouat, Jeremy. *Metal Mining in Canada, 1840–1950*. Ottawa: National Museum of Science and Technology, 2000.

———. *Roaring Days: Rossland's Mines and the History of British Columbia*. Vancouver: UBC Press, 1995.

Mulholland, James A. *A History of Metals in Colonial America*. Tuscaloosa: University of Alabama Press, 1981.

Mumford, Lewis. *Technics and Civilization*. Reprint, Chicago: University of Chicago Press, 2010.

Murillo, Dana Velasco. *Urban Indians in a Silver City: Zacatecas, Mexico, 1546–1810*. Stanford, CA: Stanford University Press, 2016.

Murphy, Lucy Eldersveld. *A Gathering of Rivers: Indians, Métis, and Mining in the Western Great Lakes, 1737–1832*. Lincoln: University of Nebraska Press, 2000.

———. "To Live Among Us: Accommodation, Gender, and Conflicts in the Western Great Lakes Region, 1760–1832." In *Contact Points: American Frontiers from the Mohawk Valley to the Mississippi, 1750–1830*." Edited by Andrew R. L. Cayton and Fredrika J. Teute, 270–303. Chapel Hill: University of North Carolina Press, 1998.

Murphy, Michelle. "'The Elsewhere within Here' and Environmental Illness; or, How to Build Yourself a Body in a Safe Space." *Configurations* 8 (2000): 87–120.

Nash, Gary. *The Urban Crucible: The Northern Seaports and the Origins of the American Revolution*. Cambridge, MA: Harvard University Press, 1979.

Nash, Linda. "Finishing Nature: Harmonizing Bodies and Environments in Late Nineteenth-Century California." *Environmental History* 8 (2003): 25–52.

Nelson, Daniel. *Farm and Factory: Workers in the Midwest, 1880–1990*. Bloomington: Indiana University Press, 1995.

———. *Managers and Workers: Origins of the Twentieth-Century Factory System in the United States, 1880–1920*. Madison: University of Wisconsin Press, 1995.

Newton, Huey P. *Revolutionary Suicide* (New York: Harcourt Brace Jovanovich, 1973),

Noble, David F. *America by Design: Science, Technology, and the Rise of Corporate Capitalism*. New York: Knopf, 1977.

———. *Forces of Production: A Social History of Industrial Automation*. New York: Knopf, 1984.

Norwood, Vera. *Made from This Earth: American Women and Nature*. Chapel Hill: University of North Carolina Press, 1993.

Nriagu, Jerome O. *Lead and Lead Poisoning in Antiquity*. New York: John Wiley and Sons, 1983.

———. "Paleoenvironmental Research: 'Tales Told in Lead.'" *Science* 281, no. 5383 (September 11, 1998): 1622–23.

Nye, David. *America as Second Creation: Technology and Narratives of New Beginnings*. Cambridge: MIT Press, 2003.

Oleson, Alexandra, and Sanborn Conner Brown. *The Pursuit of Knowledge in the Early American Republic: American Scientific and Learned Societies from Colonial Times to the Civil War*. Ann Arbor: U.M.I, 1994.

Opie, John. *Nature's Nation: An Environmental History of the United States*. Fort Worth, TX: Harcourt Brace College Publishers, 1998.

Overman, Frederick. *A Treatise on Metallurgy: Comprising Mining, and General and Particular Metallurgical Operations, with a Description of Charcoal, Coke, and Anthracite Furnaces, Blast Machines, Hot Blast, Forge Hammers, Rolling Mills, Etc., Etc*. New York: D. Appleton, 1887.

Paquette, Gabriel. *Enlightened Reform in Southern Europe and Its Atlantic Colonies, c. 1750–1830*. Farnham, UK: Ashgate, 2009.

———. *Enlightenment, Governance and Reform in Spain and Its Empire, 1759–1808*. Basingstoke, UK: Palgrave Macmillan, 2011.

Parsons, James J. "Raised Field Farmers as Pre-Columbian Landscape Engineers: Looking North from the San Jorge, Colombia," in *Prehistoric Intensive Agriculture in the Tropics*, 2 vols., ed. I. S. Farrington, International Series 232, Oxford: British Archaeological Reports, 1985.

Pascoe, Peggy. "Western Women at the Cultural Crossroads." In *Trails: Toward a New Western History*, edited by Patricia Nelson Limerick, Clyde A. Milner, and Charles E. Rankin, 40–58. Lawrence: University Press of Kansas, 1991.

Paul, Rodman W. *Mining Frontiers of the Far West, 1848–1880*. Albuquerque: University of New Mexico Press, 1963.

Paul, Rodman W., and Elliott West. *Mining Frontiers of the Far West, 1848–1880*. Albuquerque: University of New Mexico Press, 2001.

Peck, Gunther. *Reinventing Free Labor: Padrones and Immigrant Workers in the North American West, 1880–1930*. New York: Cambridge University Press, 2000.

Penman, J. T. and J. N. Gundersen, "Pipestone Artifacts from Upper Mississippi Valley Sites" *Plains Anthropologist* 44, no. 167 (February 1999): 47–57. https://www.jstor.org/stable/25669585;

Perino, Gregory. "The Krueger Site, Monroe County, Illinois." *Mississippian Site Archaeology in Illinois I: Site Reports from the St. Louis and Chicago Areas.* Illinois Archaeological Survey, Bulletin no. 8 (1971): 1–148.

Peskin, Lawrence. *Manufacturing Revolution: The Intellectual Origins of Early American Industry.* Baltimore: Johns Hopkins University Press, 2003.

Petrik, Paula. *No Step Backward: Women and Family on the Rocky Mountain Mining Frontier, Helena, Montana, 1865–1900.* Helena: Montana Historical Society Press, 1987.

Phillips, W. A. "Aboriginal Quarries and Shops at Mill Creek, Illinois." *American Anthropologist,* new series, vol. 2, no. 1 (1900): 37–52.

Polemon, John. *Second Part of the Book of Battalias, Faught in Our Age* [. . .], London: 1587; Reprint Amsterdam: Theatrum Orbis Terrarum / New York: Da Capo Press, 1972.

Prude, Jonathan. *The Coming of Industrial Order: Town and Factory Life in Rural Massachusetts, 1810–1860.* Cambridge: Cambridge University Press, 1983.

Pulsifer, William. *Notes for the History of Lead and an Inquiry into the Development of the Manufacture of White Lead and Lead Oxides.* New York: D. Van Nostrand, 1888.

Pyne, Stephen J. "Firestick History." *Journal of American History* 76 (March1990): 1132–41.

Quimby, George I., Jr. "Indian Trade Objects in Michigan and Louisiana." *Michigan Academy of Science, Arts, and Letters* 27 (1941): 543–51.

Quinn, M. L. "Industry and Environment in the Appalachian Copper Basin, 1890–1930." *Technology and Culture* 34 (1993): 575–612.

Quivik, Fredric L. "The Historic Landscape of Butte and Anaconda, Montana." In *Images of an American Land: Vernacular Architecture in the Western United States,* edited by Thomas Carter, 267–290. Albuquerque: University of New Mexico Press, 1997.

———. "Smoke and Tailings: An Environmental History of Copper Smelting Technologies in Montana, 1880-1930." PhD diss., University of Pennsylvania, 1998.

Raymond, Robert. *Out of the Fiery Furnace: The Impact of Metals on the History of Mankind.* University Park: Pennsylvania State University Press, 1986.

Reyling, August Reyling. *Historical Kaskaskia,* 1963, Illinois History and Lincoln Collections, Special Collections Division of the University of Illinois Library, Urbana-Champaign, https://libsysdigi.library.uiuc.edu/OCA/Books200906/historical kaskasooreyl/historicalkaskasooreyl.pdf.

Rice, Stephen. *Minding the Machine: Languages of Class in Early Industrial America.* Berkeley: University of California, 2004.

Richard, Thomas A. *A History of American Mining.* New York: McGraw-Hill, 1932.

Richter, Daniel. *Facing East From Indian Country: A Native History of Early America.* Cambridge, MA: Harvard University Press, 2001.

———. *The Ordeal of the Longhouse: The Peoples of the Iroquois League in the Era of European Colonization:* Chapel Hill: University of North Carolina Press, 1993.

Rickard, T. A. *Man and Metals: A History of Mining in Relation to the Development of Civilization*. 1932: Reprint, Arno Press, 1974.

Rilling, Donna. *Making Houses/Crafting Capitalism: Builders in Philadelphia, 1790–1850*. Philadelphia: University of Pennsylvania Press, 2001.

Rock, Howard B. *Artisans of the New Republic: The Tradesmen of New York City*. New York: New York University Press, 1979.

Roediger, David R. *The Wages of Whiteness: Race and the Making of the American Working Class*. Revised Edition. London: Verso, 1999.

Rome, Adam. *The Bulldozer in the Countryside: Suburban Sprawl and the Rise of American Environmentalism*. Cambridge: Cambridge University Press, 2001.

Ronda, James P. *Jefferson's West: A Journey with Lewis and Clark*. Charlottesville, VA: Thomas Jefferson Foundation, 2000.

———. *Lewis and Clark among the Indians*. Lincoln: University of Nebraska Press, 1984.

Rosenberg, Nathan. "Technology and the Environment." *Technology and Culture* 12 (October 1971), 543–61.

Rothensteiner, John E. "Earliest History of Mine La Motte." *Missouri Historical Review* 20, no. 2 (January 1926): 199–213.

Rowe, D. J. *Lead Manufacturing in Britain: A History*. London: Routledge, 2017.

Rowland, Dunbar, and Albert. G. Sanders, eds. and trans. *Mississippi Provincial Archives, 1702–1729*, vol. 2. Jackson: Press of Mississippi Department of Archives and History, 1927.

Rudwick, Martin J. S. "The Emergence of a Visual Language for Geological Science, 1760–1840." *History of Science* 14 (1976): 149–95.

Rule, John C. "Jean-Frédéric Phélypeaux, Comte de Pontchartrain et Maurepas: Reflections on His Life and His Papers," *Louisiana History: The Journal of the Louisiana Historical Association* 6, no. 4 (1965): 365–77. http://www.jstor.org/stable/4230863.

Salisbury, Neal. *Manitou and Providence: Indians, Europeans, and the Making of New England, 1500–1643*. New York: Oxford University Press, 1984.

Schama, Simon. *Landscape and Memory*. New York: A. A. Knopf, 1995.

Schlarman, Joseph H. *From Quebec to New Orleans: The Story of the French in America, Illustrated; Fort de Chartres*. Belleville, IL: Buechler, 1929.

Schockel, Bernard H. "History of Development of Jo Daviess County." In *Geography of the Galena and Elizabeth Quadrangles*, edited by Arthur Trowbridge and Eugene Shaw. Illinois State Geological Survey, Bulletin no. 26. Urbana: State of Illinois and University of Illinois, 1916.

Schroeder, Walter A. "Environmental Setting of the St. Louis Region," in *Common Fields: An Environmental History of St. Louis*, ed. Andrew Hurley, St. Louis: Missouri Historical Society Press, 1997.

———. *Opening the Ozarks: A Historical Geography of Missouri's Ste. Geneviève District 1760–1830*. Columbia: University of Missouri Press, 2002.

Scott, James. *Seeing Like A State: How Certain Schemes to Improve the Human Condition Have Failed*. New Haven, CT: Yale University Press, 1998.

Scranton, Philip. *Endless Novelty: Specialty Production and American Industrialization, 1865–1925*. Princeton, NJ: Princeton University Press, 1997.

———. *Proprietary Capitalism: The Textile Manufacture at Philadelphia, 1800–1885*. New York: Cambridge University Press, 1983.

Seed, Patricia. *Ceremonies of Possession in Europe's Conquest of the New World, 1492–1640*. Cambridge: Cambridge University Press, 1995.

Sellers, Charles Grier. *The Market Revolution: Jacksonian America, 1815–1846*. New York: Oxford University Press, 1994.

Sellers, Christopher C. *Hazards of the Job: From Industrial Disease to Environmental Health Science*. Chapel Hill: University of North Carolina Press, 1997.

———. "To Place or Not to Place: Toward an Environmental History of Modern Medicine," *Bulletin of the History of Medicine* 92, no. 1 (Spring 2018): 1–45.

Shactman, Tom. *Gentlemen Scientists and Revolutionaries: The Founding Fathers in the Age of Enlightenment*. New York: St. Martin's Press, 2014.

Shelton, Cynthia J. *The Mills of Manayunk: Industrialization and Social Conflict in the Philadelphia Region, 1787–1837*. Baltimore: Johns Hopkins University Press, 1986.

Shoemaker, Nancy. *Negotiators of Change: Historical Perspectives on Native American Women*. New York: Routledge, 1995.

Shy, John. *A People Numerous and Armed: Reflections on the Military Struggle for American Independence*. Ann Arbor: University of Michigan Press, 1990.

Sigstad, John S. "A Field Test for Catlinite." *American Antiquity* 35, no. 3 (1970): 377–82.

Silliman, Benjamin. "The Turquoise of New Mexico." *American Journal of Science* 22 (July 1881): 67–71.

Silver, Beverly J. *Forces of Labor: Workers' Movements and Globalization since 1870*. Cambridge: Cambridge University Press, 2003.

Slaughter, Thomas P. *The Whiskey Rebellion: Frontier Epilogue to the American Revolution*. New York: Oxford University Press, 1986.

Slotkin, Richard. *The Fatal Environment: The Myth of the Frontier in the Age of Industrialization, 1800–1890*. New York: Athenaeum, 1985.

Smith, Duane A. *Mining America: The Industry and the Environment, 1800–1980*. Boulder: University of Colorado Press, 1993.

———. *Rocky Mountain Mining Camps: The Urban Frontier*. Bloomington: Indiana University Press, 1967.

Smith, Henry Nash. *Virgin Land: The American West as Symbol and Myth*. Cambridge, MA: Harvard University Press, 1971.

Smith, Merritt Roe. *Harpers Ferry Armory and the New Technology: The Challenge of Change*. Ithaca, NY: Cornell University Press, 1977.

Snyder, Terri L. "Suicide, Slavery, and Memory in North America," *Journal of American History* 97, no. 1 (2010): 39–62, http://www.jstor.org/stable/40662817.

Sobel, Mechal. *The World They Made Together: Black and White Values in Eighteenth Century Virginia*. Princeton, NJ: Princeton University Press, 1987.

Spanagel, David I. "Great Convulsions and Parallel Scratches: The Era of Romantic Geology in Upstate New York." *Northeastern Geology and Environmental Sciences* 17, no. 2 (1995), 179–82.

Squier, Ephriam G., and Edwin G. Davis. *Ancient Monuments of the Mississippi Valley.* Washington, DC: Smithsonian Institution, 1848.

Starobin, Robert. *Industrial Slavery in the Old South.* New York: Oxford University Press, 1970.

Steinberg, Theodore. *Nature Incorporated: Industrialization and the Waters of New England.* Cambridge: Cambridge University Press, 1991.

Stewart, John. "A Walking Tour in Old Ste. Geneviève," *Missouri Life* 6, no. 3 (July-August 1978), 50–57.

Stewart, Mart A. *What Nature Suffers to Groe: Life, Labor, and Landscape on the Georgia Coast.* Athens: University of Georgia Press, 1996.

Stilgoe, John R. *Common Landscape of America, 1580 to 1845.* New Haven, CT: Yale University Press, 1982.

Stiller, David. *Wounding the West: Montana, Mining, and the Environment.* Lincoln: University of Nebraska Press, 2000.

Stine, Jeffrey K., and Joel A. Tarr. "At the Intersection of Histories: Technology and the Environment." *Technology and Culture* 39, no. 4 (October 1998): 601–40.

Stokes, Melvyn, and Stephen Conway, eds. *The Market Revolution in America: Social, Political, and Religious Expressions, 1800–1880.* Charlottesville: University of Virginia, 1996.

Stoll, Steven. *Larding the Lean Earth: Soil and Society in Nineteenth-Century America.* New York: Hill and Wang, 2002.

Stoltman, James B, ed. *New Perspectives on Cahokia: Views from the Periphery.* Madison: University of Wisconsin, 1991.

Stott, Richard B. *Workers in the Metropolis: Class, Ethnicity, and Youth in Antebellum New York City.* Ithaca, NY: Cornell University Press, 1990.

Stubbs, Tristan. *Masters of Violence: The Plantation Overseers of Eighteenth-Century Virginia, South Carolina, and Georgia.* Columbia: University of South Carolina Press, 2018.

Studnicki-Gizbert, Daviken. "Exhausting the Sierra Madre: Mining Ecologies in Mexico over the Lougue Duree," in *Mining North America: An Environmental History Since 1522,* eds. John Robert McNeill and George Vrtis, Oakland: University of California Press, 2017.

Sullivan, William A. *The Industrial Worker in Pennsylvania, 1800–1840.* Harrisburg, PA: Pennsylvania Historical and Museum Commission, 1955.

Surrey, N. M. Miller. *Calendar of Manuscripts in Paris Archives and Libraries Relating to the History of the Mississippi Valley to 1803.* Washington, DC: Carnegie Institution of Washington, Department of Historical Research, 1926–28.

———. *The Commerce of Louisiana during the French Regime, 1699–1763.* New York: Columbia University, 1916.

Swartzlow, Ruby. "The Early History of Lead Mining in Missouri, Part I." *Missouri Historical Review* 28, no. 3 (April 1934): 184–94.

———. "The Early History of Lead Mining in Missouri, Part II." *Missouri Historical Review* 28, no. 4 (July 1934): 287–95.

———. "The Early History of Lead Mining in Missouri, Part III." *Missouri Historical Review* 29, no. 1 (October 1934): 27–34.

———. "The Early History of Lead Mining in Missouri, Part IV." *Missouri Historical Review* 29, no. 2 (January 1935): 109–14.

———. "The Early History of Lead Mining in Missouri, Part V." *Missouri Historical Review* 29, no. 3 (April 1935): 195–205.

Taylor, Alan. *Liberty Men and Great Proprietors.* Chapel Hill: University of North Carolina Press, 1990.

———. "Unnatural Inequalities: Social and Environmental Histories." *Environmental History* 1 (1996): 6–19.

———. "'Wasty Ways': Stories of American Settlement." *Environmental History* 3 (July 1998): 291–310.

Taylor, George Rogers. *The Transportation Revolution, 1815–1865: Volume 4 of the Economic History of the United States.* New York: Rinehart, 1951.

Thomas, Cyrus. *Report on the Mound Explorations of the Bureau of American Ethnology, Twelfth Annual Report of the Bureau of Ethnology 1890–91.* Washington, DC: GPO, 1894.

Thompson, E. P. *The Making of the English Working Class.* New York: Vintage, 1966.

Thompson, Henry Clay, II. *"A History of Madison County Missouri." Democrat News* (Fredricktown, MO). Serially published ca. 1940.

Thornton, John. *Africa and Africans in the Making of the Modern World, 1400–1680.* Cambridge: Cambridge University Press, 1998.

Tocqueville, Alexis de. "A Fortnight in the Wilds," in *Journey to America*, ed. J. P. Mayer, trans. George Lawrence (New Haven: Yale University Press, 1959), http://www.iwu.edu/~matthews/journey1.html.

Todorov, Tzevtan. *The Conquest of America: The Question of the Other.* Translated by Richard Howard. New York: Harper Perennial Publishers, 1984.

Tomlins, Christopher. *Labor, Law and Ideology in Nineteenth-Century America.* Cambridge, MA: Harvard University Press, 1993.

Tomlins, Christopher L., and Bruce H. Mann, eds. *The Many Legalities of Early America.* Omohundro Institute of Early America History and Culture. Chapel Hill: University of North Carolina Press, 2001.

Trawick, Paul. *The Struggle for Water in Peru: Comedy and Tragedy in the Andean Commons*: Stanford, CA: Stanford University Press, 2003.

Trosper, Ronald. "That Other Discipline: Economics and American Indian History." In *New Directions in American Indian History*, edited by Colin G. Calloway, 199–222. Norman: University of Oklahoma Press, 1988.

Tucker, Barbara M. *Samuel Slater and the Origins of the American Textile Industry, 1790–1860*. Ithaca, NY: Cornell University Press, 1984.

Turner, Frederick Jackson. *The Frontier in American History*. New York: Henry Holt, 1920.

———. "The Significance of Frontier in American History." Annual Report of the American Historical Association, 1893.

Ulrich, Laurel Thatcher. *Good Wives: Image and Reality in the Lives of Women in Northern New England 1650–1750*. New York: Vintage Books, 1991.

Usner, Daniel H., Jr. "An American Indian Gateway: Some Thoughts on the Migration and Settlement of Eastern Indians Around Early St. Louis." *Gateway Heritage* 11, no. 3 (Winter 1990–91): 42–51.

———. *Indians, Settlers, and Slaves in a Frontier Exchange Economy: The Lower Mississippi Valley before 1783*. Chapel Hill: Published for the Institute of Early American History and Culture, Williamsburg, Virginia, by the University of North Carolina Press, 1992.

———. "Weaving Material Objects and Political Alliances: The Chitimacha Indian Pursuit of Federal Recognition" in *Native American and Indigenous Studies* 1, no. 1 (2014): 25–48.

Valenčius, Conevery Bolton. *The Health of the Country: How American Settlers Understood Themselves and Their Land*. New York: Basic Books, 2002.

Van Zandt, Cynthia J. *Brothers among Nations: The Pursuit of Intercultural Alliances in Early America, 1580–1660*. Oxford: Oxford University Press, 2008.

Vickers, Daniel. *Farmers and Fishermen: Two Centuries of Work in Essex County, Massachusetts, 1630–1850*. Published for the Omohundro Institute of Early American History and Culture, Williamsburg, Virginia. Chapel Hill: University of North Carolina Press, 1994.

Wallace, Anthony, and Anthony F. C. Rockdale. *The Growth of an American Village in the Early Industrial Revolution*. New York: Alfred A. Knopf, 1978.

Walthall, John A. *Galena and Aboriginal Trade in Eastern North America*. Springfield: Illinois State Museum, 1981.

Warren, Christian. *Brush with Death: A Social History of Lead Poisoning*. Baltimore: Johns Hopkins University Press, 2000.

Warren, Karen J., ed. *Ecofeminism: Women, Culture, Nature*. Bloomington: Indiana University Press, 1997.

Warren, Louis S. *The Hunter's Game: Poachers and Conservationists in Twentieth-Century America*. New Haven, CT: Yale University Press, 1997.

Watson, Harry. *Liberty and Power: The Politics of Jacksonian America*. Oxford: Oxford University Press, 1981.

Way, Peter. *Common Labour: Workers and the Digging of North American Canals, 1780–1860*. Cambridge: Cambridge University Press, 1993.

Weaver, John E. *North American Prairie*. Lincoln: Johnsen, 1954.

Weber, David J. "Turner, the Boltonians, and the Borderlands." *American Historical Review* 91 (February 1986): 66–81.

Weber, David J., and Jane M. Rauch, eds. *Where Cultures Meet: Frontiers in Latin American History.* Wilmington, DE: Scholarly Resources, 1994.

Werner, Abraham G. *On the External Characters of Minerals,* trans. Albert V. Carozzi, Urbana: University of Illinois Press, 1962.

West, Elliott. *The Contested Plains: Indians, Goldseekers, and the Rush to Colorado.* Lawrence: University Press of Kansas, 1998.

Whitaker, Arthur P. "The Elhuyar Mining Missions and the Enlightenment." *Hispanic American Historical Review* 31, no. 4 (1951): 557–85.

White, George W., ed. *The American Mineralogical Journal. Archibald Bruce, M.D. Contributions to the History of Geology.* vol. 1. New York: Hafner, 1968.

———. "Early Geological Observations in the American Midwest," in *Toward a History of Geology: Proceedings,* ed. Cecil J. Schneer, Cambridge, MA: Massachusetts Institute of Technology, 1970.

White, Richard. "American Environmental History: The Development of a New Historical Field." *Pacific Historical Review* 54 (1985): 297–335.

———. "Environmental History, Ecology, and Meaning." *Journal of American History* 76 (March 1990): 1111–16.

———. *"It's Your Misfortune and None of My Own": A History of the American West.* Norman: University of Oklahoma Press, 1991.

———. *Land Use, Environment, and Social Change: The Shaping of Island County, Washington.* Seattle: University of Washington Press, 1992.

———. *The Middle Ground: Indians, Empires, and Republics in the Great Lakes Region, 1650–1815.* Cambridge: Cambridge University Press, 1991.

———. *The Organic Machine: The Remaking of the Columbia River.* New York: Hill and Wang, 1995.

White, Richard, and William Cronon. "Ecological Change and Indian-White Relations." In *History of Indian-White Relations,* edited by Wilcomb E. Washburn, 417–29. Vol. 4 of *Handbook of North American Indians.* Washington, DC: Smithsonian Institution, 1988.

Whites, Leeann, Mary Neth, and Gary R. Kremer. *Women in Missouri History: In Search of Power and Influence.* Columbia: University of Missouri Press, 2004.

Whitman, Stephen T. *The Price of Freedom: Slavery and Manumission in Baltimore and Early National Maryland.* Louisville: University of Kentucky Press, 1997.

Wilentz, Sean. *Chants Democratic: New York City and the Rise of the American Working Class, 1788–1850.* New York: Oxford University Press, 1984.

Wilkinson, Richard G. "The English Industrial Revolution." In *The Ends of the Earth: Perspectives on Modern Environmental History,* edited by Donald Worster, 80–102. New York: Cambridge University Press, 1988.

Williams, Carol J. *Framing the West: Race, Gender, and the Photographic Frontier in the Pacific Northwest.* New York: Oxford University Press, 2003.
Williams, Michael. *Americans and Their Forests: A Historical Geography.* New York: Cambridge University Press, 1989.
———. *Deforesting the Earth: From Prehistory to Global Crisis.* Chicago: University of Chicago Press, 2003.
Wilson, Douglas L. "Thomas Jefferson's Library and the French Connection." *Eighteenth-Century Studies* 26, no. 4 (1993): 669. https://doi.org/10.2307/2739489.
Winslow, A. "Lead and Zinc Deposits." *Missouri Geological Survey* 7 (1894): 477–87.
Winsor, Justin. *Cartier to Frontenac: Geographical Discovery in the Interior of North America in its Historical Relations, 1534–1700.* Originally published, Boston: Houghton, Mifflin, 1894; Reprint, New York: Cooper Square, 1970.
Wirth, John D. "The Trail Smelter Dispute: Canadians and Americans Confront Transboundary Pollution, 1927–41." *Environmental History* 1 (1996): 34–51.
Wisseman, Sarah Wisseman. Randall Hughes, Thomas Emerson, and Kenneth Farnsworth, "Refining the Identification of Native American Pipestone Quarries in the Midcontinental United States," *Journal of Archaeological Science* 39 (2012): 2496–505.
Woods, W. Y. "A Strange Pre-Historic Find." *The Wisconsin Naturalist* 1, no. 1 (1890).
Worster, Donald. *Dust Bowl: The Southern Plains in the 1930s.* Oxford: Oxford University Press, 1979.
———. *Nature's Economy: A History of Ecological Ideas.* 2nd Edition. Cambridge: Cambridge University Press, 1994.
———. *Rivers of Empire: Water, Aridity, and the Growth of the American West.* New York: Oxford University Press, 1992.
———. "Seeing Beyond Culture." *Journal of American History* 76 (March 1990): 1142–47.
———. "Transformations of the Earth: Toward and Agroecological Perspective in History." *Journal of American History* 76 (March 1990): 1087–1106.
———. *The Wealth of Nature: Environmental History and the Ecological Imagination.* New York: Oxford University Press, 1993.
Wyman, Mark. *Hard Rock Epic: Western Miners and the Industrial Revolution, 1860–1910.* Berkeley: University of California Press, 1979.
York, Neil L. *Mechanical Metamorphosis: Technological Change in Revolutionary America.* Contributions in American Studies, vol. 78. Westport, CT: Greenwood Press, 1985.
Young, Otis E., Jr. *Western Mining: An Informal Account of Precious-Metals Prospecting, Placering, Lode Mining, and Milling on the American Frontier from Spanish Times to 1893.* Norman: University of Oklahoma Press, 1970.
Zakim, Michael. *Ready-Made Democracy: A History of Men's Dress in the American Republic, 1760–1860.* Chicago: University of Chicago Press, 2003.

INDEX

Page numbers in *italics* refer to illustrations.

accidents, 161–62
Adam, Joseph, 35–36
Agricola, Georgius: on assaying, 21, 33–34; authority of, on mining, 10, 20–22, 45–46, 115; background on, 20; *De Natura Fossilium*, 10, 21; *De Re Metallica*, 20–21, 45; on environmental impacts of mining, 82; on European mining techniques, 20–22, 32, 39, 42; as founder of mineralogy, 10; on furnaces, 41–42, 80, 81; on health concerns related to mining, 45–46, 82; on mining labor, 95, 97; on prospecting, 29, 30; on pumps, 54; on smelting, 40–42
Aikin, Arthur and C. R., *A Dictionary of Chemistry and Mineralogy*, 128
air quality, 46, 47, 48, 82–83, 157
Albert, J. J., 144
Alembert, Jean le Rond d', 1
Amelung, Johann Friedrich, 97
American Mineralogical Journal, 104
American system of mining, 139–69; depictions of, 148–51; environmental and health impacts of, 155–64; as part of civilizing project, 138–41, 151, 163; practices in, 142–46; promotion of, 137–40, 145, 156–58, 163. *See also* European mining practices
animals, health impacts of mining on, 45, 47, 82–83, 156–57
Aristotle, 10, 21

arsenic, 14, 63, 82, 97, 130
Ashley, William, 125
assaying, 21, 33–34, 113, 118–23
Assyrians, 19
Austin, Moses: assays and experiments of, 109–11, 113, 118–23; background on, 89; environmental concerns of, 59, 63, 82–84; and Herculaneum, 98, 124–25; immigration to Spanish Louisiana, 57, 59, 63, 69, 77, 79–81, 93; introduction of European mining methods by, 8, 11, 57, 80–81, 87, 89–97, 105, 107–12, 114, 117–18, 140; and Mine à Breton, 82–84, 87, 91–100, 103, 112–15, 140, 149; as mineralogist, 9–12, 101, 104–7, 109–13, 116, 118, 121, 123, 135–36; products made by, 88, 97, 99, 151–52; study and analysis of mining techniques by, 11, 69, 79–80, 91, 93, 110, 112–17, 139; *A Summary Description of the Lead Mines in Upper Louisiana* (prospectus requested by Jefferson), 8–9, 11–12, 14, 101–18, 121–24; transportation of lead by, 98–99
Austin, Stephen, 89, 149

Baker, George, 134
baling seals, 20, 26, 28, 57
Barre, Esteban, 71
Bates, Elias, 93
Beauvais, St. Gemme, 83
Bell, Josiah, 80, 89, 91–93, 103, 110, 111–13, 117, 120–21, 136
Bellin, Jacques-Nicolas, 24

251

Beltrami, Giacomo, 81
Bernard, Antoine, 71
Bienville, Jean-Baptiste Le Moyne de, 24
Billeron, Marianne, 64, 76
Billon, Frederic, 124–25
blacksmiths, 77, 93, 105–6
blasting, 78
blowpipes, 120–21
Bourbon Reforms, 13, 59–61, 73, 76, 79, 84
Brackenridge, Henry Marie, 9, 134, 138–41, 148, 155–57, 159
Brading, David, 60
Brandrams, Templeman & Co., 126, 152–53
Breton, François Azau, 114
Brickey, John S., 93, 94
British (later American) Illinois, 7–8
British mining practices, 6, 9, 10, 57, 91–92, 99, 141–42
Brown, Joseph, 123
Bruce, Archibald, 104
buddles, 14, 95, 129–30. *See also* troughs
Buffon, Comte de, *Natural History*, 1
business associations, 10, 61, 65, 70, 73–75, 83, 86, 90–91, 98, 105

Cabeza de Vaca, Álvar Núñez, 4
Cadillac, Antoine de la Mothe, 16, 75, 87, 165
Cahokia Mission, 37
canals, 154–55
carbonate of lead, 117–18
Carpentier, Henri, 71, 73
Carpentier, Marie, 64
cartography. *See* maps
Castor Vein, 12, 62, 64, 65, 68
catlinite (pipestone) molds, 20, 26, 28
Charity Hospital, Paris, 129–30, 133
Chevalier, Peter, 75–78, 98
children, employed in mining, 95
Chisel Mines, 87, 89, 92
cider, 134
civilizing project, 138–41, 151, 163
Clark, William, 102, 160
Clemson, Thomas, 138, 144–48, 155–56;
Observations of the La Motte Mines, 145, 157–58

colic, 47, 129–30, 133–34, 156, 158. *See also* lead poisoning
color grinding, 130–33
Company of the Indies (France), 27
Company of the West, 27, 34, 36
compasses, 143, 145
copper, 3, 106
cords, 143
Corps of Discovery, 102
Coxe, Tench, 124
Craddock, Paul, 39
Creoles, 59–61, 64, 68–70, 75, 86–87, 93, 108. *See also* European mining practices
currency, lead as, 12, 35–36, 62, 63, 66, 73

Datchurut, Jean Baptiste, 62, 64–66, 68, 73, 83, 90
Decorah, Spoon, 165–67
deforestation, 9, 59, 62, 82–84
Delassus, Charles DeHault, 83–84, 87, 188n9
Delaware Indians, 149, 151
Delisle, Guillaume, 24
Devon colic, 134
De Witt, Benjamin, 112
Diderot, Denis, 1
division of labor, 76–77, 94, 95
doctors, 138–39
Drake, Daniel, *A Systematic Treatise*, 163
Duval, Jean, 65–66

Egyptians, 19
Embargo Act (1807), 124
Encyclopedia (Diderot and d'Alembert), 1
engagés (indentured servants): health impacts on, 82–83; housing for, 59, 68, 70, 73; as mine workers, 49, 59, 64–70, 75–76; wages paid to, 62, 67
environmental impacts: air quality, 46, 47, 48, 82–83, 157; contaminated water, 45, 129–30, 157, 159; deforestation, 9, 22, 59, 62, 82–84; of early Native Americans, 17–18; in Europe, 82; of furnaces, 9, 45, 62–63, 83, 85, 129–30; reclamation project to address, 84–87; reduction of, at turn of nineteenth century, 63, 79, 83, 93; waste products, 63, 84. *See also* health

impacts; settlement conditions, near mining operations
European mining practices: Agricola on, 20–22, 32, 39, 42; Austin's introduction of, 8, 11, 57, 80–81, 87, 89–97, 105, 107–12; financing of, 90; Platte's account of, 74; prospecting, 29–30, 94; replacement of Native American by, 6, 8, 11, 12, 61, 76, 78–79, 90–91, 94, 100, 103–5, 107–9, 111–12, 118–19, 136–38, 141, 148–49, 151; similarity of Native American to, 29–33; smelting, 40–42; technologies of, 32, 54, 68, 80, 81, 87, 91, 140, 143; year-round operations, 51. *See also* American system of mining; lead mining: amalgamation of Native American and French practices in extraction, 31–34

farming: environmental threats to, 47, 157–58; mining practiced in conjunction with, 14, 19, 50, 51, 69, 77, 79, 134, 147–48
Featherstonhaugh, George W., 138, 144–45, 148
fire setting, for lead extraction from limestone, 78
flat-veins, 115–16
Flint, Timothy, 49
flint glass, 117, 152
food, for mining operations, 77
Fort de Chartres, 36, 37
Fortier, Diego, 71
Fox Indians, 55, 80–81
Freiberg Mining Academy, Germany, 86
French and Fox Wars (1730s), 55
French settlers: health impacts on, 44–49; at Kaskaskia, 16; and lead mining, 2, 4, 16; Native American cultural exchanges with, 149, 151; and Native American mining methods, 6–7, 13, 23, 40, 42–44, 50, 54, 57, 59, 91, 100, 107–9, 118–19, 135–39, 148–49, 151, 165–69; settlement possibilities for, 49–55; and slavery, 35–37
furnaces: amalgamated Native American-French, 41–44, 62, 64, 69, 80; of ancient world, 40–41; environmental impacts of, 9, 45, 62–63, 83, 85; European, 41–42, 80, 81; hillside trench, 80; for lead ash waste smelting, 85–86, 95–96; limestone ash, 96; log, 43–44, 54, 59, 62, 64, 65, 69, 79–80, 83, 85, 96, 146; log hearth, 80, 135; masonry, 9, 42, 44, 59, 61, 62, 74, 81, 96,

140; reverberatory, 62, 80, 91, 94, 95–96, 106, 135, 140, 146, 157; Scotch hearth, 80, 146–47; traditional Native American, 41–42, 80–81. *See also* smelting
fur trading, 4, 26, 55, 63

Gadobert, Pierre and Ann, 62, 64, 66–68, 70–71, 73, 83, 90
Galena (village), 80, 81
galena: appearance of, 116, 117, 144; early colonists' discovery of, 4; lead content of, 40, 106, 117, 122, 179n45; Native American uses of, 3, 19; silver content of, 87, 144, 179n45; smelting of, 40–41, 179n45. *See also* lead
gangue minerals, 118–19
Geological Survey of Virginia, 145
glassmaking, 89, 91, 97, 117, 123, 127–28, 152
glaze paints, 3, 18–20
gold, 2, 7
Gravier, Jacques, 2–4, 15–16, 21–22, 24, 34, 114, 165
Green, Drury, 93
Gregoire, Charles, 145, 146–48
Gruy, Antoine Valentin de, 9, 10, 22–23, 26, 36–37, 39–44, 48, 50–54, 56–57, 59, 61, 65, 68, 99, 107, 115–16, 138
guides. *See* Native Americans: as guides
Guire, Paul de, 75–78, 98
gunpowder, 78

Hammond, Samuel, 124
Harris, E. B., 161
health impacts: from accidents, 161–62; Agricola on, 45–46, 82; air quality issues, 46, 47, 48, 82–83, 157; from American system, 139; on animals, 45, 47, 82–83, 156–57; broken bones, 161–62; from contaminated water, 45, 129–30, 157, 159; doctors' commentary on, 138–39, 158–64; lead poisoning, 13, 47, 129–34, 138, 156–64; legislation addressing, 156–57; on mine workers, 45–46, 82–83; official recording of, 13; proactive concerns with, 47–48; of smelting, 47–48, 129–30, 155–64; village siting in relation to, 45, 48
Henderson, Joseph, 128, 131–33, 136
Henriod, Henry, 142–48, 161–63
Henry, Claude, 131

Herculaneum, 98, 124–25, *126*, 147, 154–55
Hertzog, Joseph, 127–28, 131
hillside trench furnaces, 80
Hippocrates, 129
Ho-Chunk peoples. *See* Winnebago peoples
Houghton, Thomas, 52
Humboldt, Alexander von, 83
hunting, as specified job, 77
Hutchins, Thomas, 203n94

Iberville, Pierre Le Moyne d', 24
Illinois Country of the Province of Louisiana, 36–37
Illinois tribes, 3, 15, 17
indentured servants. *See* engagés
Indian Removal Act (1830), 151
iron, 3, 105–6, 204n124

Jefferson, Thomas: Humboldt and, 83; and lead mining, 1–2, 4, 8, 12, 101–9; and Louisiana Territory, 8, 10–12, 101–9, 123; as natural philosopher, 1–2, 10; *Notes on the State of Virginia*, 1–2, 8, 101, 103
Jesuits, 12, 36
Johnson, Esau, 29–30
Jolliet, Louis, 25
Joseph Lead Company, 168
Judd, Richard, 8–9

Kalm, Peter, 203n94
Kant, Immanuel, 76
Kaskaskia (village), 2–4, 8, 36, 45, 52
Kaskaskia Indians, 4, 7, 15–16, 36
Kendall, Jonathan, 125
Kipp, John, 154
Kirwan, Richard, 111
Kitchin, Thomas, *Louisiana Map*, 25, *25*

laboratories, for assays and experiments, 109–11, 113, 118–21, 123
La Chance, Joseph, 75–78
La Chance, Nicolas, 76–77, 79, 98
Lagarciniere, Daniel Fagot, 71
La Gautrain, Karpen de, 55–56
La Malice (hunter), 77
lamp furnaces, 121

Lane, Hardage, 13, 158–60, 163–64
La Renaudière, Philippe de, 9, 10, 22–23, 26–35, 39–40, 48–53, 75, 82, 99, 116, 119, 137, 138
La Rose, Nicolas Noel dit, 61–62, 64–66, 68, 70–71, 73, 83, 90
La Salle, René-Robert Cavelier, Sieur de, 3, 15, 25
lead: amalgamation of Native American and French practices in, 68–70; as currency, 35–36; early American uses of, 3, 11, 26, 55–56, 61, 80, 84, 88–89, 97, 106, 151–52; European uses of, 20; French uses of, 28; Native American uses of, 3, 18–19, 28, 33, 116; physical properties of, 20; sources of, 25–31. *See also* assaying; galena; smelting
lead ashes, 44, 63, 70, 81, 85–87, 95–96, 130, 152. *See also* slag
lead coloring, 130–34, 159
lead crystals, 33, 39
lead mining: amalgamation of Native American and French practices in, 6–7, 13, 23, 40, 42–44, 50, 54, 57, 59, 91, 100, 107–9, 118–19, 135–39, 148–49, 151, 165–69; and American self-sufficiency, 11; dominance of European methods in, 6, 8, 12, 61, 76, 78–79, 90–91, 94, 100, 103–5, 107–9, 111–12, 118–19, 136–38, 141, 148–49, 151; extraction practices, 31–34; farming practiced in conjunction with, 14, 19, 50, 51, 69, 77, 79, 134, 147–48; Jefferson's interest in, 1–2, 4, 8, 12, 101–9; by Kaskaskia Indians, 15–16; large-scale techniques, 94–95; Native American traditions of, 16, 18–19, 22–23; phases in early America, 6; productivity of, 54, 56, 70–71, 91, 100, 122, 134–35, 147; reclamation project in, 84–87, 95–96; scholarship on, 4–6; seasonal approach to, 22, 35, 37, *38*, 50–52, 59, 70, 142; source documents on, 11–12; year-round, 51, 74, 75, 90–91, 148. *See also* American system of mining; European mining practices
lead poisoning, 13, 47, 129–34, 138, 156–64
Lead to Metal (documentary), 168
lead washing, 45, 48, 95, 118, 121, 130
lead weed, 29
Leonardo da Vinci, 46
Lesueur, Charles-Alexandre, 148–49; View of Mine La Motte Village with twelve houses and a clothes line, 148–49, *150*; View of Mine

La Motte with three miners among the lead pits and windlasses, 149, *150*
Lewis, Meriwether, 102
limestone ash furnaces, 96
Linn, Lewis Fields, 138, 141–42, 144; *Observations of the La Motte Mines*, 145
log furnaces, 43–44, 54, 59, 62, 64, 65, 69, 79–80, 83, 85, 96, 146
log hearth furnaces, 80, 135
Loisel, Antoine, 36
Lorimier, Louis, 149
Louisiana Purchase, 3, 8, 84, 101–2
Louisiana Territory: Austin's prospectus on, 8–9, 11–12, 14, 101–13, 117–18, 121–24; establishment of American government in, 101; French settlers in, 15–16; Jefferson and, 8, 10–12, 101–9, 123; maps of, 23–25, *25*; Spanish and British control of, 58; Upper, 64. *See also* Missouri Territory; *pays de Illinois*; Spanish Louisiana
Lowell, Francis Cabot, 92
Luna, Francisco, 85–87, 96, 193n98
Luzières, Pierre-Charles de Hault de Lassus de, 60–61, 63, 78–80, 84–91, 98–100, 139

Maclot, John, 125
magnifiers, 120–21
managers, 93, 122, 130, 141. *See also* overseers
Manning, Major, 146
manuals, 10, 54, 96, 109, 111–13, 115, 120, 202n75
maps, 23–25
Mark, Jacob, 90
Marquette, Jacques, 25
Martin, Susan, *Wonderful Power*, 6
Maryland Chemical Works, 131
Mason, Amable Partenay, 129–30
masonry furnaces, 9, 40, 42, 44, 59, 61, 62, 74, 81, 96, 140
Matis, Jerome, 75–78, 98
Maurepas, Jean-Frédéric Phélypeau, Comte de, 51–55
McCready, Benjamin, 162–63
McKim and Sons Chemical Works, 131
Meade, William, *Description and Analysis of an Ore of Lead from Louisiana*, 104–5, 111
Medical Society of Missouri, 158
Medina, Bartolomé de, 86

Meramec mine, 15
Merat, François Victory, 129–30, 133; *Dissertation sur la colique metallique*, 129; *Traite de la colique metallique*, 129
mercury, 86
Merieult, John, 99
Mexico, 60, 83, 86
middle ground, 7, 23, 40, 49–50, 61, 107, 167
Miller, Samuel, *A Brief Retrospect of the Eighteenth Century*, 111, 135–36
mills. *See* sawmills; stamping mills; water mills
Mine à Breton, 8, 10, 12, 14, 59, 77, 80–81, 87, 90–101, 103, 111–15, 122, 129–30, 134–35, 140, 142, 146, 149, 154, 157, 159
Mine La Motte, 12, 14, 16, 22, 26, 27, 34, 37, 41, 44–45, 48, 50, 52, 54, 56, 59–62, 64–71, 73–81, 83–84, 86–87, 91, 93, 98–100, 108, 112, 114–17, 122, 127, 134–35, 137, 139–42, 144–49, 151, 153–54, 156, 158, 163, 165, 208n59
Miner, Joseph, 85
mineralogy: areas of study, 10; early practitioners of, 10; emerging science of, 9, 203n94; knowledge gained from, 9, 11, 21, 102–6, 109, 135; of Louisiana/Missouri Territory, 103–5, 109–13, 116, 118, 121, 123, 144–45
miners. *See* workers
mineshafts, 32–33, 68, 91, 94–95, 117, 142–45, 161–62
mine waste, 62–63, 84, 130
mining associations. *See* business associations
Mississippians, 18–19
Mississippi River, 2, 3, 7–8, 58, 98, 155
Missouri Gazette (newspaper), 123–25, 127, 134
Missouri Territory: Austin's prospectus on, 8–9, 11–12, 14, 101–18, 121–24; growth and development of, 123–27, 142, 147; lead mining in, 16, 106, 110–12, 118, 122–26, 135, 138–39, 146, 152; and slavery, 132. *See also* Spanish Louisiana
Moody, Paul, 92
Morales, Juan Bonaventura, 84
Morgan, Benjamin, 154
Mullins, Matthew, 93, 110, 111–13, 117, 120–21, 136
Mullins, Timothy, 93, 110, 111–13, 117, 120–21, 136
Murphy, Lucy Eldersveld, 6
musket balls, 3, 11, 16, 20, 26, 55–56, 61, 80, 84, 88–89, 106

Native Americans: extraction practices of, 31–34; French cultural exchanges with, 149, 151; and French mining methods, 6–7, 13, 23, 40, 42–44, 50, 54, 57, 59, 91, 100, 107–9, 118–19, 135–39, 148–49, 151, 165–69; as guides, 3–6, 9, 12, 15–16, 22, 26–27, 30–31, 37, 50, 52, 75, 87, 165; knowledge gained from, 9; lead mining history and practices of, 16, 18–19, 22–23, 165–66; lead uses of, 3, 18–19, 28, 33; manufacturing processes of, 28; seasonal living patterns of, 14, 19, 22, 37, 49–50; settlement patterns and environmental impact of, 17–18; smelting practices of, 6, 7, 22, 37, 39–44, 80

natural philosophers, 1–2, 8–10, 12–13, 61, 86, 102, 110–11, 113, 115, 117, 119, 120, 135, 138, 145

nature, scientific attempts to master, 14, 106, 108, 109, 119

New Bourbon, 78–79

New Jersey Copper Mines Association, 90

New Orleans: founding of, 36; and the lead trade, 28, 53, 55, 71, 98–99, 151–52, 155; Spanish administration of, 58, 106; as trading center, 58, 106

Nicholson, William, *A Dictionary of Chemistry*, 128, 208n76

Nicollet, Gabriel, 75–78, 98

Non-Importation Act (1806), 124

O'Hara, James, 97

Old Mines, 15, 45, 48, 50, 59, 61, 129–30, 144, 146, 165

Old Mines Creek, 57

optics, 120–21

overseers, 76–77. *See also* managers

oxides, 157

paint, lead used in manufacture of, 3, 19, 31, 106, 127–28, 152, 162. *See also* glaze paints

Panic of 1796–1797, 88–89

patio process, 86

Paul, Rodman Wilson, *Mining Frontiers of the Far West*, 5

pays de Illinois: cultural amalgamation in, 6; defined, 3, 15; French loss of, 58; French settlement in, 35–37; Kaskaskia Indians in, 15; lead mining in, 2, 16, 25–44; maps of, 23–25, 25, 38;

as middle ground of cultural interchange, 7–8, 49–50; settlement possibilities in, 28, 49–55, 75

Perkins, John, 161

Perry Mines, 161

pig lead: as currency, 12, 61–63, 66, 73; disks made of, 44; manufacturing based on, 126–27, 131–32, 152–54; musket balls made from, 61, 84, 88–89; transformation of ash waste into, 85; transportation of, 44, 63, 67, 71, 73, 74, 98, 154–55

pipestone. *See* catlinite

Pittsburgh Glass Works, 97

Pivert, Perrine, 27

Planches, Louis Tanquerel des, 130, 133

Plattes, Gabriel, *A Discovery of Subterraneal Treasures*, 74, 76, 78, 92

Pliny the Elder, 46

prairie shoestring, 29

prisoners, as labor, 54

productivity, 54, 56, 70–71, 91, 100, 122, 134–35, 147

prospecting, 28–31, 94, 107–9, 114–16, 139, 145

Pryor, Nathaniel, 80

Pueblo peoples, 18–19

pumps, 53–54

rake-veins, 115–16

Ramazzini, Bernardino, 47, 129

reclamation, 84–87, 95–96

red lead, 3, 106, 117, 123, 126–28, 130–33, 152–53, 155

Renault, Philippe, 35–37, 40, 44–45, 75

respiratory protective devices, 46

reverberatory furnaces, 62, 80, 91, 94, 95–96, 106, 135, 140, 146, 157

roasting, 43

Rogers, Robert, 203n94

Rondeau, Marie Anne, 45

Roosevelt, Nicholas J., 90

Royal Mining School, Mexico City, 86

Rozier, Ferdinand, 142–48, 161–63

Samuel Wetherill & Sons, 152–55. *See also* Wetherill Brothers

Santa Fe, 85

sawmills, 93

Schoolcraft, Henry Rowe, 9, 81, 95–96, 98, 117, 122, 135, 137–41, 144, 148, 155–57, 159; *A View of the Lead Mines of Missouri*, 117–18

Schuyler, Philip A., 90
scoria. *See* slag
Scotch hearth furnaces, 80, 146–47
seals. *See* baling seals
seasonal living patterns, 14, 19, 22, 35, 37, 49–52, 70, 142
Segon, Pablo, 71
settlement conditions, near mining operations, 28, 49–55, 70–71, 74–75, 78–79, 89–90, 93, 98, 138–39, 148–49, 157–58. *See also* environmental impacts
shafts. *See* mineshafts
Shawnee Indians, 149, 151
sheet lead, 3, 11, 88–89, 91, 93, 99, 106, 151–52
shot, 3, 11, 39, 55, 88–89, 97–98, 99, 106, 124–25, 127, 135, 151
shot towers, 88–89, 93, 97, 124–25, *126*
silver, 2, 7, 15, 27, 59–60, 85–87, 144
Skeel, Stephen, 160–64
slag (scoria), 40, 43, 44, 63, 84–85. *See also* lead ashes
Slater, Samuel, 92, 153
slaves: depictions of, 149; French attitude toward, 36; health impacts on, 82–83; housing for, 59, 66, 68, 70, 73–74; in iron production, 204n124; laws pertaining to, 67–68; in lead coloring industry, 131–32; Louisiana Territory and, 132; as mine workers, 28, 35–37, 59, 62, 64–70, 75–76, 94, 95, 146, 149; other work done by, 68, 74, 77; overseers of, 76; seasonal labor by, 65, 66; in Spanish Illinois, 61, 65, 68; suicide of, 158–60. *See also* engagés
smelting: Agricola on European, 40–42; amalgamated process of, 40–44, 60, 79; British and American methods of, 6, 11–13, 80, 90–97, 107, 122, 137–38, 140, 146–47; environmental impacts of, 9, 45, 62–63, 83, 85, 129–30; health impacts of, 47–48, 129–30, 155–64; Native American methods of, 6, 7, 22, 37, 39–44, 80; origins of, 39; preparation of ore for, 53, 69, 95, 118–19, 130, 146; reclamation project in, 84–87; timber required for, 9, 33, 51, 59, 62, 82–84. *See also* furnaces
Smith, Adam, 76
Snyder, Terri, 159
sorting, 48
Soto, Hernando de, 4

Spanish Louisiana: administration of, 59–60; environmental concerns in, 82–87; establishment of, 7, 58; European practices in, 88–101; French and Creole domination of mining business in, 59, 61–62, 64; immigration policy of, 59, 60–61, 88; lead mining in, 59–100; mining-related settlements in, 70–71, 73–76, 78–79, 93; population boom in, 88–89; reclamation project in, 84–87, 95–96; villages and land routes in, *72*. *See also* Missouri Territory
Sparke, John, 128
stamping mills, 69
Ste. Geneviève: as administrative center, 12, 58–59; economic activities of, 77; Jefferson's interest in, 103; mining practices around, 12, 59, 70, 74, 108; settlements in or near, 8, 18, 54–55, 60–61, 64, 68, 78–79, 88, 147; slave population of, 61, 65, 68; as trading center, 71, 73–75, 98, 125, 147, 154–55
St. Louis, 58, 132
St. Louis Medical and Surgical Journal, 158
St. Michel (St. Michael), 75, 79, 142, 147
Stockhausen, Samuel, 47, 129
Stoddard, Amos, 8, 14, 101–3, 113, 135
stone furnaces. *See* masonry furnaces
Storch, John A., 123
Storts, John, 93
St. Philippe, 35–37
suicide, 158–60
sulphur, 14, 33, 43, 46, 82, 84, 122, 129–30
sulphuret of lead, 33, 144. *See also* galena

tailings, 63, 84, 130
Tamaroa Indians, 15, 22, 75
technologies: in European mining, 32, 54, 68, 80, 81, 87, 91, 140, 143; Europe-to-America transfers of, 92, 97; introduction of new, into Louisiana Territory, 88–90; for large-scale operations, 94–95. *See also* furnaces; tools
Thompson, Henry, 154
Thwaites, Reuben Gold, 165–66
timber: for building construction, 51; for furnace construction, 42–43, 54, 59; in *pays de Illinois*, 9, 54; for shaft construction, 32; for smelting, 9, 33, 40, 51, 59, 62, 79, 82–84. *See also* deforestation

Index | 257

tin, 3
Tissot, Samuel, 47–48
Tocqueville, Alexis de, 137
tools, 77, 108–10, 119–21, 142–43. *See also* technologies
transportation of lead, 44, 52–53, 56, 63, 67, 71, 73, 74, 98, 142, 154–55
Treaty of Paris (1763), 58
Treaty of Prairie du Chien (1835), 151
trenches, 22, 28–33, 68, 69, 108–9, 135
troughs, 121. *See also* buddles
Trudeau, Zenon, 88
Turner, Frederick Jackson, 7

Ulloa, Antonio, 60, 68
Ure, Andrew, *Dictionary of Arts, Manufactures and Mines*, 132
Ursin, Marc Antonine de la Loire des, 9, 10, 22–23, 25–26, 28–37, 40, 50–51, 53–54, 75, 99, 116, 119, 137, 138
Usner, Daniel, 7
US Topographical Engineers, 144

Valenčius, Conevery Bolton, 45
Vallé, François, 12, 41, 56, 59, 61, 64–66, 68, 71, 73–79, 83–84, 86, 90–91, 93–94, 98, 100, 151
Vallé, François, Jr., 64–65, 89–90
Vallé, Jean Baptiste, 86–87
Variat, Pierre, 75–78, 98
Vasquez, Benito, 71
veins, 115–18
Vial, Pedro, 63, 85–87, 96, 193n95

waste products: environmental impacts of, 63, 84; reduction of, at turn of nineteenth century, 63, 79

water: contaminated water, 159; contamination of, 45, 129–30, 157; as discovery aid in prospecting, 30; health effects of, 45, 48; uses of, in mining operations, 48–49, 51–54, 93, 146
water mills, 53, 146, 158
Watts, William, 88, 97
Weaver, John, 29
Werner, Abraham Gottlob, 10, 110, 111
West, Elliott, *Mining Frontiers of the Far West*, 5
West, mining in, 5
Wetherill, John, 152, 154
Wetherill, Samuel, 128, 152–54
Wetherill, Samuel, Jr., 152, 154
Wetherill Brothers, 126–27. *See also* Samuel Wetherill & Sons
White, Richard, *The Middle Ground*, 7
white lead, 3, 106, 117, 126–28, 130, 152–53, 155, 162
Wilt, Andrew, 131–34, 136, 159
Wilt, Christian, 127–28, 131–34
windlasses, 68–69, 94–95, 109, 140, 143, 149
Winnebago peoples, 29–30, 165–66
Wivarenne, Pierre, 45
women: employed in mining, 95; roles and status of, in Louisiana Territory, 67
Woodhouse, James, 119–21
Woods, W. Y., 42–43
workers: behavior of, 141; health problems of, 46–47, 82–83, 129–34; in mines, 28; skilled, 54, 55–56, 89–91, 141–42, 157. *See also* engagés

Yarnall, Sarah, 152

zane, 89, 91, 96, 97, 99, 152
Zuni peoples, 18–19

www.ingramcontent.com/pod-product-compliance
Lightning Source LLC
LaVergne TN
LVHW090054080526
838200LV00082B/7